Statistical Methods in Counterterrorism

Game Theory, Modeling, Syndromic Surveillance, and Biometric Authentication

Alyson G. Wilson
Gregory D. Wilson
David H. Olwell
Editors

Statistical Methods in Counterterrorism

Game Theory, Modeling, Syndromic Surveillance, and Biometric Authentication

Alyson G. Wilson
Los Alamos National Laboratory
Los Alamos, NM 87545
USA
agw@lanl.gov

Gregory D. Wilson
Los Alamos National Laboratory
Los Alamos, NM 87545
USA
gdwilson@lanl.gov

David H. Olwell
Department of Systems Engineering
Naval Postgraduate School
Monterey, CA 93943
USA
dholwell@nps.edu

Library of Congress Control Number: 2006922769

ISBN-10: 0-387-32904-8 e-ISBN: 0-387-35209-0
ISBN-13: 978-0387-32904-8

Printed on acid-free paper.

Printed in the United States of America. (EB)

9 8 7 6 5 4 3 2 1

springer.com

For those who lost their lives on September 11, 2001, and the men and women fighting the war on terror

Preface

In the months after September 11, 2001, in the aftermath of the attacks on the World Trade Center in New York, counterterrorism became a research interest for a broad range of Western scholars, statisticians among them. At the same time, the U.S. government, still in shock, repeated the same question during multiple hearings in Washington, D.C.: "All the data was out there to warn us of this impending attack, why didn't we see it?" Data became a large part of the response to 9/11 as Americans tried to regain a rational grip on their world. Data from flight recorders was collected and analyzed, timelines were assembled to parse out explanations of what happened, sensitive data was removed from government websites, and the White House debated what data to release to investigators and the American public. "Data" was a frequently heard term in the popular media, one of the many things that we had to protect from the terrorists, and one of the most important things that we could use to defeat them.

In the statistical community, professionals wondered how they could help the government prevent terror attacks in the future by developing and applying advanced statistical methods. The federal government is a sizable consumer and producer of statistical data, as the 9/11 commission report noted.

> The U.S. government has access to a vast amount of information. When databases not usually thought of as "intelligence," such as customs or immigration information, are included, the storehouse is immense. But the U.S. government has a weak system for processing and using what it has. [KH04, pp. 416–417]

Additionally, government decision-makers are often skeptical about statistics. Understanding that the Washington audience wasn't always receptive, the statistical community pondered how to put what they knew to work for the country. They felt specially qualified to help decision-makers see the important patterns in the oceans of data and detect the important anomalies in the seemingly homogeneous populations. At a round-table luncheon at the Joint Statistical Meetings in San Francisco in 2003, almost two years after

9/11, a dozen statisticians ate and pondered the same questions. "How do we get in the door?" "How do we get someone to let us help?"

It was hard to get in the door, because Washington was still trying to figure out what a response to terrorism in the homeland would begin to look like. The threat paradigm had shifted enough that no one quite knew what the appropriate questions were, let alone the appropriate responses. Potential bioterrorism is a case in point. Dread diseases like smallpox had been conceptualized and studied as diseases, as public health problems, and as potential battlefield weapons, but had not been extensively studied as agents terrorists might set loose in a major population center. When a set of anthrax mailings followed close on the heels of the World Trade Center bombings, it was as if our world-view had been fractured. Many old questions of interest faded away, many new ones appeared, others were yet to be discovered. Biologists, epidemiologists, biostatisticians, public health experts, and government decision-makers woke up the next day wondering where to begin. The same was true across many fronts and many lines of inquiry in those months. The U.S. government wound up organizing an entirely new Department of Homeland Security to address the raft of new problems that emerged after 9/11. In the decision-maker's estimation, the new problems were different enough that existing structures like the Federal Bureau of Investigation, Centers for Disease Control and Prevention, and Immigration and Naturalization Services were not sufficient or appropriately specialized to address this new threat.

At the time of this writing, the science of counterterrorism is also still unfolding. The government has begun to engage the country's research community through grants and collaborative opportunities, but across the sciences, and in statistics, the interesting problems and viable methodologies are still in a very speculative stage. Speculative is also exciting, though. Researchers feel lucky to be able to help define the landscape of a new research enterprise. This book encompasses a range of approaches to new problems and new problem spaces. The book is divided into four sections pertinent to counterterrorism: game theory, biometric authentication, syndromic surveillance, and modeling. Some of the chapters take a broad approach to defining issues in the specific research area, providing a more general overview. Other chapters provide detailed case studies and applications. Together they represent the current state of statistical sciences in the area of counterterrorism.

Game theory has long been seen as a valuable tool for understanding possible outcomes between adversaries. It played an important role in cold war decision and policymaking, but the opening section of this book rethinks game theory for the age of terrorism. In a world of asymmetric warfare, where your adversary is not a country with national assets and citizens at risk in the event of retaliation, the stakes are different. The section on game theory presented in this text provides an overview of statistical research issues in game theory and two articles that look specifically at game theory and risk analysis.

Biometric authentication has become a more prominent research area since 9/11 because of increased interest in security measures at border entry stations and other locations. Authentication of fingerprints, faces, retinal scans, etc., is usually an issue in the context of identity verification, i.e., does this passport match the person in front of me who is trying to use it? Beyond the logistics of collecting the information on everyone who applies for a passport or visa, storing it on the identity documents in a retrievable form, upgrading the computer equipment at all border crossings, and training border police to use the new technology, the issues of accurate identification are still to be worked out. Security agencies would also like to be able to use face recognition to pick known terrorists or criminals out of crowds using video cameras and real-time analysis software. The stakes for false positives are high — a man suspected as a potential terrorist bomber was held down by police and shot in the head in the London subway in 2005, and many individuals have wound up in long-term detention under the mere suspicion that they were members of terrorist organizations. Current technological shortcomings also have strong cultural implications: fingerprint authentication works less well with laborers who have worn skin and calluses on their hands; retinal scans work better with blue eyes than with brown. The section on biometric authentication in this book provides an overview of the history of its use with law enforcement and the courts and outlines some of the challenges faced by statisticians developing methods in this area. The two papers both address reducing error rates, specifically for authentication, although there are a myriad of other applications.

Syndromic surveillance has long been an issue of interest for biostatisticians, epidemiologists, and public health experts. After 9/11, however, more government funding became available to study issues related to sudden outbreaks of infectious diseases that might be the result of bioterrorism. Traditionally, research in this area would have looked at things like normal seasonal influenza cases, perhaps with an eye to preparing for possible flu pandemics caused by more virulent strains. But in the case of a bioterrorist incident, the concerns are a little different. For example, you want to be able to detect an outbreak of smallpox or cluster of anthrax infections as soon as possible so you can begin to respond. This may involve collecting and monitoring new data sources in near real-time: hospital admissions of patients with unusual symptoms, spikes in over-the-counter sales of cold medicines, etc. Collecting, integrating, and analyzing such new types of data involves the creation of new infrastructure and new methodologies. The section in this book on syndromic surveillance provides an overview of challenges and research issues in this growing area and includes articles on monitoring multiple data streams, evaluating statistical surveillance methods, and the spatiotemporal syndromic analysis.

Modeling is the bread and butter for many working statisticians and naturally is being applied to address issues in counterterrorism. Many of the speculative questions researchers and decision-makers have about terrorism can be more practically and efficiently tested in computer models as opposed to actual physical experiments. As the section overview points out, "we cannot expose a population to a disease or chemical attack and see what happens." This overview highlights the main issues addressed in the section and suggests future research directions. The section includes articles on developing large disease simulations, analyzing distributed databases, modeling of the concentration field in a building following release of a contaminant, and modeling the sensitivity of radiation detectors that might be deployed to screen cargo.

We would like to thank David Banks for suggesting this monograph, Sallie Keller-McNulty and Nancy Spruill for their ongoing support, and Hazel Kutac for her tireless editorial and production work.

Reference

[KH04] Kean, T. H., and L. Hamilton. 2004. *The 9/11 commission report.* Washington, DC: National Commission on Terrorist Attacks upon the United States.

Los Alamos, NM — *Alyson Wilson*
Los Alamos, NM — *Gregory Wilson*
Monterey, CA — *David Olwell*

January 2006

Contents

Part I

Game Theory

Game Theory in an Age of Terrorism: How Can Statisticians Contribute?

Ronald D. Fricker, Jr.

Department of Operations Research, Naval Postgraduate School,
rdfricker@nps.edu

In *The Law of Loopholes in Action* [Gel05], David Gelernter argues that "every loophole will eventually be exploited; every loophole will eventually be closed." His thesis applied to terrorism means that terrorists will find security loopholes via continual exploration and that, once discovered, specific defensive measures have to be put in place to close each loophole.

The net effect of the Law of Loopholes, as anyone who flies regularly today knows, is an ever-expanding set of security rules and requirements. Such rules and requirements are useful for helping prevent the reoccurrence of a particular type of incident. But, when a determined adversary's focus is on causing general destruction and mayhem, then as one loophole is plugged, the adversary simply shifts its attention and energies to looking for and trying to exploit a different loophole.

The problem, of course, is that it is impossible to defend all potential targets (and their associated loopholes) against all threats all of the time. While it *is* important to implement certain new and improved defensive tactics, precisely because it is impossible to protect everything at all times, it is equally as important (and arguably more important) to implement offensive strategies to deter and disrupt these adversaries.

The question is, how to identify effective offensive and defensive strategies and tactics?

One approach is through the use of *game theory*, the mathematically based study and analysis of adversarial conflicts. The classic text *The Compleat Strategyst* [Wil66] characterizes *games of strategy* as having the following characteristics:

- A conflict: the participants (e.g., individuals, organizations, countries; known as "players" in game theory parlance) are at cross-purposes or have opposing interests.
- Adversarial reaction and interaction: each player has some control over the course of the conflict or its outcome via one or more decisions.

- Outside forces: some aspects of the conflict are outside of the players' control and may be governed by chance or are unknown.

These characteristics clearly apply to the problem of thwarting terrorists and defeating terrorism.

The first extensive treatment of game theory was *Theory of Games and Economic Behavior* by John von Neumann and Oskar Morgenstern [VM44] in 1944. The seminal work on the subject, "Zur Theorie der Gesellschaftsspiele" by von Neumann [von28], was written in 1928. John von Neumann characterized the difference between games such as chess and games of strategy by saying "Chess is not a game. Chess is a well-defined form of computation. You may not be able to work out the answers, but in theory there must be a solution, a right procedure in any position. Now real games are not like that at all. Real life is not like that. Real life consists of bluffing, of little tactics of deception, of asking yourself what is the other man going to think I meant to do. And that is what games are about in my theory" [Pou92].

Game theoretic methods provide a structured way to examine how two adversaries will interact under various conflict scenarios. The results often provide insight into why real-world adversaries behave the way they do. In the middle and late 20th century, a great deal of game theoretic research focused on analyzing the arms race, nuclear brinkmanship, and Cold War strategies [Pou92]. While in the pre-9/11 era, game theory was also applied to terrorism, post-9/11 this work has expanded [SA03].

1 Game Theory Applied to Terrorism

In what is surely a gross oversimplification of the field (apologies to game theorists in advance), there are three broad categories of game theoretic methods applicable to the analysis of terrorism:

1. *Classic games* can generally be illustrated in a tabular form in which the players, their strategies, and their "payoffs" are completely specified. These types of games are often studied to determine whether there are a pair of strategies that result in an equilibrium between the two players (a "saddle point") and how the players will behave given the existence or absence of a saddle point.
2. *Repetitive (or repeated) games*, which are games that occur over time and the opponents repeatedly interact in a series of conflicts. These games are studied to gain insight into how players behave and react to their opponent's behavior and which behavioral strategies result in favorable or unfavorable final outcomes.
3. *Tabletop games* consisting of the simulation of an adversarial interaction with two or more actual (human) players using rules, data, and procedures designed to depict a conflict. "Tabletop" refers to the manner of older war games in which a battle was played out using miniature markers and

maps on a table, much like the board game Risk. These types of games are generally less structured than the previous types, meaning the players have a much larger set of strategies available than can be easily tabularized.

Recent applications of game theoretic methods to the study of terrorism include: assessing strategies for how nations allocate expenditures for terrorism deterrence and the resulting implications for being attacked [AST87, SL68]; measures evaluating how various military employment policies/strategies encourage or discourage states from sponsoring terrorism [Art04]; assessing insurance risks via models that explicitly account for malicious terrorist intent [Maj02]; determining whether or not a stated policy of nonnegotiation with terrorist hostage-takers deters such behavior and under what conditions [LS88]; and evaluating the effects of focusing national antiterrorism policy on deterrence or prevention [SA03].

2 Statistics and Game Theory

In the parlance of game theory, much of classical statistics is a "one-person game" because there is no adversary. Classic statistical problems, particularly inferential problems, concern the estimation of an unobserved parameter or parameters. In these problems, the "adversary" is nature, manifested as randomness in some form or another, not as a willful opponent.

A frequent assumption in statistical methods, analyses, and models is that the parameter or population under study is fixed and the most important uncertainty to quantify is that which comes from sampling variability. Even in those problems where the parameter may change over time, the usual assumption is that the underlying mechanism that generates an outcome is unaffected by that outcome. (For example, in a regression model we assume the dependent variable does not or cannot affect the independent variable.) Neither of these assumptions is likely to be true in a game theory problem, where the population of interest is an intelligent adversary capable of changing its form, tactics, and responses.

The upshot is that most statisticians are not used to thinking about problems such as those addressed by game theory. However, statisticians *are* used to addressing problems in which uncertainty is either a natural component or must be quantified, and there is a lot of uncertainty in game theoretic models about deterring, detecting, and thwarting terrorists.

3 How Can Statisticians Contribute?

Game theoretic models tend to be fairly abstract models of reality. This has not prevented the models from providing useful insights into strategies for addressing certain types of conflicts, but it does lead to two specific questions:

1. How well do the models fit observed data?
2. How can model uncertainty be quantified?

Both are questions that statisticians are well-suited to help address.

Possible ways statisticians could contribute to the further development of game theoretic methods, both in general and for terrorist problems in particular, include the following.

- Game theory models, including the strategies and their payoffs, are often defined in an ad hoc manner using expert judgment. A relevant statistical question is, how might data from past incidents and other knowledge be used to *infer* either the terrorist's "game" or the strategies they perceive or prefer? That is, how might a game be "fit" to observed data?
- The payoffs in game theory are utilities representing the desirability of the various outcomes to the players. In the absence of information, the utilities are often simply rankings of the various outcomes. A better methodology would be to elicit utilities from policymakers or subject-matter experts, much like one might elicit prior probabilities for a Bayesian analysis. Relevant questions include, what is (are) the best way(s) to elicit the utilities and how should utilities from multiple experts be combined?
- Once the payoffs are specified, the analysis of a game often treats them as fixed and known. How might the games be created, analyzed, and evaluated so that the uncertainty in payoffs is accounted for in the results, including the specification of the optimal strategy?
- Tabletop games are often useful for developing new insights and/or out-of-the-box potential strategies, but they also often can only explore a small portion of the "game space." Relevant questions include how to characterize and account for the uncertainty in game design (e.g., a terrorist opponent's capabilities) and how statistical methods might be used to help design a series of games to best explore the "capabilities/strategy space."
- Finally, for new types of games that incorporate uncertainty, as well as for a set or series of more traditional games, how can graphical methods be employed to best display important game results, including appropriate depictions of uncertainty and variability?

The two chapters that follow this one discuss and examine how risk analysis can be combined with game theory. In "Combining Game Theory and Risk Analysis in Counterterrorism: A Smallpox Example," Banks and Anderson describe how to use risk analysis to generate random payoff matrices, which are then used to estimate the probability that a given strategy is optimal. In "Game-Theoretic and Reliability Methods in Counterterrorism and Security," Bier discusses the literature on reliability and risk analytic methods for rare events, game theory, and approaches for combining the two methods for defending complex systems against terrorist attack.

These two efforts represent a promising start towards addressing some of the problems described above. Yet more remains to be done.

References

[Art04] Arthur, K. 2004. "Understanding the military's role in ending state-sponsored terrorism." Master's thesis, Naval Postgraduate School.

[AST87] Atkinson, S. E., T. Sandler, and J. T. Tschirhart. 1987. "Terrorism in a bargaining framework." *Journal of Law and Economics* 30:1–21.

[Gel05] Gelernter, D. 2005. "The law of loopholes in action." *Los Angeles Times*, B13, May 6.

[LS88] Lapan, H. E., and T. Sandler. 1988. "To bargain or not to bargain: That is the question." *American Economic Review* 78:16–20.

[Maj02] Major, J. A. 2002. Advanced techniques for modeling terrorism risk. National Bureau of Economic Research Insurance Group Conference, February 1, 2002. http://www.guycarp.com/portal/extranet/pdf/major_terrorism.pdf, downloaded on June 30, 2005.

[Pou92] Poundstone, W. 1992. *Prisoner's dilemma.* New York: Doubleday.

[SA03] Sandler, T., and D. G. Arce M. 2003. "Terrorism and game theory." *Simulation and Gaming* 34:319–337.

[SL68] Sandler, T., and H. E. Lapan. 1968. "The calculus of dissent: An analysis of terrorists' choice of targets." *Synthese* 76:245–261.

[von28] von Neumann, J. 1928. "Zur Theorie der Gesellschaftsspiele." *Mathematische Annalen* 100:295–300.

[VM44] von Neumann, J., and O. Morgenstern. 1944. *Theory of games and economic behavior.* Princeton, NJ: Princeton University Press.

[Wil66] Williams, J. D. 1966. *The compleat strategyst*, rev. ed. New York: McGraw-Hill Book Company.

Combining Game Theory and Risk Analysis in Counterterrorism: A Smallpox Example

David L. Banks[1] and Steven Anderson[2]

[1] Institute of Statistics and Decision Sciences, Duke University, banks@stat.duke.edu
[2] Center for Biologics Evaluation and Research, U. S. Food and Drug Administration, AndersonSt@cber.fda.gov

1 Introduction

The U.S. government wishes to invest its resources as wisely as possible in defense. Each wasted dollar diverts money that could be used to harden crucial vulnerabilities, prevents investment in future economic growth, and increases taxpayer burden. This is a classic conflict situation; a good strategy for the player with fewer resources is to leverage disproportionate resource investment by its wealthy opponent. That strategy rarely wins, but it makes the conflict sufficiently debilitating that the wealthy opponent may be forced to consider significant compromises.

Game theory is a traditional method for choosing resource investments in conflict situations. The standard approach requires strong assumptions about the availability of mutual information and the rationality of both opponents. Empirical research by many people [KT72] shows that these assumptions fail in practice, leading to the development of modified theories with weaker assumptions or the use of prior probabilities in the spirit of Bayesian decision theory. This paper considers both traditional game theory (minimax solution for a two-person, zero-sum game in normal form) and also a minimum expected loss criterion appropriate for extensive-form games with prior probabilities. However, we emphasize that for terrorism, the zero-sum model is at best an approximation; the valuation of the wins and the losses is likely to differ between the opponents.

Game theory requires numerical measures of payoffs (or losses) that correspond to particular sets of decisions. In practice, those payoffs are rarely known. Statistical risk analysis allows experts to determine reasonable probability distributions for the random payoffs. This paper shows how risk analysis can support game theory solutions and how Monte Carlo methods provide insight into the optimal game theory solutions in the presence of uncertainty about payoffs.

Our methodology is demonstrated in the context of risk management for a potential terrorist attack using the smallpox virus. The analysis we present here is a simplified version that aims at methodological explanation rather than analysis or justification of specific healthcare policies. As a tabletop exercise, the primary aim is only to provide a blueprint for a more rigorous statistical risk analysis. The underlying assumptions, modeling methods used here, and any results or discussion of the modeling are based on preliminary and unvalidated data and do not represent the opinion of the Food and Drug Administration (FDA), the Department of Health and Human Services, or any branch of the U.S. government.

2 Game Theory for Smallpox

The smallpox debate in the United States has focused upon three kinds of attack and four kinds of defense. The three attack scenarios suppose that there might be:

- No smallpox attack,
- A lone terrorist attack on a small area (similar to the likely scenario for the anthrax letters), or
- A coordinated terrorist attack upon multiple population centers.

The four defense scenarios that have been publicly considered by U.S. agency officials are:

- Stockpile smallpox vaccine,
- Stockpile vaccine and develop biosurveillance capabilities,
- Stockpile vaccine, develop biosurveillance, and inoculate key personnel, and
- Provide mass vaccination to nonimmunocompromised citizens in advance.

Although there are many refinements that can be considered for both the attack and the defense scenarios, these represent the possibilities discussed in the public meetings held in May and June 2002 [McK02].

Suppose that analysts used game theory as one tool to evaluate potential defense strategies. Then the three kinds of attack and four kinds of defense determine a classic normal-form payoff matrix for the game (Table 1).

Table 1. Attack—defense cost matrix

	No Attack	Single Attack	Multiple Attack
Stockpile Vaccine	C_{11}	C_{12}	C_{13}
Biosurveillance	C_{21}	C_{22}	C_{23}
Key Personnel	C_{31}	C_{32}	C_{33}
Everyone	C_{41}	C_{42}	C_{43}

The C_{ij} entries are the costs (or payoffs) associated with each combination of attack and defense, and we have used abbreviated row and column labels to identify the defenses and attacks, respectively, as described before.

For each of the 12 attack–defense combinations, there is an associated cost. These costs may include dollars, human lives, time, and other resources. For our calculation, all of these costs are monetized, according to principles detailed in Sect. 3. The monetized value of a human life is set to $2.86 million, following the Department of Transportation's figures for cost–benefit analyses of safety equipment.

Note that there is very large uncertainty in the C_{ij} values. Portions of the cost (e.g., those associated with expenses already entailed) may be known, but the total cost in each cell is a random variable. These random variables are not independent, since components of the total cost are common to multiple cells. Thus it is appropriate to regard the entire game theory table as a multivariate random variable whose joint distribution is required for a satisfactory analysis that propagates uncertainty in the costs through to uncertainty about best play.

Classical game theory [Mye91, Chap. 3] determines the optimal strategies for the antagonists via the minimax theorem. This theorem asserts that for any two-person cost matrix in a strictly competitive game (which is the situation for our example), there is an equilibrium strategy such that neither player can improve their expected payoff by adopting a different attack or defense. This equilibrium strategy may be a pure strategy, in which case optimal play is a specific attack–defense pair. This happens when the attack that maximizes the minimum damage and the defense that minimizes the maximum damage coincide in the same cell. Otherwise, the solution is a mixed strategy, in which case the antagonists pick attacks and defenses according to a probability distribution that must be calculated from the cost matrix. There may be multiple equilibria that achieve the same expected payoff, and for large matrices it can be difficult to solve the game.

Alternatively, one can use Bayesian decision theory to solve the game. Here a player puts a probability distribution over the actions of the opponent, and then chooses their own action so as to minimize the expected cost [Mye91, Chap. 2]. Essentially, one just multiplies the cost in each row by the corresponding probability, sums these by row, and picks the defense with the smallest sum. This formulation is easier to solve, but it requires one to know or approximate the opponent's probability distribution, and it does not take full account of the mutual strategic aspects of adversarial games (i.e., the assigned probabilities need not correspond to any kind of "if I do this, then he'll do that" reasoning). Bayesian methods are often used in extensive-form games, where players make their choices over time, conditional on the actions of their opponent.

In developing our analysis of the smallpox example we make two assumptions about time. First, we use only the information available by June 1, 2002; subsequent information on the emerging program costs is not included. This keeps the analysis faithful in spirit to the decision problem actually faced by U.S. government policymakers in the spring of 2002 (their initial plan was universal vaccination, but ultimately they chose the third scenario with stockpiling, biosurveillance, and very limited vaccination of some first responders). Second, all of the estimated cost forecasts run to October 1, 2007. The likelihood of changing geopolitical circumstances makes it unrealistic to attempt cost estimates beyond that fiscal year.

3 Risk Analysis for Smallpox

Statistical risk analysis is used to estimate the probability of undesirable situations and their associated costs. In the same way that it is used in engineering (e.g., for assessing nuclear reactor safety [Spe85]) or the insurance industry (e.g., for estimating the financial costs associated with earthquakes in a specific area [Bri93]), this paper uses risk analysis to estimate the costs associated with different kinds of smallpox attack/defense combinations.

Risk analysis involves careful discussions with domain experts and structured elicitation of their judgments about probabilities and costs. For smallpox planning, this requires input from physicians, public health experts, mathematical epidemiologists, economists, emergency response administrators, government accountants, and other kinds of experts. We have not conducted the in-depth elicitation from multiple experts in each area that is needed for a fully rigorous risk analysis; however, we have discussed the cost issues with representatives from each area, and we believe that the estimates in this section are sufficiently reasonable to illustrate, qualitatively, the case for combining statistical risk analysis with game theory for threat management in the context of terrorism.

Expert opinion was typically elicited in the following way. Each expert was given a written document with background on smallpox epidemiology and a short description of the attacks and defenses considered in this paper. The expert often had questions; these were discussed orally with one of the authors and, to the extent possible, resolved on the basis of the best available information. Then the expert was asked to provide a point estimate of the relevant cost or outcome and the range in which that value would be expected to fall in 95% of similar realizations of the future. If these values disagreed with those from other experts, then the expert was told of the discrepancy and invited to alter their opinion. Based on point estimate and the range, the authors and the expert chose a distribution function with those parameters, which also respected real-world requirements for positivity, integer values, known skew, or other properties. As the last step in the interview, the expert was given access to all the other expert opinions obtained to that point and asked if there were any that seemed questionable; this led, in one case, to an expert being recontacted and a subsequent revision of the elicitation. But it should be emphasized that these interviews were intended to be short and did not use the full range of probes, challenges, and checks that are part of serious elicitation work.

The next three subsections describe the risk analysis assumptions used to develop the random costs for the first three cells (C_{11}, C_{21}, C_{31}) in the game theory payoff matrix. Details for developing the costs in the other cells are available from the authors. These assumptions are intended to be representative, realistic, and plausible, but additional input by experts could surely improve upon them. Many of the same costs arise in multiple cells, introducing statistical dependency among the entries. (That is, if a given random payoff matrix assumes an unusually large cost for stockpiling in one cell of the random table, then the same high value should appear in all other cells in which stockpiling occurs.)

3.1 Cell (1,1): Stockpile Vaccine/No Attack Scenario

Consider the problem of trying to estimate the costs associated with the (1,1) cell of the payoff matrix, which corresponds to no smallpox attack and the stockpiling of vaccine. This estimate involves combining costs with very different levels of uncertainty.

At the conceptual level, the cost C_{11} is the sum of four terms:

$$C_{11} = \mathrm{ET}_{\mathrm{dry}} + \mathrm{ET}_{\mathrm{Avent}} + \mathrm{ET}_{\mathrm{Acamb}} + \mathrm{VIG} + \mathrm{PHIS},$$

where $\mathrm{ET}_{\mathrm{dry}}$ and $\mathrm{ET}_{\mathrm{Avent}}$ are the costs of efficacy and safety testing for the Dryvax and Aventis vaccines, respectively; $\mathrm{ET}_{\mathrm{Acamb}}$ is the cost of new vaccine production and testing from Acambis; VIG is the cost of producing sufficient doses of vaccinia immune globulin to treat adverse reactions and possible exposures; and PHIS is the cost of establishing the public healthcare infrastructure needed to manage this stockpiling effort.

There is no uncertainty about $\mathrm{ET}_{\mathrm{Acamb}}$; the contract fixes this cost at \$512 million. But there is substantial uncertainty about $\mathrm{ET}_{\mathrm{dry}}$ and $\mathrm{ET}_{\mathrm{Avent}}$ since these entail clinical trials and may require follow-on studies; based on discussions with experts, we believe these costs may be realistically modeled as independent uniform random variables, each ranging between \$2 and \$5 million. There is also large uncertainty about the cost for producing and testing sufficient doses of VIG to be prepared for a smallpox attack; our discussions suggest this is qualitatively described by a normal random variable with mean \$100 million and a standard deviation of \$20 million. There is great uncertainty about PHIS (which includes production of bifurcated inoculation needles, training, storage costs, shipment readiness costs, etc.). Based on the five-year operating budget of other government offices with analogous missions, we assume this cost is normally distributed with mean \$940 million and standard deviation \$100 million.

3.2 Cell (2,1): Biosurveillance/No Attack Scenario

Biosurveillance programs are being piloted in several major metropolitan areas. These programs track data, on a daily basis, from emergency room admission records to quickly discover clusters of disease symptoms that suggest bioterrorist attack. Our cost estimates are based upon discussions with the scientists working in the Boston area [RKD02] and with the Pittsburgh team that developed monitoring procedures for the Salt Lake City Olympic games.

The cost C_{21} includes the cost C_{11} since this defense strategy uses both stockpiling of vaccine and increased biosurveillance. Thus

$$C_{21} = C_{11} + \text{PHIB} + \text{PHM} + \text{NFA} \times \text{FA},$$

where PHIB is the cost of the public health infrastructure needed for biosurveillance, including the data input requirements and software; PHM is the cost of a public health monitoring center, presumably at the Centers for Disease Control and Prevention, that reviews the biosurveillance information on a daily basis; NFA is the number of false alarms from the biosurveillance system over five years of operation; and FA is the cost of a false alarm.

For this exercise, we assume that PHIB is normally distributed with mean \$900 million and standard deviation \$100 million (for a five-year funding horizon); this is exclusive of the storage, training, and other infrastructure costs in PHIS, and it includes the cost of hospital nursing-staff time to enter daily reports on emergency room patients with a range of disease symptoms (not just those related to smallpox). PHM is modeled as a normal random variable with mean \$20 million and standard deviation \$4 million (this standard deviation was proposed by a federal administrator and may understate the real uncertainty). False alarms are a major problem for monitoring systems; it is difficult to distinguish natural contagious processes from terrorist attacks. We expect about one false alarm per month over five years in a national system of adequate sensitivity, and thus FA is taken to be a Poisson random variable with mean 60. The cost for a single false alarm is modeled as a normal random variable with mean \$500,000 and standard deviation \$100,000.

3.3 Cell (3,1): Key Personnel/No Attack Scenario

One option, among several possible policies that have been discussed, is for the United States to inoculate about 500,000 key personnel, most of whom would be first-responders in major cities (i.e., emergency room staff, police, and public health investigators who would be used to trace people who have come in contact with carriers). If chosen, this number is sufficiently large that severe adverse reactions become a statistical certainty.

The cost of this scenario subsumes the cost C_{21} of the previous scenario, and thus

$$C_{31} = C_{21} + \frac{\text{NKP} \times \text{IM}}{25,000} + \text{PAE} \times \text{NKP} \times \text{AEC},$$

where NKP is the number of key personnel; IM is the cost of the time and resources needed to inoculate 25,000 key personnel and monitor them for adverse events; PAE is the probability of an adverse event; and AEC is the average cost of one adverse event.

We assume that NKP is uniformly distributed between 400,000 and 600,000 (this reflects uncertainty about how many personnel would be designated as "key"). The IM is tied to units of 25,000 people, since this is a one-time cost and represents the number of people that a single nurse might reasonably inoculate and maintain records upon in a year. Using salary tables, we approximate this cost as a normal random variable with mean $60,000 and standard deviation $10,000.

The probability of an adverse event is taken from Anderson [And02], which is based upon Lane et al. [LRN70]; the point estimate for all adverse events is 0.293, but since there is considerable variation and new vaccines are coming into production, we have been conservative about our uncertainty and assumed that the probability of an adverse event is uniformly distributed between 0.15 and 0.45. Of course, most of these events will be quite minor (such as local soreness) and would not entail any real economic costs.

The AEC is extremely difficult to estimate. For purposes of calculation, we have taken the value of a human life to be $2.86 million (the amount used by the National Highway Transportation Safety Administration in cost–benefit analyses of safety equipment). But most of the events involve no cost, or perhaps a missed day of work that has little measurable impact on productivity. After several calculations and consultations, this analysis assumes that AEC can be approximated as a gamma random variable with mean $40 and standard deviation $100 (this distribution has a long right tail).

4 Analysis

The statistical risk analysis used in Sect. 3, albeit crude, shows how expert judgment can generate the random payoff matrices. The values in the cells of such tables are not independent, since many of the cost components are shared between cells. In fact, it is appropriate to view the table as a matrix-valued random variable with a complex joint distribution.

Random tables from this joint distribution can be generated by simulation. For each table, one can apply either the minimax criterion to determine an optimal strategy in the sense of von Neumann and Morgenstern [VM44], or a minimum expected loss criterion to determine an optimal solution in the sense of Bayesian decision theory [Mye91, Chap. 2]. By doing this repeatedly, for many different random tables, one can estimate the proportion of time that each defense strategy is superior.

Additionally, it seems appropriate to track not just the number of times a defense strategy is optimal, but also weight this count by some measure of the difference between the costs of the game under competing defenses. For example, if two defenses yield game payoffs that differ only by an insignificant amount, it seems unrealistic to give no credit to the second-best strategy. For this reason we also use a scoring algorithm in which the score a strategy receives depends upon how well separated it is from the optimal strategy. Specifically, suppose that defense strategy i has value V_i on a given table. Then the score S_i that strategy i receives is

$$S_i = 1 - \frac{V_i}{\max V_j},$$

and this ensures that strategies are weighted to reflect the magnitude of the monetized savings that accrue from using them. The final rating of the strategies is obtained by averaging their scores from many random tables.

4.1 Minimax Criterion

We performed the simulation experiment described above 100 times and compared the four defense strategies in terms of the minimax criterion. Although one could certainly do more runs, we believe that the approximations in the cost modeling are so uncertain that additional simulation would only generate spurious accuracy.

Among the 100 runs, we found that the Stockpile strategy won 9 times, the Biosurveillance strategy won 24 times, the Key Personnel strategy won 26 times, and the Vaccinate Everyone strategy won 41 times. This lack of a clear winner may be, at some intuitive level, the cause of the widely different views that have been expressed in the public debate on preparing for a smallpox attack.

If one uses scores, the results are even more ambiguous. The average score for the four defense strategies ranged between 0.191 and 0.326, indicating that the expected performances were, on average, quite similar.

From public policy standpoint, this may be a fortunate result. It indicates that in terms of the minimax criterion, any decision is about equally defensible. This gives managers flexibility to incorporate their own judgment and to respond to extra scientific considerations.

4.2 Minimum Expected Loss Criterion

The minimax criterion may not be realistic for the game theory situation presented by the threat of smallpox. In particular, the normal-form game assumes that both players are ignorant of the decision made by their opponent until committed to a course of action. For the smallpox threat, there has been a vigorous public discussion on what preparations the United States should make. Terrorists know what the United States has decided to do, and presumably this will affect their choice of attack. Therefore the extensive-form version of game theory seems preferable. This form can be thought of as a decision tree, in which players alternate their moves. At each stage, the player can use probabilistic assessments about the likely future play of the opponent.

The minimum expected loss criterion requires more information than does the minimax criterion. The analyst needs to know the probabilities of a successful smallpox attack conditional on the United States selecting each of the four possible defenses. This is difficult to determine, but we illustrate how one can do a small sensitivity analysis that explores a range of probabilities for smallpox attack.

Table 2 shows a set of probabilities that we treat as the baseline case. We believe it accords with a prudently cautious estimate of the threat of a smallpox attack.

Table 2. Baseline probabilities of attack for different defenses

	No Attack	Single Attack	Multiple Attack
Stockpile Vaccine	0.95	0.040	0.010
Biosurveillance	0.96	0.035	0.005
Key Personnel	0.96	0.039	0.001
Everyone	0.99	0.005	0.005

To interpret Table 2, it says that if the United States were to only stockpile vaccine, then the probability of no smallpox attack is 0.95, the probability of a single attack is 0.04, and the probability of multiple attacks is 0.01. Similarly, one reads the attack probabilities for other defenses across the row. All rows must sum to one.

The minimum expected loss criterion multiplies the probabilities in each row of Table 2 by the corresponding costs in the same row of Table 1, and then sums across the columns. The criterion selects the defense that has the smallest sum.

As with the minimax criterion, one can simulate many payoff tables and then apply the minimum expected loss criterion to each. In 100 repetitions, Stockpile won 96 times, Biosurveillance won 2 times, and Vaccinate Everyone won twice. The scores showed roughly the same pattern, strongly favoring the Stockpile defense.

We now consider two alternative sets of probabilities shown in Tables 3 and 4. Table 3 is more pessimistic and has larger attack probabilities. Table 4 is more optimistic and has smaller attack probabilities. A serious sensitivity analysis would investigate many more tables, but our purpose is illustration and we doubt that the quality of the assessments that underlie the cost matrix can warrant further detail.

For Table 3, 100 simulation runs found that Stockpile won 15 times, Biosurveillance won 29 times, Key Personnel won 40 times, and Vaccinate Everyone won 16 times. In contrast, for Table 4, the Stockpile strategy won 100 times in 100 runs. The scores for Table 3 ranged from 18.2 to 38.8, which are

Table 3. Pessimistic probabilities of attack for different defenses

	No Attack	Single Attack	Multiple Attack
Stockpile Vaccine	0.70	0.20	0.10
Biosurveillance	0.80	0.15	0.05
Key Personnel	0.85	0.10	0.05
Everyone	0.90	0.05	0.05

Table 4. Optimistic probabilities of attack for different defenses

	No Attack	Single Attack	Multiple Attack
Stockpile Vaccine	0.980	0.0100	0.0100
Biosurveillance	0.990	0.0050	0.0050
Key Personnel	0.990	0.0050	0.0050
Everyone	0.999	0.0005	0.0005

quite similar. In contrast, for Table 4 nearly all the weight of the score was on the Stockpile defense.

These results show that the optimal strategy is sensitive to the choice of probabilities used in the analysis. Determining those probabilities requires input from the intelligence community and the judgment of senior policymakers.

5 Conclusions

This paper has outlined an approach combining statistical risk analysis with game theory to evaluate defense strategies that have been considered for the threat of smallpox. We believe that this approach may offer a useful way of structuring generic problems in resource investment for counterterrorism.

The analysis in this paper is incomplete.

1. We have focused upon smallpox, because the problem has been framed rather narrowly and quite definitively by public discussion. But a proper game theory analysis would not artificially restrict the options of the terrorists, and should consider other attacks, such as truck bombs, chemical weapons, other diseases, and so forth (which would get difficult, but there may be ways to approximate). It can be completely misleading to seek a local solution, as we have done.
2. Similarly, we have not fully treated the options of the defenders. For example, heavy investment in intelligence sources is a strategy that protects against many different kinds of attacks and might well be the superior solution in a less local formulation of the problem.
3. We have not considered constraints on the resources of the terrorists. The terrorists have limited resources and can invest in a portfolio of different kinds of attacks. Symmetrically, the United States can invest in a portfolio

of defenses. This aspect of the problem is not addressed — we assume that both parties can fund any of the choices without sacrificing other goals.
4. The risk analysis presented here, as discussed previously, is not adequate to support public policy formulation.

Nonetheless, despite these limitations, the methodology has attractive features. First, it is easy to improve the quality of the result through better risk analysis. Second, it automatically raises issues that have regularly emerged in policy discussions. Third, it captures facets of the problem that are not amenable to either game theory or risk analysis on their own, because classical risk analysis is not used in adversarial situations and because classical game theory does not use random costs.

Appendix: Background on Smallpox

Although the probability that the smallpox virus (*Variola major*) might be used against the United States is thought to be small, the public health and economic impact of even a limited release would be tremendous. Any serious attack would probably force mass vaccination programs, causing additional loss of life due to adverse reactions. Other economic consequences could easily be comparable to those of the attacks of September 11, 2001.

A smallpox attack could potentially be initiated through infected humans or through an aerosol [HIB99]. In 12 to 14 days after natural exposure patients experience fever, malaise, body aches, and a body rash [FHA88]. During the symptomatic stages of the disease the patient can have vesicles in the mouth, throat, and nose that rupture to spread the virus during a cough or sneeze. Person-to-person spread usually occurs through inhalation of virus-containing droplets or from close contact with an infected person. As the disease progresses, the rash spreads to the head and extremities and evolves into painful, scarring vesicles and pustules. Smallpox has a mortality rate of approximately 30%, based on data from the 1960s and 1970s [Hen99].

Various mathematical models of smallpox spread exist and have been used to forecast the number of people infected under different exposure conditions and different public health responses [KCW02, MDL01]. There is considerable variation in the predictions from these models, partly because of differing assumptions about the success of the "ring vaccination" strategy that has been planned by the Centers for Disease Control and Prevention (CDC) [CDC02], and this is reflected in the public debate on the value of preemptive inoculation versus wait-and-see preparation. However, the models are in essential agreement that a major determinant of the size of the epidemic is the number of people who are exposed in the first attack or attacks.

The current vaccine consists of live vaccinia or cowpox virus and is effective at preventing the disease. Also, vaccination can be performed within the first 2 to 4 days postexposure to reduce the severity or prevent the occurrence of the disease [Hen99].

Vaccination is not without risk; the major complications are serious infections and skin disease such as progressive vaccinia, eczema vaccinatum, generalized vaccinia, and encephalitis. Approximately 12 people per million have severe adverse reactions that require extensive hospitalization, and about one-third of these die — vaccinia immune globulin (VIG) is the recommended therapy for all of these reactions except encephalitis. Using data from Lane et al. [LRN70], we estimate that 1 in 71,429 people suffer postvaccinial encephalitis, 1 in 588,235 suffer progressive vaccinia, 1 in 22,727 suffer eczema vaccinatum, and 1 in 3,623 suffer generalized vaccinia. Additionally, 1 in 1,656 people suffer accidental infection (usually to the eye) and 1 in 3,289 suffer some other kind of mild adverse event, typically requiring a person to miss a few days of work. Other studies give somewhat different numbers [NLP67a, NLL67b]. People who have previously been successfully vaccinated for smallpox are less likely to have adverse reactions, and people who are immunocompromised (e.g., transplant patients, those with AIDS) are at greater risk for adverse reactions [CDC02, Guide B, parts 3, 5, and 6].

Because the risk of smallpox waned in the 1960s, vaccination of the U.S. population was discontinued in 1972. It is believed that the effectiveness of a smallpox vaccination diminishes after about 7 years, but residual resistance persists even decades later. It has been suggested that people who were vaccinated before 1972 may be substantially protected against death, if not strongly protected against contracting the disease [Coh01].

The United States currently has about 15 million doses of the Wyeth Dryvax smallpox vaccine available. The vaccine was made by scarification of calves with the New York City Board of Health strain and fluid containing the vaccinia virus was harvested by scraping [RMK01]. Recent clinical trials on the efficacy of diluted vaccine indicate that both the five- and ten-fold dilutions of Dryvax achieve a take rate (i.e., a blister forms at the inoculation site, which is believed to be a reliable indicator of immunization) of at least 95%, so the available vaccine could be administered to as many as 150 million people should the need arise [FCT02, NIA02].

The disclosure by the pharmaceutical company Aventis [Ens02] of the existence in storage of 80 to 90 million doses of smallpox vaccine that were produced more than 30 years ago has added to the current stockpile. Testing is being done on the efficacy of the Aventis vaccine stock, including whether it, too, could be diluted if needed.

Contracts to make new batches of smallpox vaccine using cell culture techniques have been awarded to Acambis. The CDC amended a previous contract with Acambis in September 2001 to ensure production of 54 million doses by late 2002. Another contract for the production of an additional 155 million doses was awarded to Acambis in late November 2001, and the total cost of these contracts is $512 million. After production, additional time may be needed to further test the safety and efficacy of the new vaccine [RMK01].

References

[And02] Anderson, S. 2002. "A risk-benefit assessment of smallpox and smallpox vaccination." Technical Report, Office of Biostatistics and Epidemiology, Center for Biologics Evaluation and Research, U.S. Food and Drug Administration, Rockville, MD.

[Bri93] Brillinger, D. R. 1993. "Earthquake risk and insurance." *EnviroMetrics* 4:1–21.

[CDC02] Centers for Disease Control and Prevention. 2002. Smallpox response plan and guidelines version 3.0. http://www.bt.cdc.gov/agent/smallpox/response-plan/index.asp.

[Coh01] Cohen, J. 2001. "Smallpox vaccinations: How much protection remains?" *Science* 294:985.

[Ens02] Enserink, M. 2002. "New cache eases shortage worries." *Science* 296:25–26.

[FHA88] Fenner, F., D. Henderson, I. Arita, Z. Jezek, and I. Ladnyi. 1988. *Smallpox and its eradication.* Geneva: World Health Organization.

[FCT02] Frey, S. E., R. B. Couch, C. O. Tacket, J. J. Treanor, M. Wolff, F. K. Newman, R. L. Atmar, R. Edelman, C. M. Nolan, and R. B. Belshe. 2002. "Clinical responses to undiluted and diluted smallpox vaccine." *New England Journal of Medicine* 346 (17): 1265–1274.

[Hen99] Henderson, D. A. 1999. "Smallpox: Clinical and epidemiological features." *Emerging Infectious Diseases* 5:537–539.

[HIB99] Henderson, D. A., T. V. Inglesby, J. G. Bartlett, M. S. Ascher, E. Eitzen, P. B. Jahrling, J. Hauer, M. Layton, J. McDade, M. T. Osterholm, T. O'Toole, G. Parker, T. Perl, P. K. Russell, and K. Tonat. 1999. "Smallpox as a biological weapon — Medical and public health management." *Journal of the American Medical Association* 281 (22): 2127–2137.

[KT72] Kahneman, D., and A. Tversky. 1972. "Subjective probability: A judgment of representativeness." *Cognitive Psychology* 3:430–454.

[KCW02] Kaplan, E., D. Craft, and L. Wein. 2002. "Emergency response to a smallpox attack: the case for mass vaccination." *Proceedings of the National Academy of Sciences* 99:10935–10940.

[LRN70] Lane, J. M., F. L. Ruben, J. M. Neff, and J. D. Millar. 1970. "Complications of smallpox vaccination, 1968: Results of ten statewide surveys." *Journal of Infectious Diseases* 122:303–309.

[McK02] McKenna, M. A. 2002. "No mass smallpox vaccinations, panel recommends." *Atlanta Journal-Constitution*, June 21, p. 1.

[MDL01] Meltzer, M. I., I. Damon, J. W. LeDuc, and J. D. Millar. 2001. "Modeling potential responses to smallpox as a bioterrorist weapon." *Emerging Infectious Diseases* 7:959–969.

[Mye91] Myerson, R. B. 1991. *Game theory: Analysis of conflict.* Cambridge, MA: Harvard University Press.

[NIA02] National Institute of Allergy and Infectious Diseases. March 28, 2002. "NIAID study results support diluting smallpox vaccine stockpile to stretch supply." *NIAID News.* http://www.niaid.nih.gov/newsroom/releases/smallpox.htm.

[NLP67a] Neff, J. M., J. M. Lane, J. P. Pert, H. Moore, and J. Millar. 1967. "Complications of smallpox vaccination, I: National survey in the United States, 1963." *New England Journal of Medicine* 276:1–8.

[NLL67b] Neff, J. M., R. H. Levine, J. M. Lane, E. Ager, H. Moore, B. Rosenstein, and J. Millar. 1967. "Complications of smallpox vaccination, United States, 1963, II: Results obtained from four statewide surveys." *Pediatrics* 39:916–923.

[RMK01] Rosenthal, S. R., M. Merchilinsky, C. Kleppinger, and K. L. Goldenthal. 2001. "Developing new smallpox vaccines." *Emerging Infectious Diseases* 7:920–926.

[RKD02] Ross, L., K. Kleinman, I. Dashevsky, C. Adams, A. DeMaria, Jr., and R. Platt. 2002. "Use of automated ambulatory-care encounter records for detection of acute illness clusters, including potential bioterrorism events." *Emerging Infectious Diseases* 8:753–760.

[Spe85] Speed, T. P. 1985. "Probabilistic risk assessment in the nuclear industry: WASH-1400 and beyond." Edited by L. LeCam and R. Olshen, *Proceedings of the Berkeley Conference in Honor of Jerzy Neyman and Jack Kiefer*, Volume 2. Pacific Grove, CA: Wadsworth, 173–200.

[VM44] von Neumann, J., and O. Morgenstern. 1944. *Theory of games and economic behavior.* Princeton, NJ: Princeton University Press.

Game-Theoretic and Reliability Methods in Counterterrorism and Security*

Vicki Bier

Center for Human Performance and Risk Analysis, University of Wisconsin-Madison, bier@engr.wisc.edu

1 Introduction

After the September 11, 2001, terrorist attacks on the World Trade Center and the Pentagon, and the subsequent anthrax attacks in the United States, there has been an increased interest in methods for use in security and counter-terrorism. However, the development of such methods poses two challenges to the application of conventional statistical methods. One is the relative scarcity (fortunately) of empirical data on severe terrorist attacks. The second is the intentional nature of such attacks [BS98].

In dealing with extreme events (i.e., "events that are both rare and severe" [BHL99], such as disasters or failures of highly redundant engineered systems), for which empirical data are likely to be sparse, classical statistical methods have been of relatively little use. Instead, methods such as reliability analysis were developed, using decomposition [Rai68, Arm85] to break complex systems down into their individual components (such as pumps and valves) for which larger amounts of empirical failure-rate data may be available. Reliability analysis has become an important tool for analyzing and protecting against threats to the operability of complex engineered systems. Beginning with the Reactor Safety Study [NRC75], modern risk-analysis methods [Bie97, BC01] built on the techniques of reliability analysis, adding consequence-analysis models to allow for estimation of health and safety impacts as well as loss of functionality. Quantification of risk-analysis models generally relies on some combination of expert judgment [Coo91] and Bayesian rather than classical statistics to estimate the parameters of interest in the face of sparse data [BHL99, BFH04]. Zimmerman and Bier [ZB02] argue that "Risk assessment in its current form (as a systems-oriented method that is flexible enough to

* This article is a revision of an article originally published in *Modern Statistical and Mathematical Methods in Reliability* (2005), World Scientific Publishing Company [Bie05].

handle a variety of alternative conditions) is a vital tool for dealing with extreme events."

However, the routine application of reliability and risk analysis by itself is not adequate in the security domain. Protecting against intentional attacks is fundamentally different from protecting against accidents or acts of nature (which have been the more usual focus of engineering risk analysis). In particular, an intelligent and adaptable adversary may adopt a different offensive strategy to circumvent or disable our protective security measures. Game theory [Dre61, FT91] provides a way of taking this into account analytically.

Thus, security and counterterrorism can benefit from a combination of techniques that have not usually been used in tandem. This paper discusses approaches for applying risk and reliability analysis and game theory to the problem of defending complex systems against attacks by knowledgeable and adaptable adversaries.

2 Applications of Risk and Reliability Analysis to Security

Early applications of engineering risk analysis to counterterrorism and security include Martz and Johnson [MJ87] and Cox [Cox90]. More recently (following September 11), numerous risk analysts have proposed its use for homeland security [PG02, Gar03, Zeb03, ZHN04]. Because security threats can span such a wide range, the emphasis has been mainly on risk-based decision-making (i.e., using the results of risk analyses to target security investments at the most likely and most severe threats) rather than on detailed models of particular types of threats.

Much of this work [HML98, EFW00a, EFW00b, EHL01] has been directed specifically towards threats against critical infrastructure, beginning even before September 11, due to concerns raised by the President's Commission on Critical Infrastructure Protection [Pre97], among others. In particular, Haimes et al. [HML98] provide a useful taxonomy of methods for protecting infrastructure systems from terrorism, grouping countermeasures into four categories: (1) security — restricting access to key sites or facilities; (2) redundancy — providing alternate means for performing key functions; (3) robustness — making systems stronger or less sensitive to upset; and (4) resilience — ensuring that key systems and/or functions can be restored quickly. For more recent applications of risk analysis to critical infrastructure, see Haimes [Hai02a, Hai02b], Haimes and Longstaff [HL02], and Lemon and Apostolakis [LA04]. There has also been some effort to adapt earthquake risk modeling techniques to security problems [Ise04, Wer04].

In the reliability area, Levitin and colleagues have by now amassed a large body of work applying reliability analysis to problems of security [Lev02, Lev03a, Lev03b, LL00, LL01, LL03, LDX03]. Much of this work combines reliability analysis with optimization, to identify the most cost-effective risk reduction strategies. Examples include determining the optimal physical separation of components that are functionally in parallel with each other and the optimal allocation of physical protection to various hierarchies of a system (e.g., hardening the system as a whole, or its subsystems, or individual components).

Risk and reliability analyses (and the concomitant approach of risk-based decision-making) have a great deal to contribute to ensuring the safety (risk analysis) and functionality (reliability analysis) of complex engineered systems. In particular, they provide "a systematic approach to organizing and analyzing scientific knowledge and information for potentially hazardous activities or for substances that might pose risks under specified circumstances" [NRC94], by integrating information on a wide variety of possible threats within a single framework and quantifying the frequency and severity of those threats. Furthermore, basic reliability analysis results can be useful in security analysis even if they were not initially developed with that in mind. For example, results on least-cost diagnosis of coherent systems [Ben81, CQK89, CCS96] can be readily adapted to yield results on optimal attack strategies. Moreover, the recognition that risk may often be unknown to within an order of magnitude or more (almost certainly more than that, in the case of security threats) imposes some discipline on the analysis process, suggesting that the majority of the modeling effort should probably be devoted to analyzing the dominant contributors to risk. This means that complex models of phenomena that are unlikely to contribute much to the overall level of risk can often be replaced by much simpler approximations.

However, unlike in applications of risk analysis to problems such as the risk of nuclear power accidents, the relationship of recommended risk-reduction actions to the dominant risks emerging from the analysis is not straightforward. In most applications of risk analysis, risk reduction actions follow the usual "80/20 rule" (originally due to Pareto) — the decision-maker can review a list of possible actions, ranked based on the magnitude of risk reduction per unit cost, and choose the most cost-effective, typically getting something on the order of 80% of the benefit for perhaps 20% of the cost. This does not work so well in the security context (especially if the potential attacker can readily observe system defenses), since the effectiveness of investments in defending one component can depend critically on whether other components have also been hardened (or, conversely, if the attacker can easily identify alternative targets that have not yet been hardened).

Risk and reliability analyses are clearly important in identifying the most significant security threats, particularly in complex engineered systems (whose vulnerabilities may depend on networks of interdependencies that cannot be readily identified without detailed analysis). However, in the security context, the results of such analyses do not lead in a straightforward manner to an application of the Pareto principle. In particular, risk and reliability analyses generally assume that the threat or hazard is static, whereas in the case of security, the threat is adaptive and can change in response to the defenses that have been implemented. Therefore, simply rerunning an analysis with the same postulated threat but assuming that some candidate security improvements have been implemented will in general significantly overestimate the effectiveness of the candidate improvements — in many cases, by an extremely large margin. For example, in an analysis of asymmetric warfare against civilian targets, Ravid [Rav02] argues that since the adversary can change targets in response to defensive investments, "investment in defensive measures, unlike investment in safety measures, saves a lower number of lives (or other sort of damages) than the apparent direct contribution of those measures." Game theory provides one natural way of addressing this limitation in the applicability of risk and reliability analyses to security.

3 Applications of Game Theory to Security

Due to its value in understanding and modeling conflict, game theory has a long history of being applied to security-related problems, beginning with military applications [Dre61]; for specific examples see Haywood [Hay54], Berkovitz and Dresher [BD59, BD60], and Leibowitz and Lieberman [LL60]. It has also been extensively used in political science [Bra75, Bra85, BK88]; e.g., in the context of arms control. Recently, there have also been exploratory applications of game theory and related ideas to computer security, Anderson [And01], Burke [Bur99], Chaturvedi et al. [CGM00], Cohen [Coh99], Schneier [Sch00, Sch01].

With respect to applications of game theory and related methods to security in general, there is a large body of work already, much of it by economists [BR00, FL02, FL03, AS01, SA03, ES04, SE04, KZ03, LZ05]. Much of the work in this area until now has been designed to provide "policy insights" [SA03]; i.e., to inform policy-level decisions such as public versus private funding of defensive investments [LZ05], or the relative merits of deterrence and other protective measures [FL02, FL03, KZ03, SE04].

Of course, the events of September 11 have resulted in greater interest in this type of work. Perhaps more significantly with respect to the topic of this paper, there has also recently been interest in using game theory not only to explore the effects of different policies, but also to generate detailed guidance in support of operational-level decisions; e.g., determining which assets to protect or how much to charge for terrorism insurance.

For example, Enders and Sandler [ES04] study substitution effects in terrorism, observing that "installation of screening devices in U.S. airports in January 1973 made skyjackings more difficult, thus encouraging terrorists to substitute into other kinds of hostage missions." Similarly, they note that: "If the government were to secure its embassies or military bases, then attacks against such facilities would become more costly on a per-unit basis. If, moreover, the government were not at the same time to increase the security for embassy and military personnel when outside their facilities, then attacks directed at these individuals (e.g., assassinations) would become relatively cheaper" (and hence presumably more frequent).

Clearly, security improvements that appear to be cost-justified without taking into account the fact that attacks may be deflected to other targets may turn out to be wasteful (at least from a public perspective) if they merely deflect attacks to other targets of comparable value. For example, anthrax sterilization equipment installed in every post office in the country (which was considered, but fortunately never implemented), if publicly known, might never sterilize a single anthrax spore, since terrorists could deliver anthrax just as effectively by Federal Express, United Parcel Service, or even bicycle courier. In fact, such deflection of risk to other targets was not considered by the U.S. Postal Service in evaluating the cost-effectiveness of proposed security-improvement measures [CKS03].

In an application of what Shubik [Shu87] (see also Smith and von Winterfeldt [SV04]) calls "conversational" game theory, the Brookings Institution [OOD02] has recommended that "policymakers should focus primarily on those targets at which an attack would involve a large number of casualties, would entail significant economic costs, or would critically damage sites of high national significance." While game theory is not explicitly mentioned in the Brookings report, game-theoretic thinking clearly underlies this recommendation.

The Brookings recommendation constitutes a reasonable "zero-order" suggestion about how to prioritize targets for investment. Under this type of "weakest-link" model, defensive investment is allocated only to the target(s) that would cause the most damage if attacked. Importantly, though, such weakest-link models tend to be unrealistic in practice. For example, Arce M. and Sandler [AS01] note that the extreme solutions associated with weakest-link models "are not commonly observed among the global and transnational collective action problems confronting humankind." In particular, real-world decision-makers will generally want to "hedge" by investing in defense of additional targets to cover contingencies such as whether they have failed to correctly estimate which targets will be the most attractive to the attackers.

Moreover, it is important to go beyond the zero-order heuristic of protecting only the most valuable assets (or those that would do the most damage if successfully attacked), to also take into account the success probabilities of attacks against various possible targets. This can be important, since terrorists appear to take the probability of success into account in their choice of targets; for example, Woo [Woo03] has observed that "al-Qaeda is sensitive to target hardening," and that "Osama bin Laden has expected very high levels of reliability for martyrdom operations." Thus, even if one choice of target is potentially more damaging than another, it may not merit as much defensive investment as a target that is less valuable but more vulnerable (and hence may have a greater likelihood of being attacked).

Models that take the success probabilities of potential attacks into account include Bier and Abhichandani [BA03], Bier et al. [BNA05], Major

[Maj02], and Woo [Woo02, Woo03]. The results by Bier and colleagues represent weakest-link models, in which at equilibrium the defender invests in only those components that are most vulnerable (i.e., have the highest probability of being successfully attacked), or would cause the highest expected damage given an attack (taking into account both the success probability of an attack and the value of the component). In particular, in a series system, we assume that the attacker chooses which target to attack to maximize either the success probability of the attack [BA03], or more generally the expected damage of the attack [BNA05]. Therefore, it is never worthwhile for the defender to invest in any components other than those perceived as most attractive by the attacker.

By contrast, the models proposed by Major [Maj02] and Woo [Woo02, Woo03] achieve the more realistic result of hedging at optimality. In particular, Major [Maj02] assumes that the defender allocates defensive resources optimally, but that the attacker does not know this and randomizes the choice of targets to protect against the possibility that the allocation of defensive resources was suboptimal. The result is that the attacker's probability of choosing any particular target is "inversely proportional to the marginal effectiveness of defense ... at that target." Moreover, since the attacker randomizes in choosing which asset to target, the optimal defensive investment involves hedging (i.e., positive defensive investment even in assets that are not "weakest links").

Woo [Woo02, Woo03] extends the model introduced by Major and provides one possible strategy for estimating the values of the parameters in the resulting model. In particular, Major [Maj02] treats the values of the various potential targets as being exogenous. While O'Hanlon et al. [OOD02] were able to estimate the values of various potential targets to roughly an order of magnitude quite soon after September 11, obtaining more accurate estimates of asset values can itself be a difficult and time-consuming task. Therefore, Woo [Woo02, Woo03] assumes simply that the various types of targets have been rank-ordered in value by terrorism experts. He then converts these ordinal rankings into cardinal estimates of the targets' attractiveness to terrorists using Fechner's law [WE92, Fec1860]. This is a concept from the early days of psychophysics according to which "an arithmetic progression in perceptions requires a geometrical progression in their stimuli" [Woo02]. Fechner's logarithmic relationship of perception to stimulus is certainly plausible in this application, but is a somewhat ad hoc assumption; moreover, it may be more plausible to assume that perceived attractiveness is a logarithmic function of one or more cardinal measures of damage (such as lives lost or economic impact) than of ordinal rankings.

The basic game analyzed by Major and Woo involves simultaneous play by attackers and defenders, so the assumption that the attacker cannot readily observe the defender's investment makes sense in that context. (Note that it is still somewhat heroic to assume that the attacker can observe the marginal effectiveness of possible defensive investments. In practice, even the defender may not know the effectiveness of investments that have not been seriously considered for implementation!) However, many types of defenses (such as security guards) are either public knowledge or readily observable by attackers; moreover, some defenses (such as hardening of large buildings or installation of anthrax sterilization equipment in post offices) are not only observable, but also involve large capital outlays, and hence are difficult to change in response to evolving defender perceptions about likely attacker strategies. In such situations, it seems counterintuitive to assume that the attacker can observe the marginal effectiveness of defensive investments in each possible target, but cannot observe which defenses the defender has actually implemented. Presumably, in cases of sequential play (in which the defender commits to a particular choice of defensive investments, and the attacker observes these before selecting an attack strategy), those defenses that have already been implemented should be easier to observe than the hypothetical effectiveness of defenses that have not been implemented.

Since the basic premise underlying the models of Major and Woo is of questionable applicability in cases of sequential play (which are likely to be commonly encountered in practice), recent work [BOS05] achieves the same goal (i.e., an optimal defensive strategy that allows for hedging in equilibrium) in a different manner. In particular, in this model, attackers and defenders are assumed to have different valuations for the various potential targets. This is reasonable given the observation by Woo [Woo03] that "If a strike against America is to be inspirational [to al-Qaeda], the target should be recognizable in the Middle East"; thus, for example, attacks against iconic targets such as the Statue of Liberty or the Sleeping Beauty Castle at Disneyland may be disproportionately attractive to attackers relative to the economic damage and loss of life that they would cause. Moreover, while defenders may well prefer that attackers choose targets that are difficult and costly to attack, the attackers most likely care more about such factors than the defenders do. In addition to allowing attacker and defender valuations to differ, the proposed model assumes that attackers can observe defensive investments perfectly (which is conservative, but perhaps not overly so), but that defenders are uncertain about the attractiveness of each possible target to the attackers. This last assumption is reasonable in light of the fact that lack of knowledge about attacker values, goals, and motivations is precisely one of the reasons for gathering intelligence about potential attackers.

The model of Bier et al. [BOS05] has several interesting features in addition to the possibility of defensive hedging at equilibrium. First, it is interesting to note that such hedging does not always occur. In particular, it will often be optimal for the defender to put no investment at all into some targets even if they have a nonzero probability of being attacked — especially when the defender is highly budget constrained, and the various potential targets differ greatly in their values (both of which seem likely to be the case in practice). Moreover, in this model, if the allocation of defensive resources is suboptimal, defending one set of targets could in principle deflect attacks to alternative targets that are simultaneously less attractive a priori to the attackers, but also more damaging to the defenders. For example, making particular targets less vulnerable to attack could lead terrorists to adopt attack strategies that are more costly or difficult for them to implement, or would yield less publicity benefit to the attackers, but are also more lethal. This could be an important consideration, in light of the past substitution effects documented by Enders and Sandler [ES04].

3.1 Security as a Game between Defenders

The work discussed above has primarily viewed security as a game between an attacker and a defender, focusing on anticipating the effects of defensive actions on possible attackers — although Anderson [And01] views information security in part as a game between the providers of information security products. However, it also makes sense to consider the effects of defensive strategies adopted by one agent on the incentives faced by other defenders. Some types of defensive actions (such as installation of visible burglar alarms or car alarms) may actually increase risk to other potential victims. This type of situation can lead to overinvestment in security when viewed from the perspective of society as a whole, because the payoff to any one individual or organization from investing in security is greater than the net payoff to the entire society. Conversely, other types of defensive actions — such as vaccination [HAT94, Phi00], fire protection [OS02], installation of vehicle tracking systems (if their installation in a particular vehicle is not readily observable by potential car thieves) [AL98], or use of antivirus protection software [And01] — decrease the risk to other potential victims. This type of situation can be expected to result in underinvestment in security, since defenders may attempt to "free ride" on the investments of others, and in any case are unlikely to take positive externalities affecting other agents into account in their decision-making.

To better account for situations in which security investment confers positive externalities, Kunreuther and Heal [KH03, HK03] proposed a model of interdependent security where agents are vulnerable to "infection" from other agents. For example, consider the supply chain for food and agricultural products, in which companies could be vulnerable to contamination introduced upstream in the supply chain, and hence are vulnerable to the security weaknesses of other companies; Kunreuther and Heal have applied similar models to airlines that are vulnerable to threats in checked baggage transferred from partner airlines. In this context, not only will defensive investment on the part of one agent benefit other agents, it may also be extremely costly or difficult for agents to defend their own systems against infection spread (however unintentionally) by their partners, and they may therefore need to rely on their partners to protect them against such threats. Kunreuther and Heal consider in particular the case where even a single successful attack can be catastrophic — in other words, where the consequences of a successful attack (e.g., business failure) are "so serious that it is difficult to imagine an alternative event with greater consequences." In the context of this model, they show that failure of one agent to invest in security can make it unprofitable for other agents to invest in security, even when they would normally find it profitable to do so. Moreover, they show that this game can in some cases have multiple equilibrium solutions (e.g., an equilibrium in which all players invest and another in which no players invest). Kunreuther and Heal [KH03] discuss numerous possible coordinating mechanisms that can help to ensure that all players arrive at the socially optimal level of defensive investment, such as voluntary standards [And01], like those put forth by the International Organization for Standardization, or contracts.

Recent work [BG05, ZB05] has extended these results to the case of attacks occurring over time (according to a Poisson process), rather than the static model assumed in the original analysis. In this model, differences in discount rates among agents can lead some agents with low discount rates not to invest in security when it would otherwise be in their interests to do so, if other agents (e.g., with higher discount rates) choose not to invest in security. In particular, when an agent has a high discount rate, future losses due to attacks will have a low present value, so the agent will not find it worthwhile to invest in security. When the agent has a moderately small discount rate, the losses due to future attacks will tend to loom relatively large, so the agent will find investing in security to be worthwhile. When the discount rate of the agent is in the intermediate range, the agent will effectively be ambivalent about whether to invest, and will prefer to invest only when other agents also invest. Finally, when an agent's discount rate is extremely small, then investing will again be worthwhile only when other agents also invest. The reason for this last (somewhat counterintuitive) result is that investing in security is assumed to eliminate only the risk from direct attack, not the risk of "infection" from other agents, and hence merely postpones rather than eliminates the loss from an attack; at extremely low discount rates, merely postponing the loss is of little value.

Differences in discount rates can arise for a variety of reasons, ranging from participation in different industries with different typical rates of return, to risk of impending bankruptcy causing some agents to have extremely short-time horizons, to myopia (adopting a higher discount rate than is in the agent's enlightened self-interest). As in the simpler model, coordinating mechanisms (as well as efforts to counteract myopia) can be important here in ensuring that the socially optimal level of investment is achieved when multiple equilibrium solutions are possible. Thus, heterogeneous time preferences can complicate the task of achieving security in an interdependent world, but an understanding of this phenomenon can help in identifying promising solutions.

4 Combining Reliability Analysis and Game Theory

We have seen that many of the recent applications of risk and reliability analysis to security do not explicitly model the adaptive response of potential attackers to defensive investments, and hence may vastly overstate both the effectiveness and the cost-effectiveness of those investments. Similarly, much of the existing game-theoretic security work focuses on nonprobabilistic games. Moreover, even those models that explicitly consider the success probabilities of potential attacks (e.g., [Maj02, Woo02, Woo03, KH03, HK03]) generally consider individual assets or components in isolation, and fail to consider the effect that disabling one or several components can have on the functionality of the larger system of which they may be a part. Combining the techniques of risk and reliability analysis with game theory could therefore be a fruitful way of studying and protecting against intentional threats to complex systems such as critical infrastructure.

Hausken [Hau02], an economist, has integrated probabilistic risk analysis and game theory (although not in the security context), by interpreting system reliability as a public good and elucidating the incentives of different players responsible for maintaining particular components of a larger system. In particular, he views security as a game between defenders responsible for different portions of an overall system, and elucidates the relationships between the series or parallel structure of the system and classic games such as the coordination game, the battle of the sexes, chicken, and the prisoner's dilemma [Hir83, Hir85].

Rowe [Row02], a risk analyst, argues that the implications of "the human variable" in terrorism risk (in particular, the fact that terrorists can adapt in response to our defenses) have yet to be adequately appreciated. He presents a simple game-theory framework for addressing the need to evaluate possible protective actions in light of terrorists' ability to "learn from experience and alter their tactics." This approach has been used in practice to provide input to prioritizing defensive investments among multiple potential targets and multiple types of threats.

Banks and Anderson [BA03a] apply similar ideas to the evaluation of options for responding to the threat of bioterrorism (in particular, intentionally introduced smallpox). The approach adopted by Banks and Anderson embeds risk analysis (quantified using expert opinion) in a game-theoretic formulation of the defender's decision problem. This enables them to account for both the adaptive nature of the threat and also the uncertainty about the costs and benefits of particular defensive actions. They conclude that this approach "captures facets of the problem that are not amenable to either game theory or risk analysis on their own."

Recent results by the author and colleagues [BA03, BNA05] use game theory to explore the nature of optimal investments in the security of simple series and parallel systems as a building block to the analysis of more complex systems. The results suggest that defending series systems against informed and determined attackers is an extremely difficult challenge. In a series system, if the attacker knows about (or can observe) the system's defenses, the defender's options for protecting a series system are extremely limited. In particular, the attacker's ability to respond strategically to the defender's investments deprives the defender of the ability to allocate defensive investments according to their cost-effectiveness; rather, if potential attackers know (or can readily learn) about the effectiveness of any defensive measures, defensive investments in series systems must essentially equalize the strength of all defended components to be economically efficient. This is consistent with the observation by Dresher [Dre61] in the military context that, for optimal allocation of defensive resources, "It is necessary that each of the defended targets yield the same payoff to the attacker."

This emphasizes the importance of redundancy as a defensive strategy. Essentially, redundancy reduces the flexibility available to the attacker in choice of targets (since the attacker must now disable multiple redundant components to disable a system) and increases the flexibility available to the defender (since the defender can now choose which of several redundant components to defend, based on the cost-effectiveness of doing so). Traditional reliability design considerations such as spatial separation and functional diversity are also important components of defensive strategy to help ensure that attacks against redundant components are likely to succeed or fail more or less independently of each other (i.e., to ensure that redundant components cannot all be disabled by the same type of attack).

It is clearly important in practice to extend the types of security models described above to more complicated system structures (including both parallel and series subsystems), such as that shown below, rather than simple parallel or series systems. Recent work [AB04] begins to address this challenge, at least under particular assumptions. However, achieving fully general results (e.g., for arbitrary system structures and more general assumptions about the effects of security investments on the costs and/or success probabilities of potential attacks) is likely to be difficult and may require heuristic approaches.

In addition, for reasons of mathematical convenience, the models developed until now have generally assumed that the success probability of an attack on a particular component is a convex function of the resources invested to defend that component. While in many contexts this is a reasonable assumption (e.g., due to declining marginal returns to defensive investments), it is clearly not fully general. For example, certain types of security improvements (such as relocating a critical facility to a more secure location) are "inherently discrete" [KZ03], in the sense that they require some minimal level of investment to be feasible. This will tend to result in step changes in the success probability of an attack as a function of the level of defensive investment. Similarly, if security investment beyond some threshold deters potential attackers from even attempting an attack, then the likelihood of a successful attack could decrease rapidly beyond that threshold. Such effects can result in the success probability of an attack being a nonconvex function of the defensive investment (at least in certain regions, e.g., when the level of investment is not too large). This makes the problem of identifying the optimal level of defensive investment more complicated and can change the nature of the optimal solutions (e.g., increasing the likelihood that there will be multiple local optima, and that not investing in security may be the optimal strategy).

Finally, it would, of course, be worthwhile to extend our models to include the dimension of time, rather than the current static or "snapshot" view of system security. This would allow us to model imperfect attacker information (including, for example, Bayesian updating of the probability that an attack will succeed based on a past history of successful and failed attacks) as well as the possibility of multiple attacks over time.

5 Conclusions

As noted above, protecting engineered systems against intentional attacks is likely to require a combination of game theory and reliability analysis. Risk and reliability analysis by itself will likely not be sufficient to address many critical security challenges, since it does not take into account the attacker's response to the implementation of reliability or security improvements. However, most current applications of game theory to security deal with individual components or assets in isolation, and hence could benefit from the use of reliability analysis tools and methods to more fully model the risks to complex networked systems such as computer systems, electricity transmission systems, or transportation systems. In the long run, approaches that embed systems reliability models in a game-theoretic framework may make it possible to take advantage of the strengths of both approaches.

Acknowledgments

This material is based upon work supported in part by the U.S. Army Research Laboratory and the U.S. Army Research Office under grant number DAAD19-01-1-0502, by the U.S. National Science Foundation under grant number DMI-0228204, by the Midwest Regional University Transportation Center under project number 04-05, and by the U.S. Department of Homeland Security through the Center for Risk and Economic Analysis of Terrorism Events (CREATE) under grant number EMW-2004-GR-0112. Any opinions, findings, and conclusions, or recommendations expressed herein are those of the author and do not necessarily reflect the views of the sponsors. I would also like to acknowledge the numerous colleagues and students who have contributed to the body of work discussed here.

References

[And01] Anderson, R. 2001. "Why information security is hard: An economic perspective." *18th Symposium on Operating Systems Principles, New Orleans, LA.* http://www.ftp.cl.cam.ac.uk/ftp/users/rja14/econ.pdf.

[AS01] Arce M., D. G., and T. Sandler. 2001. "Transnational public goods: Strategies and institutions." *European Journal of Political Economy* 17:493–516.

[Arm85] Armstrong, J. S. 1985. *Long-range forecasting: From crystal ball to computer.* New York: Wiley.

[AL98] Ayres, I., and S. Levitt. 1998. "Measuring the positive externalities from unobservable victim precaution: An empirical analysis of Lojack." *Quarterly Journal of Economics* 113:43–77.

[AB04] Azaiez, N., and V. M. Bier. 2004. "Optimal resource allocation for security in reliability systems." Submitted to *European Journal of Operational Research.*

[BA03a] Banks, D., and S. Anderson. 2003. "Game-theoretic risk management for counterterrorism." Technical Report, Office of Biostatistics and Epidemiology, Center for Biologics Evaluation and Research, U.S. Food and Drug Administration.

[BC01] Bedford, T., and R. Cooke. 2001. *Probabilistic risk analysis: Foundations and methods.* Cambridge, UK: Cambridge University Press.

[Ben81] Ben-Dov, Y. 1981. "Optimal testing procedures for special structures of coherent systems." *Management Science* 27:1410–1420.

[BD59] Berkovitz, L. D., and M. Dresher. 1959. "A game-theory analysis of tactical air war." *Operations Research* 7:599–620.

[BD60] Berkovitz, L. D., and M. Dresher. 1960. "Allocation of two types of aircraft in tactica." *Operations Research* 8:694–706.

[Bie97] Bier, V. M. 1997. "An overview of probabilistic risk analysis for complex engineered systems." In *Fundamentals of risk analysis and risk management,* edited by V. Molak, 1–18. Boca Raton, FL: Lewis Publishers.

[Bie05] Bier, V. M. 2005. "Game-theoretic and reliability methods in counterterrorism and security." In *Modern statistical and mathematical methods in reliability,* edited by A. Wilson, N. Liminios, S. Keller-McNulty, and Y. Armijo, Volume 10 of *Series on Quality, Reliability, and Engineering Statistics,* 17–38. Singapore: World Scientific Publishers.

[BA03] Bier, V. M., and V. Abhichandani. 2003. "Optimal allocation of resources for defense of simple series and parallel systems from determined adversaries." In *Risk-based decisionmaking in water resources X,* 59–76. Reston, VA: American Society of Civil Engineers.

[BFH04] Bier, V. M., S. Ferson, Y. Y. Haimes, J. H. Lambert, and M. J. Small. 2004. "Risk of extreme and rare events: Lessons from a selection of approaches." In *Risk analysis and society: Interdisciplinary perspectives,* edited by T. McDaniels and M. J. Small. Cambridge, UK: Cambridge University Press.

[BG05] Bier, V. M., and A. Gupta. 2005. "Myopia and interdependent security risks." Submitted to *The Engineering Economist.*

[BHL99] Bier, V. M., Y. Y. Haimes, J. H. Lambert, N. C. Matalas, and R. Zimmerman. 1999. "Assessing and managing the risk of extremes." *Risk Analysis* 19:83–94.

[BNA05] Bier, V. M., A. Nagaraj, and V. Abhichandani. 2005. "Protection of simple series and parallel systems with components of different values." *Reliability Engineering and System Safety* 87:313–323.

[BOS05] Bier, V. M., S. Oliveros, and L. Samuelson. 2005. "Choosing what to protect: Strategic defensive allocation against an unknown attacker." *Journal of Public Economic Theory* (in press).

[Bra75] Brams, S. J. 1975. *Game theory and politics.* New York: Free Press.

[Bra85] Brams, S. J. 1985. *Superpower games: Applying game theory to superpower conflict.* New Haven, CT: Yale University Press.

[BK88] Brams, S. J., and D. M. Kilgour. 1988. *Game theory and national security.* Oxford, UK: Basil Blackwell.

[BS98] Brannigan, V., and C. Smidts. 1998. "Performance based fire safety regulation under intentional uncertainty." *Fire and Materials* 23:341–347.

[BR00] Brauer, J., and A. Roux. 2000. "Peace as an international public good: An application to southern Africa." *Defence and Peace Economics* 11:643–659.

[Bur99] Burke, D. 1999. "Towards a game theory model of information warfare." Ph.D. diss., Air Force Institute of Technology. http://www.iwar.org.uk/iwar/resources/usaf/maxwell/students/2000/afit-gss-lal-99d-1.pdf.

[CGM00] Chaturvedi, A. M., M. Gupta, S. Mehta, and L. Valeri. 2000. "Fighting the wily hacker: Modeling information security issues for online financial institutions using the Synthetic Environments for Analysis and Simulation environment (SEAS)." *INET 2000 Proceedings.* Internet Society. http://www.isoc.org/inet2000/cdproceedings/7a/7a_4.htm.

[CKS03] Cleaves, D. J., A. E. Kuester, and D. A. Schultz. 2003. A methodology for managing the risk of terrorist acts against the U. S. Postal Service. *Society for Risk Analysis Annual Meeting, Baltimore, MD.*

[Coh99] Cohen, F. 1999. "Managing network security: Attack and defense strategies." Technical Report 9907, Fred Cohen and Associates Strategic Security and Intelligence.

[Coo91] Cooke, R. M. 1991. *Experts in uncertainty: Opinion and subjective probability in science.* Oxford, UK: Oxford University Press.

[CCS96] Cox, L., S. Chiu, and X. Sun. 1996. "Least-cost failure diagnosis in uncertain reliability systems." *Reliability Engineering and System Safety* 54:203–216.

[CQK89] Cox, L., Y. Qiu, and W. Kuehner. 1989. "Heuristic least-cost computation of discrete classification functions with uncertain argument values." *Annals of Operations Research* 21:1–30.

[Cox90] Cox, L. A. 1990. "A probabilistic risk assessment program for analyzing security risks." In *New risks: Issues and management*, edited by L. A. Cox and P. F. Ricci. New York: Plenum Press.

[Dre61] Dresher, M. 1961. *Games of strategy: Theory and applications.* Englewood Cliffs, NJ: Prentice-Hall.

[ES04] Enders, W., and T. Sandler. 2004. "What do we know about the substitution effect in transnational terrorism?" In *Researching terrorism: Trends, achievements, failures*, edited by A. Silke and G. Ilardi. London: Frank Cass. http://www-rcf.usc.edu/tsandler/substitution2ms.pdf.

[EFW00a] Ezell, B. C., J. V. Farr, and I. Wiese. 2000. "Infrastructure risk analysis model." *Journal of Infrastructure Systems* 6 (3): 114–117.

[EFW00b] Ezell, B. C., J. V. Farr, and I. Wiese. 2000. "Infrastructure risk analysis of municipal water distribution system." *Journal of Infrastructure Systems* 6 (3): 118–122.

[EHL01] Ezell, B. C., Y. Y. Haimes, and J. H. Lambert. 2001. "Cyber attack to water utility supervisory control and data acquisition SCADA systems." *Military Operations Research* 6 (2): 23–33.

[Fec1860] Fechner, G. T. 1860. *Elemente der Psychophysik*. Leipzig: Breitkopf und Haertel.

[FL02] Frey, B. S., and S. Luechinger. 2002. "Terrorism: Deterrence may backfire." Working paper 136, Institute for Empirical Research in Economics, University of Zurich.

[FL03] Frey, B. S., and S. Luechinger. 2003. "How to fight terrorism: Alternatives to deterrence." *Defence and Peace Economics* 14:237–249.

[FT91] Fudenberg, D., and J. Tirole. 1991. *Game theory.* Cambridge, MA: MIT Press.

[Gar03] Garrick, B. J. 2003. "Perspectives on the use of risk assessment to address terrorism." *Risk Analysis* 22:421–424.

[Hai02a] Haimes, Y. Y. 2002. "Roadmap for modelling risks of terrorism to the homeland." *Journal of Infrastructure Systems* 8:35–41.

[Hai02b] Haimes, Y. Y. 2002. "Strategic responses to risks of terrorism to water resources." *Journal of Water Resources Planning and Management* 128:383–389.

[HL02] Haimes, Y. Y., and T. Longstaff. 2002. "The role of risk analysis in the protection of critical infrastructure against terrorism." *Risk Analysis* 22:439–444.

[HML98] Haimes, Y. Y., N. C. Matalas, J. H. Lambert, B. A. Jackson, and J. F. R. Fellows. 1998. "Reducing vulnerability of water supply systems to attack." *Journal of Infrastructure Systems* 4: 164–177.

[Hau02] Hausken, K. 2002. "Probabilistic risk analysis and game theory." *Risk Analysis* 22:17–27.

[Hay54] Haywood, O. G. 1954. "Military decision and game theory." *Journal of the Operations Research Society of America* 2:365–385.

[HK03] Heal, G. M., and H. C. Kunreuther. 2003. You only die once: Managing discrete interdependent risks. http://ssrn.com/abstract=419240.

[HAT94] Hershey, J., D. Asch, T. Thumasathit, J. Meszaros, and V. Waters. 1994. "The roles of altruism, free riding, and bandwagoning in vaccination decisions." *Organizational Behavior and Human Decision Processes* 59:177–187.

[Hir83] Hirshleifer, J. 1983. "From weakest-link to best-shot: The voluntary provision of public goods." *Public Choice* 41:371–386.

[Hir85] Hirshleifer, J. 1985. "From weakest-link to best-shot: Correction." *Public Choice* 46:221–223.

[Ise04] Isenberg, J. 2004. Comparison of seismic and security risk criteria for bridges. *83rd Transportation Research Board Annual Meeting, Washington, DC,* January 11–15.

[KZ03] Keohane, N. O., and R. J. Zeckhauser. 2003. "The ecology of terror defense." *Journal of Risk and Uncertainty* 26:201–229.

[KH03] Kunreuther, H., and G. Heal. 2003. "Interdependent security." *Journal of Risk and Uncertainty* 26:231–249.

[LZ05] Lakdawalla, D., and G. Zanjani. 2005. "Insurance, self-protection, and the economics of terrorism." *Journal of Public Economics* 89:1891–1905.

[LL60] Leibowitz, M. L., and G. J. Lieberman. 1960. "Optimal composition and deployment of a heterogeneous local air-defense system." *Operations Research* 8:324–337.

[LA04] Lemon, D. M., and G. E. Apostolakis. 2004. "A methodology for the identification of critical locations in infrastructures." Working paper ESD-WP-2004-01, Engineering Systems Division, Massachusetts Institute of Technology. http://esd.mit.edu/WPS/esd-wp-2004-01.pdf.

[Lev02] Levitin, G. 2002. "Maximizing survivability of acyclic transmission networks with multi-state retransmitters and vulnerable nodes." *Reliability Engineering and System Safety* 77:189–199.

[Lev03a] Levitin, G. 2003. "Optimal allocation of multi-state elements in linear consecutively connected systems with vulnerable nodes." *European Journal of Operational Research* 150:406–419.

[Lev03b] Levitin, G. 2003. "Optimal multilevel protection in series-parallel systems." *Reliability Engineering and System Safety* 81:93–102.

[LDX03] Levitin, G., Y. Dai, M. Xie, and K. L. Poh. 2003. "Optimizing survivability of multi-state systems with multi-level protection by multi-processor genetic algorithm." *Reliability Engineering and System Safety* 82:93–104.

[LL00] Levitin, G., and A. Lisnianski. 2000. "Survivability maximization for vulnerable multi-state systems with bridge topology." *Reliability Engineering and System Safety* 70:125–140.

[LL01] Levitin, G., and A. Lisnianski. 2001. "Optimal separation of elements in vulnerable multi-state systems." *Reliability Engineering and System Safety* 73:55–66.

[LL03] Levitin, G., and A. Lisnianski. 2003. "Optimizing survivability of vulnerable series-parallel multi-state systems." *Reliability Engineering and System Safety* 79:319–331.

[Maj02] Major, J. 2002. "Advanced techniques for modeling terrorism risk." *Journal of Risk Finance* 4:15–24.

[MJ87] Martz, H. F., and M. E. Johnson. 1987. "Risk analysis of terrorist attacks." *Risk Analysis* 7:35–47.

[NRC94] National Research Council. 1994. *Science and judgment in risk assessment.* Washington, DC: National Academies Press.

[NRC75] Nuclear Regulatory Commission. 1975. Reactor safety study: An assessment of accident risks in U. S. commercial nuclear power plants. WASH-1400.

[OOD02] O'Hanlon, M., P. Orszag, I. Daalder, M. Destler, D. Gunter, R. Litan, and J. Steinberg. 2002. *Protecting the American homeland.* Washington, DC: Brookings Institution.

[OS02] Orszag, P., and J. Stiglitz. 2002. "Optimal fire departments: Evaluating public policy in the face of externalities." Working paper, Brookings Institution. http://www.brookings.edu/views/papers/orszag/20020104.pdf.

[PG02] Pate-Cornell, E., and S. Guikema. December, 2002. "Probabilistic modeling of terrorist threats: A systems analysis approach to setting priorities among countermeasures." *Military Operations Research* 7:5–20.

[Phi00] Philipson, T. 2000. "Economic epidemiology and infectious diseases." In *The handbook of health economics*, edited by A. Culyer and J. Newhouse, 1762–1799. Netherlands: Elsevier Science.

[Pre97] President's Commission on Critical Infrastructure Protection. 1997. *Critical foundations: Protecting America's infrastructures.* Washington, DC: Government Printing Office.

[Rai68] Raiffa, H. 1968. *Decision analysis: Introductory lectures on choices under uncertainty.* Reading, MA: Addison-Wesley.

[Rav02] Ravid, I. 2002. Theater ballistic missiles and asymmetric war. The Military Conflict Institute. http://www.militaryconflict.org/Publications.htm.

[Row02] Rowe, W. D. 2002. "Vulnerability to terrorism: Addressing the human variables." In *Risk-based decisionmaking in water resources X*, edited by Y. Y. Haimes and D. A. Moser, 155–159. Reston, VA: American Society of Civil Engineers.

[SA03] Sandler, T., and D. G. Arce M. 2003. "Terrorism and game theory." *Simulation and Gaming* 34:319–337.

[SE04] Sandler, T., and W. Enders. 2004. "An economic perspective on transnational terrorism." *European Journal of Political Economy* 20:301–316.

[Sch00] Schneier, B. 2000. *Secrets and lies: Digital security in a networked world.* New York: Wiley.

[Sch01] Schneier, B. 2001. Managed security monitoring: Network security for the 21st century. Counterpane Internet Security, Inc. http://www.counterpane.com/msm.pdf.

[Shu87] Shubik, M. 1987. "When is an application and when is a theory a waste of time." *Management Science* 33:1511–1522.

[SV04] Smith, J. E., and D. von Winterfeldt. 2004. "Decision analysis." *Management Science* 50:561–574.

[WE92] Weber, D., and E. Englund. 1992. "Evaluation and comparison of spatial interpolators." *Mathematical Geology* 24 (4): 381–391.

[Wer04] Werner, S. January 11–15, 2004. Use of earthquake-based highway system vulnerability assessment tool for transportation security applications. *83rd Transportation Research Board Annual Meeting.*

[Woo02] Woo, G. 2002. "Quantitative terrorism risk assessment." *Journal of Risk Finance* 4:7–14.

[Woo03] Woo, G. 2003. Insuring against Al-Qaeda. Insurance Project Workshop, National Bureau of Economic Research, Inc. http://www.nber.org/~confer/2003/insurance03/woo.pdf.

[Zeb03] Zebroski, E. June 2003. "Risk management lessons from man-made catastrophes: Implications for anti-terrorism risk assessments using a simple relative risk method." *American Nuclear Society Topical Meeting on Risk Management.* San Diego.

[ZB05] Zhuang, J., and V. M. Bier. 2005. "Subsidized security and stability of equilibrium solutions in an n-player game with errors." Submitted to *Games and Economic Behavior.*

[ZHN04] Zilinskas, R. A., B. Hope, and D. W. North. 2004. "A discussion of findings and their possible implications from a workshop on bioterrorism threat assessment and risk management." *Risk Analysis* 24:901–908.

[ZB02] Zimmerman, R., and V. M. Bier. April 12–13, 2002. Risk assessment of extreme events. Columbia-Wharton/Penn Roundtable on Risk Management Strategies in an Uncertain World, Palisades, NY. http://www.ldeo.columbia.edu/CHRR/Roundtable/Zimmerman_WP.pdf.

Part II

Biometric Authentication

Biometric Authentication

David L. Banks

Institute of Statistics and Decision Sciences, Duke University,
banks@stat.duke.edu

Biometric identification is an old technology. Face recognition is a long-standing tool in law enforcement; the Wild West wanted posters are just one example. Signatures and handwriting have been accepted in United States law courts to establish authorship since 1831 [US1831]. Fingerprints also have a long history: Joao de Barros reports their use in China in the 14th century, and Sir William Hershel used them in 1856 on contracts in India to prevent subsequent repudiation. And all statisticians should know that Sir Francis Galton [Gal1892] wrote an influential book on fingerprints in 1892, which contains the taxonomy of minutia that is still in use today. Work still continues on this: Dass [Das04] applies Markov random field methods for fingerprint matching.

In the post 9/11 era, governments around the world are investigating biometric identification as a means to discourage terrorism. (There are also direct applications in preventing identity theft.) Countries are planning to provide biometric authentication on passports, and secure facilities want biometric access control. New techniques are based upon DNA samples, capillary patterns in the iris of the eye, voice prints, acceleration patterns on pressure-sensitive signature plates, and keystroke rhythms when typing passwords. But all of these ideas require statistical justification and a legal framework.

The statistical justification concerns the probability of a false match and the probability of a missed match. If the type I and type II error rates are too large, then the method has little value. Often there are not single answers for these estimates. For example, before DNA testing, blood type used to be helpful in excluding suspects, but in general was not legally useful for confirming identity. However, for some very rare blood types, it could be extremely specific and highly probative. So there are circumstances in which a method can work very much better than its average behavior. The converse is also true; biometric distinctions between twins generally show much worse performance than their average behavior.

The legal questions concern the use of the biometric technology. Biometric methods that have no standing in law have limited applicability (but are still being tested for some applications). One issue concerns the fact that suspects in the United States cannot be compelled to testify against themselves, and thus need not provide DNA samples or signature specimens. A related issue is the evolving body of privacy law — courts may decide that it is too intrusive to require identification for the routine business of daily life, such as entering a subway station or buying a car.

A more crucial issue is whether a court can even accept the biometric match as evidence. It used to be that the legal standard for such evidence was *Frye v. United States* [Fry23], a 1923 decision that held that "while courts will go a long way in admitting expert testimony deduced from a well-recognized scientific principle or discovery, the thing from which the deduction is made must be sufficiently established to have gained general acceptance in the particular field in which it belongs." This argument was used to exclude prosecutorial evidence based upon a precursor of the lie detector, but has subsequently upheld the use of handwriting, fingerprint, and DNA evidence.

The legal landscape shifted recently when the Supreme Court upheld the Ninth Circuit Court's decision in *Daubert v. Merrell Dow Pharmaceuticals Inc.* [Dau91]. That 1991 decision ruled that scientific evidence must be "subjected to verification and scrutiny by others in the field" and established five criteria that expert testimony must satisfy. In the context of biometric evidence, these are generally taken to mean that the methods must be transparent, published, and have validated estimates of type I and type II error.

One consequence of *Daubert* is a fresh skepticism of latent fingerprint evidence. This is driven by several cases in which Federal Bureau of Investigation (FBI) experts made a false match — the most conspicuous case was Brandon Mayfield, the Oregon lawyer who was arrested in connection with the train bombing in Madrid in 2004. When making a fingerprint identification, the FBI protocol requires that two experts examine the match and agree on the finding, where the determination of a match depends upon a complex set of procedures involving "points of comparison" and other features. However, the two experts do not work independently, and often have knowledge of information that may be prejudicial (in Mayfield's case, there were three experts, and they knew he was a convert to Islam).

Another recent case was Stephen Cowans, who was convicted in 1997 of shooting a police officer, but exonerated in 2004 when DNA evidence proved that the FBI had incorrectly matched his fingerprint. Both the Mayfield and the Cowans cases raise the statistical issue of *search bias*; as fingerprint libraries grow, the chance of finding a near match increases, and thus the probability of a false match must increase. Statisticians can help quantify this.

But these legal issues may not be so important for counterterrorism. Here the Holy Grail for biometric identification is face recognition, and there are reasons to believe that many applications would be juridically acceptable. For example, intelligence agencies and the Department of Homeland Security really want to have technology that allows people entering the United States to be quickly checked against a library of terrorist photos; it seems unlikely that such use would violate any protected rights.

The following two papers are outstanding examples of how statistical research has begun to address face recognition — but both papers describe methods that could apply to more general problems. They represent solid work that aims at understanding, estimating, and reducing the error rates that have made biometric identification so problematic.

Mitra's paper uses the MACE filter and builds statistical models in the spectral domain for faces images. Her approach combines ideas in data mining with traditional statistics and uses complexity penalties to determine appropriate model fits. The models allow for variation in illumination, which is one of the many hard problems in this area. In contrast, Rukhin's paper takes a more synthetic approach. Using copula theory, he studies how algorithms might be combined to achieve better error rates than any single algorithm acting alone. Some of the methodology behind this relates to work on document retrieval, which ranks the quality of match in a database. Another thread relates to the theory of nonparametric measures of correlation. Both papers demonstrate the kinds of contributions that statistical thinking can make to hard problems of national importance.

But the problems these methods face are significant. Many people look alike (Websites have sprung up to discuss the astonishing resemblance between Saruman the White, as played by Christopher Lee, and Sheik Yassin, the founder of Hamas). For counterterrorism, algorithms need to work on people who have shaved a beard or donned false eyeglasses, and the recognition software must automatically correct for differences in lighting, expression, and the angle of the photo. Even with the human eye, which is much more powerful than any existing algorithm, the false-alarm rate might be too large for border security purposes.

References

[Das04] Dass, S. D. 2004. "Markov random field models for directional field and singularity extraction in fingerprint images." *IEEE Transactions on Image Processing* 13:1358–1367.

[Dau91] Daubert v. Merrell Dow Pharmaceuticals Inc. 951 F.2d 1128 Ninth Circuit Court of Appeals. 1991.

[Fry23] Frye v. United States. 293 F. 1013 D. C. Circuit Court. 1923.

[Gal1892] Galton, F. 1892. *Finger prints.* London: Macmillan and Co.

[US1831] United States vs. Samperyac. 23 F. Cas. 932, Arkansas 1831.

Towards Statistically Rigorous Biometric Authentication Using Facial Images

Sinjini Mitra

Department of Statistics and CyLab, Carnegie Mellon University,
smitra@stat.cmu.edu

1 Introduction

In the modern electronic information age, there is an ever-growing need to authenticate and identify individuals for ensuring the security of a system. Traditional methods of authentication and identification make use of identification (ID) cards or personal identification numbers (PINs), but such identifiers can be lost, stolen, or forgotten. In addition, these methods fail to differentiate between an authorized person and an impostor who fraudulently acquires knowledge or "token" of the authorized person. Security breaches have led to losses amounting to millions of dollars in agencies like banks and telecommunication systems that depend on token-based security systems.

In traditional statistical literature, the term *biometrics* or *biometry* refers to the field of statistical methods applicable to data analysis problems in the biological sciences, such as agricultural field experiments to compare the yields of different varieties of a crop, or human clinical trials to measure the effectiveness of competing therapies. Recently the term *biometrics* has also been used to denote the unique biological traits (physical or behavioral of individuals that can be used for identification), and *biometric authentication* is the newly emerging technology devoted to verification of a person's identity based on his/her biometrics. The purpose of biometric authentication is to provide answers to questions like the following:

- Is this person authorized to enter a facility?
- Is this individual entitled to access privileged information?
- Is the given service being administered only to enrolled users?

These questions are vital for ensuring security of many business and governmental organizations. Since it relies on "something you are" rather than "something you know or possess," a biometric in principle cannot be stolen, forgotten, or duplicated and is less prone to fraud than PINs and ID cards. For all these reasons, the field of biometrics has been growing exponentially in recent years (especially after the attacks of September 11, 2001), and the rapidly evolving technology is being widely used in forensics for criminal identification in law enforcement and immigration, in experimental form in restricting access to automated teller machines (ATMs) and computer networks, as well as in various forms of e-commerce and electronic banking. Moreover, the recent practice of recording biometric information (photo and fingerprint) of foreign passengers at all U.S. airports and also the proposed inclusion of digitized photos in passports show the growing importance of biometrics in U.S. homeland security.

Typically used biometrics include face images, fingerprints, iris measurements, palm prints, hand geometry, hand veins (physical traits), and voiceprint, gait, and gesture (behavioral traits). Generally, biometric systems are composed of two parts: (1) the enrollment and (2) the identification part. The former involves the registration of a user's characteristic, which is subsequently to be used as a criterion for classification purposes. This procedure involves sample capturing with the help of digital cameras or similar devices, feature extraction for developing a sample template, and storing the template with the relevant database. The second part provides the user interface to have the end user's characteristic captured, compared to the existing templates, and verified whether he or she is authentic or an impostor.

Face recognition is probably the most popular biometric-based method because of its potential to be both accurate as well as nonintrusive and user-friendly. It analyzes facial characteristics to verify whether the image belongs to a particular person. Faces are rich in information about individual identity, mood and mental state, and position relationships between face parts, such as eyes, nose, mouth, and chin, as well as their shapes and sizes, are widely used as discriminative features for identification. Much research has been done on face recognition in the past decades in the field of computer science, and yet face authentication still poses many challenges. Several images of a single person may be dramatically different because of changes in viewpoint, color, and illumination, or simply because the person's face looks different from day to day due to appearance-related changes like makeup, facial hair, glasses, etc.

Several authentication methods based on face images have been developed for recognition and classification purposes. In face authentication, as in most image processing problems, it is necessary to extract relevant *discriminative* features that distinguish individuals. But one hardly knows in advance which possible features will be discriminative. For this reason, most of the face authentication systems today use some kind of efficient automatic feature extraction technique. Jonsson et al. [JKL99] used support vector machines (SVM) to extract relevant discriminatory information from the training data and build an efficient face authentication system, and Li et al. [LKM99] used linear discriminant analysis (LDA) for efficient face recognition and verification. Liu et al. [LCV02] applied principal components analysis (PCA) for modeling variations arising in face images from expression changes and registration errors by using the motion field between images in a video clip. Havran et al. [HHC02] performed face authentication based on independent component analysis (ICA), and Palanivel et al. [PVY03] proposed a method for video-based, real-time face authentication using neural networks. A recently developed face authentication system is the minimum average correlation energy (MACE) filter [VSV02, SVK02]. The MACE filter was originally proposed by Mahalanobis et al. [MVC87] as an effective automatic target recognition tool, and Vijaya Kumar et al. [VSV02] first used it to authenticate a facial expression database, obtaining impressive results. Savvides and Vijaya Kumar [SV03] showed that the filter-based methods produce more accurate authentication results than traditional methods based on LDA and PCA, especially in the presence of distortions such as illumination changes and partial occlusions.

The present chapter reports some initial work on establishing a firmer statistical foundation for face authentication systems and in verifying the accuracy of proposed methods in engineering and computer science, which are mostly empirical in nature. Given the sensitive nature of their applications today, it is imperative to have rigorous authentication systems where inaccurate results may have a drastic impact. The layout of the chapter is as follows. Section 2 describes some basic statistical tools that can be employed for evaluation of authentication techniques and Sect. 3 provides brief descriptions of the databases used for our study. Section 4 introduces the MACE filter authentication system along with its statistical aspects, and Sect. 5 discusses statistical model-based systems and the associated challenges and comparison with the MACE system.

2 Performance Evaluation of a Biometric System

In the design of a biometric, a primary consideration is to know how to measure the accuracy of such a system. This is critical for determining whether the system meets the requirements of a particular application and how it will respond in practice. Many statistical tools are available to help in this regard. According to Shen et al. [SSK97], two important aspects of performance evaluation that need to be addressed for any practical authentication system are:

1. To determine the reliability of error rates, and
2. To determine how the nature and quality of data influence system performance.

2.1 Decision Landscapes

Biometric identification fits squarely into the classical framework of statistical decision theory. The result of a decision-making algorithm is a match score T and a threshold τ. If $T > \tau$, the system returns a match, otherwise if $T \leq \tau$, the system decides that a match has not been made. These decisions give rise to four possible outcomes in any pattern recognition problem: either a given pattern is, or is not, the target; and in either case, the decision made by the recognition algorithm may be either correct or incorrect. These are usually referred to as (1) false accept (FA), (2) correct accept (CA), (3) false reject (FR), and (4) correct reject (CR). Obviously the first and third outcomes are errors (analogous to the type I and type II errors, respectively, that occur commonly in hypothesis testing) and are the focus of the statistical aspects of any biometric system performance.

To make this discussion more rigorous, let us denote by $f_A(x)$ and $g_I(y)$ the density of the distribution of the match scores for the authentics and the impostors, respectively. Then the false rejection rate (FRR) is defined as the probability that T is less than τ given that T comes from the distribution of the authentic user scores. The false acceptance rate (FAR), on the other hand, is defined as the probability that T is greater than τ given that T belongs to the impostor score distribution. Mathematically,

$$\text{FRR} = P(T \leq \tau | T \in \text{Authentic}) = \int_{-\infty}^{\tau} f_A(x) \mathrm{d}x, \quad (1)$$

$$\text{FAR} = P(T > \tau | T \in \text{Impostor}) = \int_{\tau}^{\infty} g_I(y) \mathrm{d}y. \quad (2)$$

When the underlying distributions are Gaussian, these probabilities have closed-form solutions in terms of the z-scores. But generally they are unknown and difficult to model. However, empirical estimates can be formed based on observed samples in the following way:

$$\hat{p}_{FRR} = \frac{\#(T \leq \tau | T \in \text{Authentic})}{\#\text{Authentic}}, \quad \hat{p}_{FAR} = \frac{\#(T > \tau | T \in \text{Impostor})}{\#\text{Impostor}}, \quad (3)$$

where $\hat{p}_{FRR}$ and $\hat{p}_{FAR}$ are respectively the estimators of FRR and FAR, #Authentic and #Impostor are respectively the total number of authentic and impostor user match scores. Often it is of interest to establish confidence intervals for these estimates of error rates and conduct hypothesis tests of whether the performance of the system under consideration meets or exceeds the system design requirement (for example, to check whether the FAR and FRR are below a prespecified threshold). Bolle et al. [BRP99, BPR00] suggest the use of binomial distributions, normal approximations, and also bootstrapping for estimating the error rate confidence intervals and developing tests of significance. An alternative that can be used when score distributions are bimodal is the beta-binomial distribution proposed by Schuckers [Sch03]. All these approaches are based on a number of assumptions, which do not hold in practice, and this calls for a much more thorough evaluation of the score distributions along with the threshold criterion τ, which can be achieved with the help of receiver operating characteristic (ROC) curves.

2.2 The Receiver Operating Characteristic Curve

The ROC curve, frequently used in engineering applications [Ega75] and in measuring effectiveness of drugs in clinical studies [HM82], is obtained by plotting the different values that the FAR and FRR take with varying τ (the decision threshold or the cutoff point). The position of the ROC on the graph reflects the accuracy of the system, independent of any decision threshold that may be used. It covers all possible thresholds, with each point on the curve denoting the performance of the system for each possible threshold, expressed in terms of the proportions of true and false positive and negative results for that threshold. The curve would be higher for authentication devices that provide greater separation of the distributions for authentics and impostors (i.e., have higher accuracy) and lower for devices that provide lesser separations (i.e., have lower accuracy). The ROC of random guessing lies on the diagonal line.

The threshold adopted for a diagnostic decision is usually chosen so as to minimize the net costs and benefits of the error rates for a given application. For example, if security is the prime consideration, then τ will be so chosen as to give a low FAR. Different thresholds thus reflect different trade-offs between FAR and FRR — as τ increases, the FRR increases and FAR decreases and vice versa as τ decreases. If all costs could be measured and expressed in the same units, then this optimal threshold could be calculated for any ROC curve.

The ROC curve yields a concise graphical summary of the performance of any biometric authentication system. Figure 1 shows a typical ROC curve. A similar ROC can be drawn for FRR, but it supplies no additional information and hence does not require separate representation. A single measure of overall performance that is often used for an authentication system is the equal error rate (EER). This is defined as the point at which the FAR equals the FRR.

Moreover, combining ROC with modeling techniques can establish a stronger statistical basis for the diagnostic evaluation of the performance of an authentication system. Ishwaran and Gatsonis [IG00] exploited the correspondence between ordinal regression models and ROC estimation technique to develop hierarchical models for analyzing clustered data (with both heterogeneity and correlations) and used a Bayesian approach based on Markov chain Monte Carlo (MCMC) methods to model fitting. Although their application involved diagnostic radiology studies with multiple interpreters, some type of variation on their approach should be adaptable to fit the authentication framework and can obviate the need for assumptions such as equality of variances and independence underlying the binomial distributions [BPR00], which seldom hold for real images.

2.3 Collection of Test Data

The quality of the test data and the conditions under which they are collected influence any practical authentication system and must be taken into consideration. Poor-quality data increases the noise variance in a model, which in turn has an adverse effect on the ROC curves. Performance figures can be very application, environment, and population dependent, and these aspects

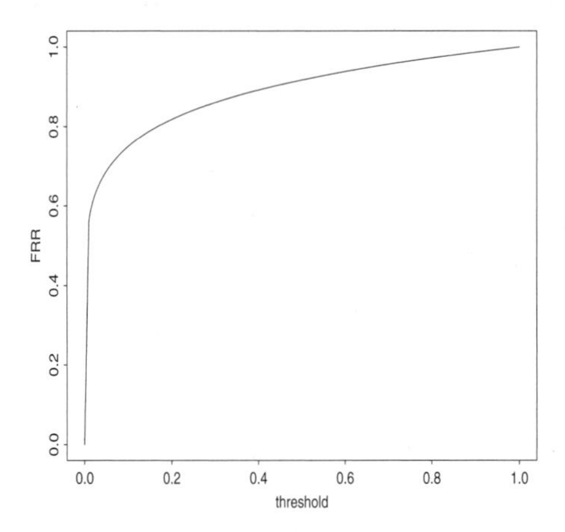

Fig. 1. An ROC curve showing the FRR.

should therefore be decided in advance. For instance, it is often helpful to know beforehand what the recommended image quality and matching decision thresholds are for a particular target application of an authentication system. These settings play a key role in determining the nature of the required database and hence in its collection procedure. Moreover, a knowledge of expected error rates (even if approximate) is greatly advantageous as it directly tells us the number of test images to use. In most situations, however, it is impossible to get hold of such prior information and may require some preliminary testing of systems to determine which factors are most significant and which may be safely ignored.

Enumerated below are a set of the factors that affect image quality and need to be considered for collecting facial images for evaluating an automated facial recognition system.

- Illumination — light intensity, light source angle, and background light.
- Pose created by camera angles.
- Movement of the subject — static, fast moving, or slow moving.
- Surroundings — crowded, empty, single subject, or multiple subjects.
- Spatial (number of pixels), gray-scale resolution, and clarity.
- Number of images collected from each individual.

All these different conditions affect any facial recognition system to a considerable extent and often methods that work well under one given situation do not work so well under other conditions. This calls for a refinement of the methods to handle all possible situations. Thus an understanding of the particular database characteristics is imperative for comparing and contrasting performance of different authentication systems.

3 Face Image Databases

We consider two different face databases here. The first one is a part of the "Cohn-Kanade AU-coded Facial Expression Database" [KCT00], consisting of images of 55 individuals expressing four different kinds of emotions — neutral, joy, anger, and disgust. Each person was asked to express one emotion at a time by starting with a neutral expression and gradually evolving into its peak form. The data thus consists of video clips of people showing an emotion, each clip being broken down into several frames. Figure 2 shows some sample images.

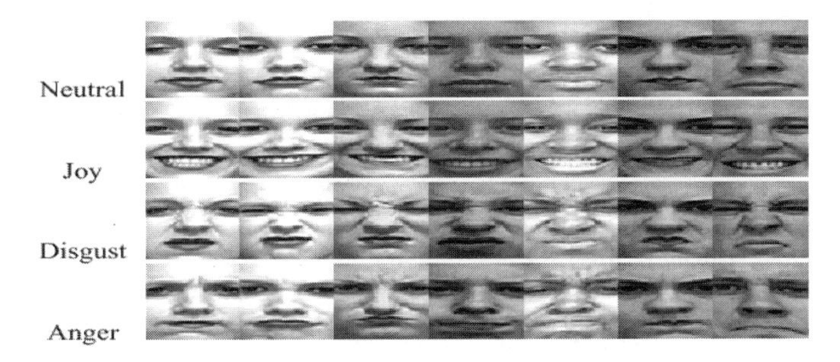

Fig. 2. Sample images of 7 subjects from the Cohn-Kanade database. Each column shows the four expressions of a subject.

The second dataset used is the publicly available "CMU-PIE Database" [SBB02], which contains 41,368 images of 68 people under 13 different poses, 43 different illumination conditions, and with 4 different expressions. This dataset is hence more diverse than the Cohn-Kanade dataset, which makes it more conducive to statistical analysis. Figure 3 shows some sample images from this dataset. We will work with only a small subset of the PIE database with neutral expressions but varying illumination.

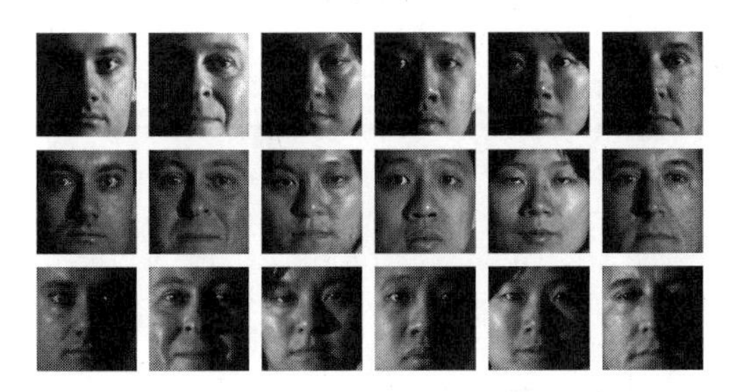

Fig. 3. Sample images of 6 subjects from the CMU-PIE database. Each column shows a subject under three different illumination conditions.

The reason for considering these two particular databases is that they represent the two most common conditions that occur in real images — expression variations and illumination variations. Hence any face authentication system faces these challenges, and we wish to study how efficient the systems we consider in this chapter are in handling them.

4 The MACE Filter

We first look at an existing authentication system called the minimum average correlation energy (MACE) filter. It is based on a simple linear filter (so called due to its application method, as is common in the engineering literature), is easy to implement, and has been reported to produce impressive results [VSV02]. We treat this only as an example of a typical face authentication system that exists today, point out the relative drawbacks from a statistical point of view, and describe simple tools to increase its statistical validity and rigor. Moreover, this system serves as a baseline for comparison with the model-based approach we propose in the next section.

The MACE filter is defined as:

$$\mathbf{h}_{MACE} = D^{-1}X(X^{+}D^{-1}X)^{-1}\mathbf{c}, \tag{4}$$

where X is a matrix of the vectorized 2D fast Fourier transforms (FFTs) of the training images of a person (X^{+} denoting the conjugate transpose), D is a diagonal matrix of the average power spectrum of the training images, and $\mathbf{c}$ is a column vector of ones. A filter is synthesized for each person in a database and applied to a test image via convolution. An inverse Fourier transform on the result yields the final output. If the test image belongs to an authentic person, a sharp spike occurs at the origin of the output plane indicating a match, while for an impostor, a flat surface is obtained suggesting a mismatch. A quantitative measure for authentication is the peak-to-sidelobe ratio (PSR), computed as PSR $= \frac{peak - mean}{\sigma}$, where *peak* is the maximum value of the final output, and the *mean* and the standard deviation σ are computed from a 20×20 sidelobe region centered at the peak (excluding a 5×5 central mask). PSR values are high for authentics and considerably lower for impostors. We do not include more details here owing to irrelevance and space constraints, but an interested reader is referred to Vijaya Kumar et al. [VSV02]. Figure 4 shows the MACE output for two images in the CMU-PIE database (using three training images per person). In both cases, the image has been so shifted as to display the origin at the center of the plane, as is conventional in most engineering applications.

4.1 Advantages and Disadvantages

The MACE filter is a non-model-based empirical methodology involving heuristics in the authentication procedure. The main drawback of the MACE

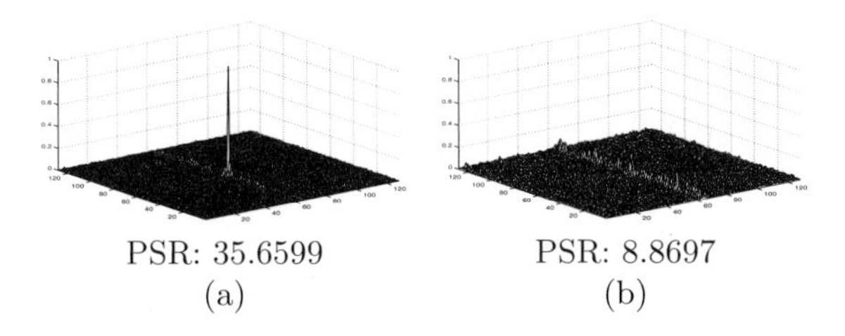

Fig. 4. MACE filter output for (a) an authentic and (b) an impostor.

authentication system is its sensitivity to distortions that occur commonly in practice. Vijaya Kumar [Vij92] describes the technique by which distortion tolerance can be built into the filter. The resulting filter is obtained by replacing D in $\mathbf{h}_{MACE}$ by $\alpha D + \sqrt{1-\alpha^2}I$, which is given by:

$$\mathbf{h} = (\alpha D + \sqrt{1-\alpha^2}I)^{-1}X[X^{+}(\alpha D + \sqrt{1-\alpha^2}I)^{-1}X]^{-1}\mathbf{c}, \tag{5}$$

with $\alpha \in [0,1]$. $\alpha = 1$ gives $\mathbf{h}_{MACE}$, while $\alpha = 0$ gives a non-MACE filter. It has been found to be effective for handling illumination changes (Savvides and Vijaya Kumar [SV03]) but deteriorates considerably in the presence of other perturbations. For example, when applied to images from the Cohn-Kanade database, the results are not satisfactory. Figure 5 shows the ROC curve obtained by plotting the FAR and the FRR for different thresholds on the PSR values for the two datasets. While we observe an EER of 0.9% for a threshold PSR value of around 20 for PIE, a relatively higher EER of 32% at a threshold PSR value of around 30 is obtained for the Cohn-Kanade database, which shows the inefficiency of MACE in the presence of expression variations. Moreover, the d' statistics in Table 1 corroborate all these findings by showing that the PIE database has a bigger separation between the authentic and impostor PSRs than the Cohn-Kanade database and hence is easier to authenticate. The d' is a simple statistical measure defined as:

$$d' = \frac{\mu_1 - \mu_2}{\sqrt{(\sigma_1^2 + \sigma_2^2)}}, \tag{6}$$

where μ_1 and μ_2 are the means of the two distributions to be compared and σ_1 and σ_2 are the respective standard deviations. The two distributions can be the distribution of the similarity measure for the authentics and the impostors, and hence a bigger d' signifies greater ease of authentication [BPR00].

Table 1. d' statistics for the two datasets

Database	# of Authentics	# of Impostors	d'
Cohn-Kanade	495	26730	1.0330
CMU-PIE	1365	87360	3.4521

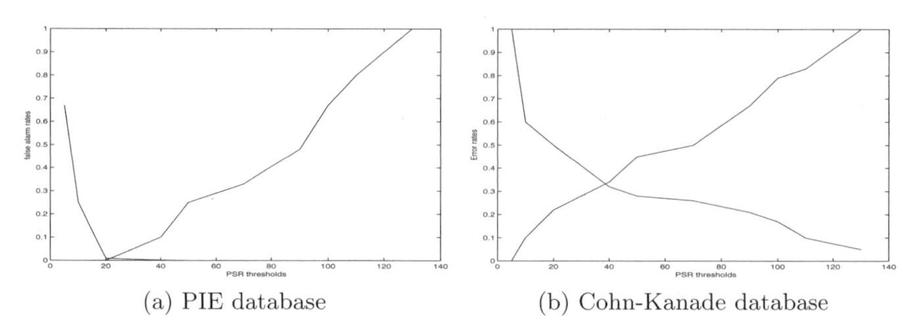

(a) PIE database (b) Cohn-Kanade database

Fig. 5. ROC curve for authenticating the two databases. The descending curve represents the FAR and the ascending one represents the FRR. The point of intersection of the two curves gives the EER.

The success of the MACE system also depends critically on selecting a suitable training set [Vij92]. A bigger N is often required to be able to represent all possible distortions, which, on the other hand, makes computations harder. So far, the choice of N has been solely based on experimental studies, and it is sensitive to the nature of the images in a database. No concrete guidelines exist to show how the number of training images affects the error rates in a given situation. Apart from this, the choices of the sidelobe dimension and α, in the case of the distortion-tolerant MACE, are also based on experimentation. So it is necessary to study their effects on the PSR values, and hence on the authentication results. Similarly, no analysis has been reported so far on how the PSR values and the results vary with the nature of the images (e.g., levels of distortions, resolution). Moreover, some associated measures of the variability in the PSR estimates like standard errors and confidence intervals should be provided so as to assess their reliability. Given the significance of PSR in the authentication process, developing its statistical aspects is expected to establish a firmer basis for the entire MACE technology.

The semblance of the distortion-tolerant version of MACE with ridge regression provides a scope for employing statistical methods like cross-validation or bootstrapping for obtaining α rigorously. In particular, a technique similar to the one described by Golub et al. [GHW79] using the generalized cross-validation method to choose a good ridge parameter can be adapted to estimate α. Alternative methods for introducing distortion tolerance into the filter include shrinkage estimators like James-Stein, stabilizing techniques, or Bayesian models. We do not explore this in depth in this chapter.

4.2 Statistical Analysis of PSR

Since PSR forms the MACE authentication score, its changes are closely related to changes in its performance and all statistical analyses should be based on those. Some particular statistical aspects of PSR that we are interested in investigating are: (1) determine the effect of different image properties (resolution, quality) and filter parameters (sidelobe dimension, number of training images, α) on the PSR value and (2) develop standard errors, confidence intervals for the PSR values and the error rates (FAR, FRR). Preliminary exploratory studies show that authentication results deteriorate with increasing resolution (Fig. 6), more training images (Fig. 7), and increasing sidelobe dimension (Fig. 8). In all these cases, the impostor PSRs get inflated thus causing an increased chance of false authentication.

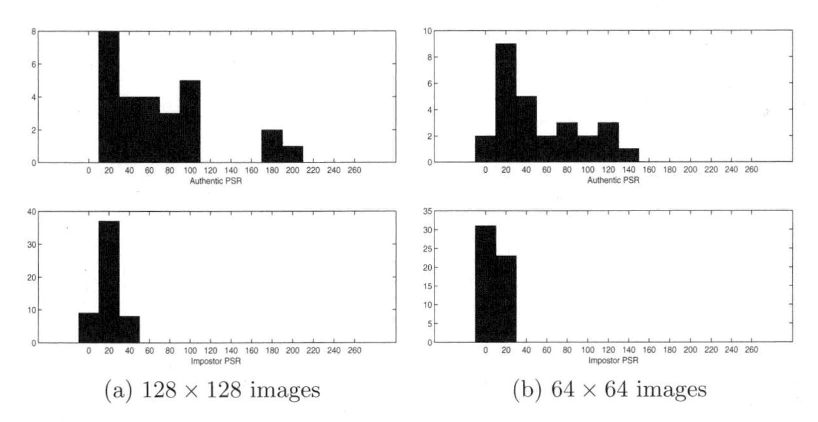

(a) 128 × 128 images (b) 64 × 64 images

Fig. 6. PSR values for an authentic and an impostor using images of different resolutions.

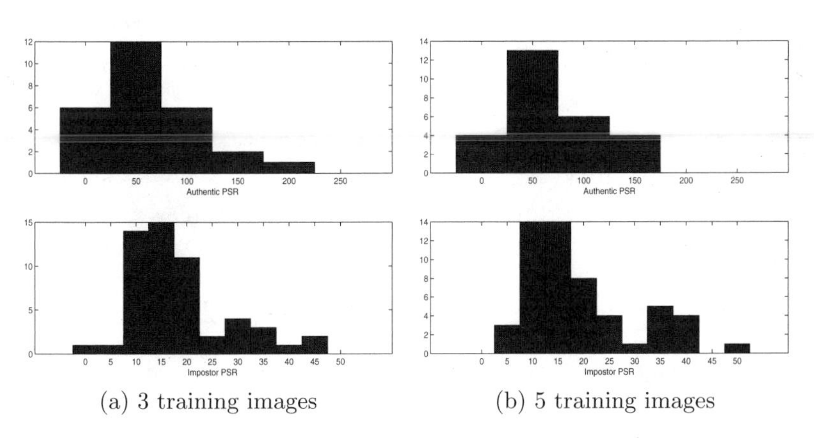

(a) 3 training images (b) 5 training images

Fig. 7. PSR values for an authentic and an impostor using different number of training images.

The above exploratory analyses show quite clearly that the authentication performance of MACE is highly influenced by certain image properties and filter parameters. The former depend on the database collection process,

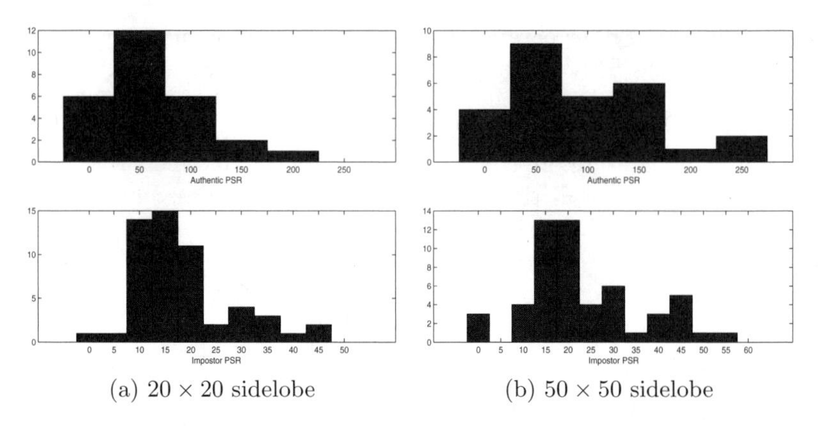

(a) 20 × 20 sidelobe (b) 50 × 50 sidelobe

Fig. 8. PSR values for an authentic and an impostor using different dimensions for the sidelobe region.

whereas the latter are chosen by a user of the system. The effect of all these factors on the PSR values (and hence on the authentication) can be studied with the help of statistical *regression models*. The regression coefficients will quantitatively determine to what extent the PSR value changes due to a change in a particular covariate value (say, when the number of training images used increases by one). Such a model can thus predict the PSR value of an observed image once its properties are known and can also help in determining the optimal levels of the filter design parameters for the best authentication results. It can therefore be used to provide guidelines for both the data collection and the filter design procedures, so that one knows exactly what to expect in a given situation.

The residuals from the fitted models can provide an estimate of the PSR distribution also. This in turn helps to compute standard errors and confidence intervals, which provide a means for assessing the reliability of the values. Although PSR forms the MACE authentication criterion, it suffers from a lack of a concrete threshold and is subjective. Moreover, the PSR distribution can be used to estimate the probabilities of false detection (FAR, FRR), also to devise statistical tests to determine if the error rates meet a specific criterion threshold, and to detect significant differences between the authentic and impostor PSR values. Alternatively, asymptotic methods [like central limit theorem (CLT)] can be used under some mild conditions for simplifying computations.

A simple linear model with the logarithm of PSR as the response is:

$$\log(\mathrm{PSR}_i) = \beta_0 + \beta_1 x_{1i} + \beta_2 x_{2i} + \ldots + \beta_p x_{pi} + \varepsilon_i, \quad (7)$$
$$\varepsilon_i \sim^{iid} N(0, \sigma^2), \ i = 1, \ldots, N,$$

where $x_1, \ldots, x_p$ are the p potential covariates, $\beta_0, \beta_1, \ldots, \beta_p$ are the regression coefficients, ε is the error or noise, and N is the number of observations. We use the logarithm transformation to adjust for nonnegative values (PSR values are nonnegative) and achieve variance stabilization to some extent. The possible covariates under consideration are enumerated in Table 2.

Table 2. The potential covariates for the regression models

Image Properties	Filter Characteristics
Authentic/impostor (binary)	Number of training images
Distortions like expression, illumination, noise, occlusion (categorical)	Noise tolerance parameter α
Image resolution	Sidelobe dimension

We fitted a simple model using the PIE database with two binary covariates, one denoting balanced or unbalanced illumination and one denoting whether the particular person is an authentic or an impostor. The model is:

$$\log(\mathrm{PSR}_i) = 1.8378 + 1.9796 \times \mathrm{authentic}_i + 0.0193 \times \mathrm{illum}_i, \; i = 1, \ldots, N. \quad (8)$$

Both the covariates have a significant effect on PSR (p-values < 0.0001), that of the variable denoting authenticity being much stronger. Some sample predictions based on this model are reported in Table 3 and a histogram of the residuals is presented in Fig. 9, which look to be approximately normal.

Table 3. Predicted PSR values for different covariate values in Model 1

Covariate	Predicted log(PSR)	Predicted PSR
Authentic & balanced illum.	3.8174	45.4858
Authentic & unbalanced illum.	3.8367	46.3722
Impostor & balanced illum.	1.8378	6.2827
Impostor & unbalanced illum.	1.8571	6.4051

Assumption Checks

A histogram of the residuals from the above model is shown in Fig. 9. They are seen to be approximately normally distributed. Other model diagnostics like Q–Q plots also did not show any major deviation from the assumption of normality.

Figure 10 shows the histogram of all PSR and log(PSR) values from the PIE database. The distribution of the raw PSR values is highly positively skewed and has large variation, whereas log(PSR) seems to be more amenable to a normal distribution (more symmetric although not perfectly). The variability is also considerably reduced and this justifies our use of log(PSR) as the outcome variable instead of the raw PSR values. These observations are

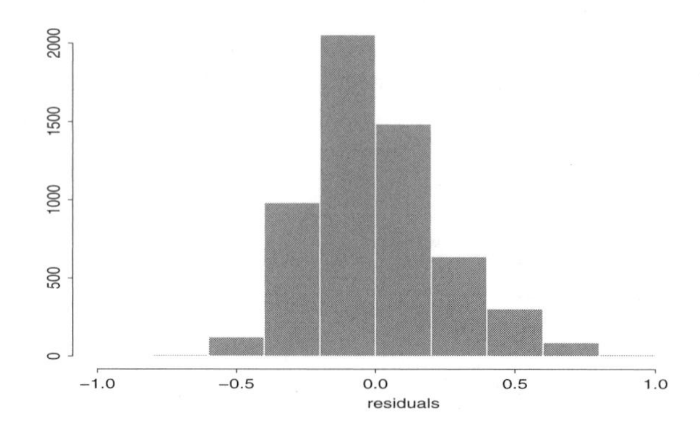

Fig. 9. Distribution of the residuals from the simple model.

Table 4. Skewness and kurtosis estimates for the PSR distributions

Variable	Measure	Authentic	Impostor	All Combined
PSR	Skewness	1.2389	-0.1254	6.9209
	Kurtosis	3.8564	3.0722	91.7597
log(PSR)	Skewness	0.3437	-0.7087	0.3954
	Kurtosis	2.8386	1.5201	4.1944

further corroborated by the sample skewness and kurtosis coefficients shown in Table 4. They show more rigorously how far removed the combined PSR distribution is from normality despite the separate authentic and impostor distributions being relatively closer to normal distributions.

The assumption of independence across the PSR values from the different images belonging to the same individual does not hold. However, the PSR values for different individuals can be safely assumed to be independent, thus representing a classical longitudinal data framework. But, on the other hand, correlations introduce more parameters and increase the complexity of the model. No drastic deviation from the linearity assumptions is observed. Moreover, nonlinear models also make parameter interpretation more complex. Thus overall, the linear models seemed to be useful and valid in providing the initial sample framework for understanding the behavior of the PSR values. The violation of the independence assumption is the most critical that needs to be addressed.

Confidence Intervals

The 95% confidence intervals for the authentic and impostor PSR and log(PSR) values computed using the regression model are shown in Table 5. Note that, by varying the values of the covariates, similar confidence intervals can be constructed for PSR values under different conditions, which will provide really helpful guidelines to users of the MACE system.

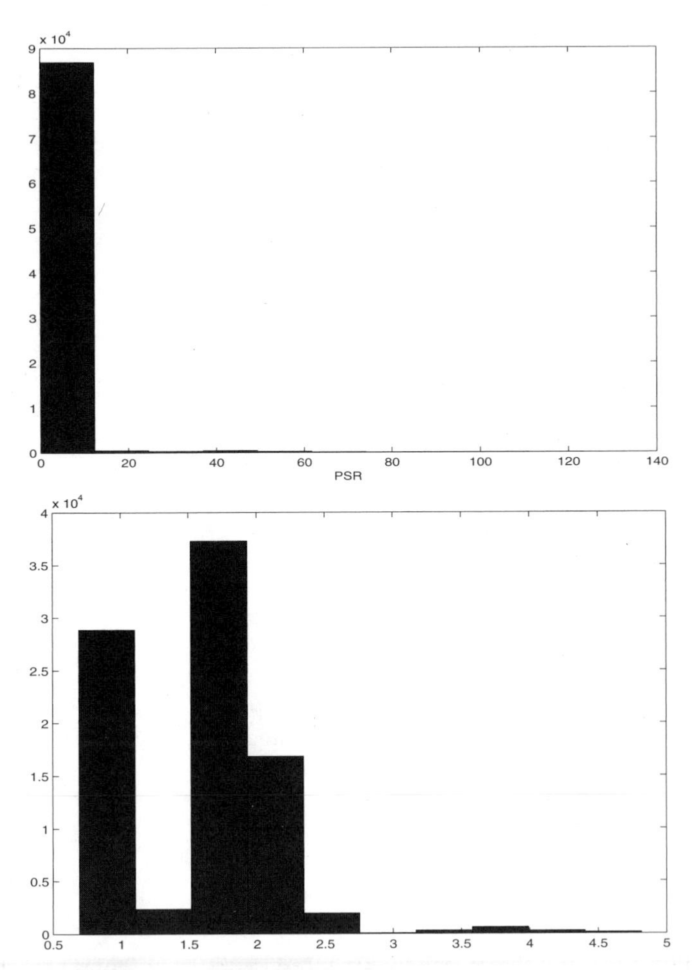

Fig. 10. Histograms of the PSR values (left) and log(PSR) values (right) from the CMU-PIE database.

Table 5. 95% confidence intervals for authentic and impostor PSR values (both raw and logarithm). The CIs for raw PSR are obtained by exponentiating those for log(PSR)

Variable	Lower 95% CI	Upper 95% CI
Authentic log(PSR)	3.1004	4.0734
Impostor log(PSR)	1.0219	1.9949
Authentic PSR	22.2068	58.7564
Impostor PSR	2.7785	7.3515

Table 6 shows the estimates of the error rates, the associated standard errors, and 95% confidence intervals for the optimal values of FAR and FRR in the CMU-PIE database (equal to the EER) computed using a binomial distribution. Similar confidence intervals can also be calculated at the different thresholds, but those at the EER are most important as it provides the overall measure of performance for an authentication device.

Table 6. Means, standard deviations, and 95% confidence intervals for the error rates from authenticating the PIE database using the MACE filter

Error rate	$\hat{p}$	$\hat{\sigma}$	n	Lower CI	Upper CI
FRR	0.009	0.0026	1365	0.0040	0.0140
FAR	0.0086	0.0003	87360	0.0080	0.0092

5 Statistical Model-Based Systems

Studying the MACE system and the statistical aspects of the PSR values obtained from it have given us an insight into the rudiments of a rigorous authentication system. We now proceed to explore options for building statistical model-based tools. Research on face modeling has so far been more or less confined to the spatial domain, some common models being Markov random fields (MRF) [LZS01] and principal components analysis (PCA) [TP91]. However, these spatial models, despite providing a good fit to face data, are inadequate for efficient classification since they largely ignore the phase component of the face image spectrum, which plays a vital role in face-based classification (discussed at length in the next section). This leads us to build models directly in the spectral domain, a novel approach to authentication as per our knowledge.

5.1 The Spectral Domain

Let $x(n_1, n_2)$ denote the original 2D image. Then the *image spectrum* X is defined by the discrete Fourier transform (DFT) [Lim90]:

$$\begin{aligned} X(j,k) &= \frac{1}{N_1 N_2} \sum_{n_1=0}^{N_1-1} \sum_{n_2=0}^{N_2-1} x(n_1, n_2) \mathrm{e}^{-i2\pi(n_1 j/N_1 + n_2 k/N_2)} \\ &\overset{\text{(polar form)}}{=} |X(j,k)| \mathrm{e}^{i\theta_x(j,k)}, \ j = 0, 1, \ldots, N_1 - 1, \\ &\quad k = 0, 1, \ldots, N_2 - 1, \end{aligned} \tag{9}$$

where $|X(j,k)|$ is called the *magnitude* and $\theta_x(j,k)$ the *phase*. For a typical image, these components are shown in Fig. 11. Many signal processing appli-

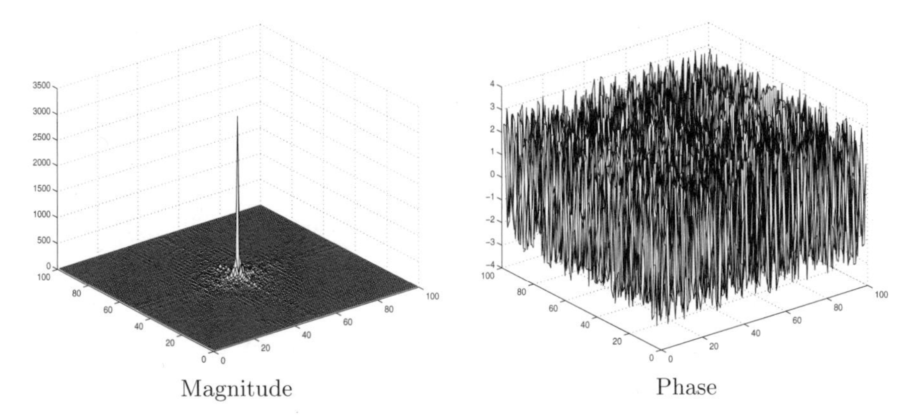

Fig. 11. The Fourier domain components of a face image.

cations in computer engineering involve this frequency-domain representation of signals, rather than the original signal itself in the spatial domain. Operations performed in one domain have corresponding operations in the other (linearity, scaling, convolution, multiplication, symmetry, etc.), thus demonstrating a link between the two domains. One of the primary reasons to prefer the spectral domain is that it often simplifies computations considerably. For example, the operation of convolution in the spatial domain is equivalent to the simple multiplication operation in the spectral domain.

Hayes [Hay82] describes an experiment that dramatically illustrates that phase captures more of the image intelligibility than magnitude. It consists of reconstructing images of two people from their Fourier coefficients by swapping their phase and magnitude components (Fig. 12). Clearly, both the reconstructed images bear more resemblance to the original image that contributed the phase. This establishes the significance of phase in face identification, hence ignoring it in modeling may have severe consequences on authentication tools based on it.

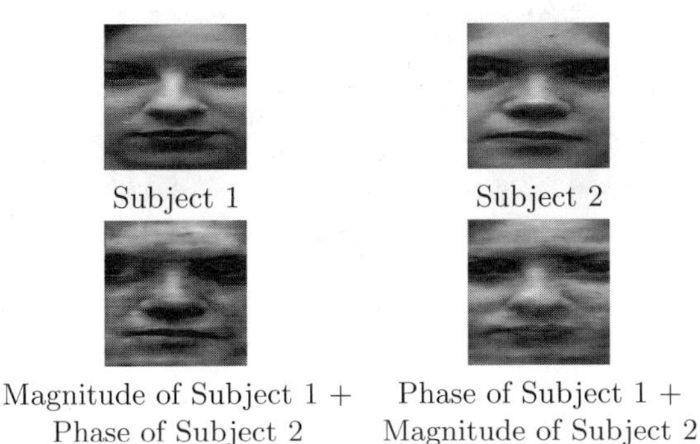

Fig. 12. The importance of phase.

The Fourier transforms of most typical images have energy (analogous to mass in the spatial domain) in the frequency domain concentrated in a small region near the origin. This is because images usually have large regions where the intensities change slowly [Lim90]. Thus any image can be modeled without significant loss of quality and intelligibility from a small fraction of the transform coefficients near the origin, a notion that is useful in any modeling strategy. Figure 13 shows some images reconstructed by setting most of the Fourier coefficients to zero. However, the high-frequency (low-energy) components represent facial structures containing discriminating information for recognition.

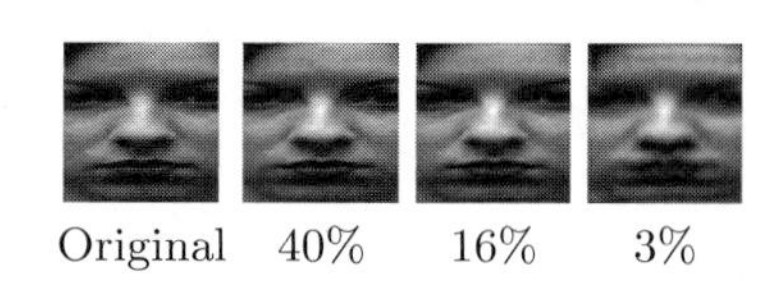

Fig. 13. Reconstruction using a few Fourier coefficients.

We thus aim at modeling the image spectrum directly to exploit the valuable information contained in it, especially in the phase. The goal therefore is to generate statistical models to adequately represent an appropriate number of frequency coefficients around the origin that retain identifiability to a reasonable extent. For example, the last face in Fig. 13 has lost some fine details and is less recognizable.

5.2 Analysis of the Image Spectrum

We will consider the PIE database for all the modeling experiments. Figure 14 shows how illumination varies over all the images of a person in this database.

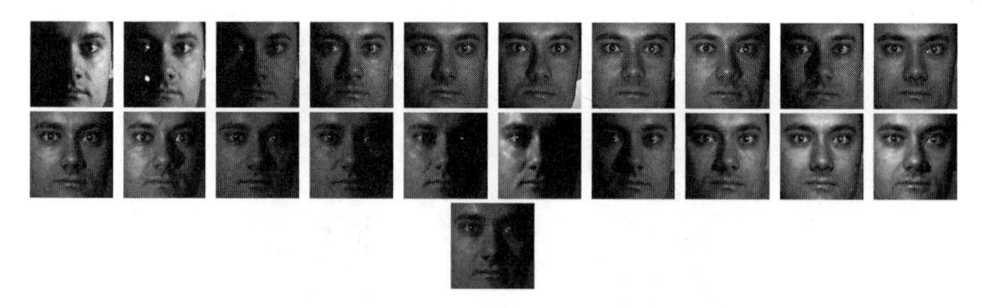

Fig. 14. The 21 images of a person with illumination variations.

In the beginning, some exploratory analyses are performed to study the behavior of the frequency components of an image. We first consider one-dimensional series of magnitude and phase components along a row or a column of the spectral plane. Without loss of generality, we chose the row that passes through the origin of the spectrum. Let us define $L_{0,j}^{n,m}$ and $P_{0,j}^{n,m}$ respectively as the log-magnitude and the phase at the jth frequency on the row through the origin of the mth neutral image from the nth person, $j = 1, 2, \ldots$, $n = 1, \ldots, 65$, $m = 1, \ldots, 21$. We chose the logarithm of the magnitude to account for the nonnegativity. Figure 15 shows the plot of the logarithm of the magnitudes for three images of a person for the chosen row. Note that these plots exhibit a dominating trend component, which, however, is not present in phase. This led us to construct the residual series for both the components

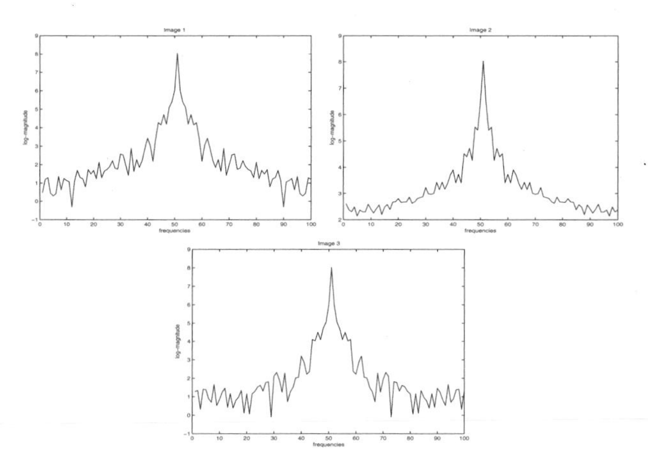

Fig. 15. Plots of the log-magnitude series for the row through the origin of the spectrum.

by subtracting the respective frequencywise means defined as:

$$\bar{L}_{0,j}^{n} = \frac{1}{21} \sum_{m=1}^{21} L_{0,j}^{n.m}, \quad \bar{P}_{0,j}^{n} = \tfrac{1}{21} \sum_{m=1}^{21} P_{0,j}^{n.m}, \quad \forall j, n. \tag{10}$$

The residual series are then computed as:

$$Y_{0,j}^{n,m} = L_{0,j}^{n,m} - \bar{L}_{0,j}^{n}, \quad Z_{0,j}^{n,m} = P_{0,j}^{n,m} - \bar{P}_{0,j}^{n}, \quad \forall j, m, n. \tag{11}$$

Figures 16 and 17 respectively show the residual log-magnitude and residual phase of the images of a person for the row through the origin of the spectrum, along with a plot of the respective means across the 21 frames of that person. For space constraints, we include here only the plots for the first 8 frames.

All these plots show that both the log-magnitude and the phase components are symmetric around the origin. The residual log-magnitude series do not have a pronounced trend component as the original series (Fig. 15).

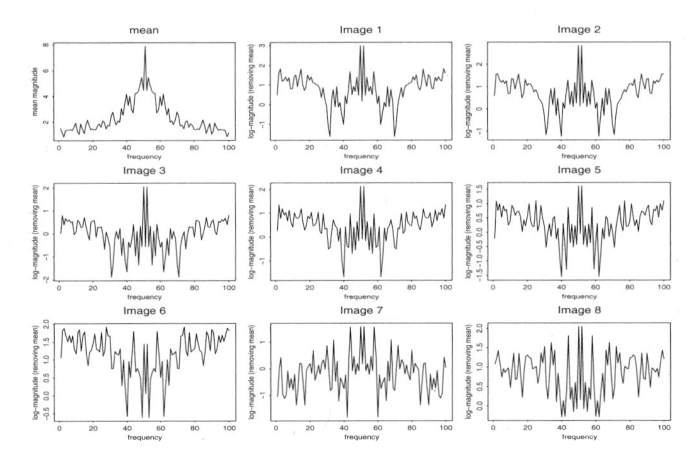

Fig. 16. Residual log-magnitude series and the mean for the row through the origin of the spectrum.

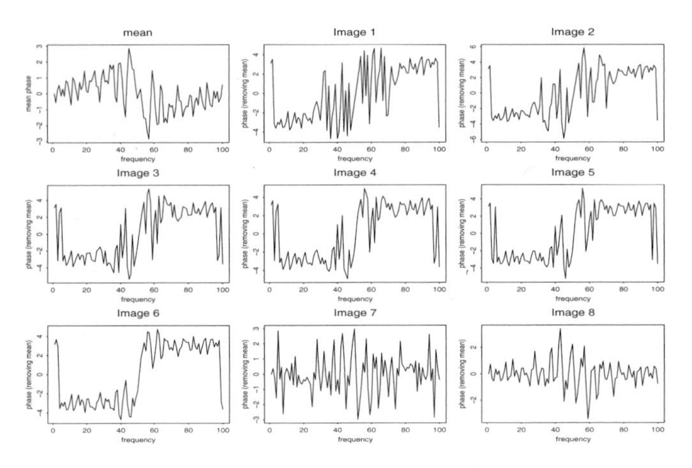

Fig. 17. Residual phase series and the mean for the row through the origin of the spectrum.

Although the magnitude series seems to have a similar structure across the different frames, the phase series seemed to differ considerably across the different frames of a person that represent different illumination conditions. All these observations suggest that the magnitude component has a clearer structure than phase, does not vary considerably with distortions, and it may be possible to capture its variation with the help of simple quantities like the mean, unlike phase. Figure 18 shows the 2D plots of the raw log-magnitude and phase (not the residuals) for these eight frames of a person.

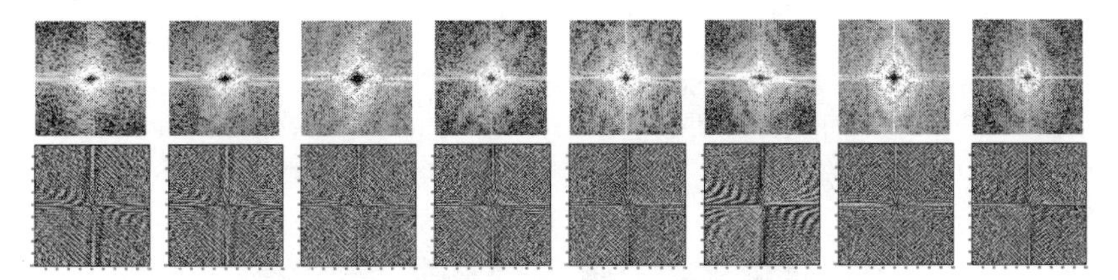

Fig. 18. 2D plots of the logarithm of the magnitude spectrum (top) and phase spectrum (bottom).

5.3 Difficulties in Phase Modeling

Perhaps the greatest challenge facing phase models is its "wrapping around" property, which is depicted in a schematic in Fig. 19. The phase angle lies in the range $(-\pi, \pi)$, and any model based on stationarity assumptions fails completely to represent this and hence loses discriminative information pertaining to identification. All these make modeling the phase angle a difficult task, but we have learned that ignoring even the slightest phase information leads to drastic results. The magnitude, on the other hand, does not suffer from these drawbacks and can be modeled using any traditional statistical approach.

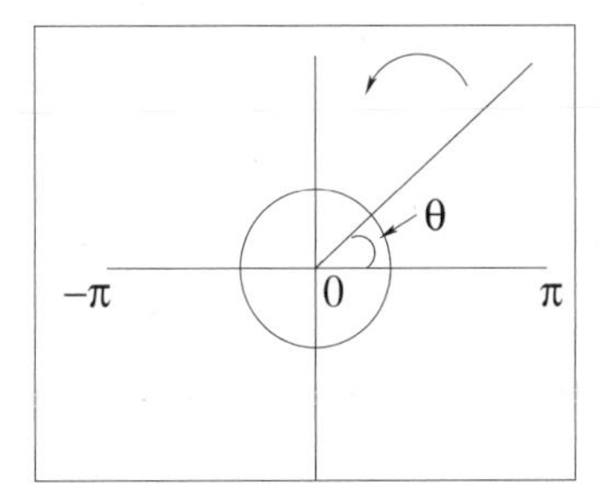

Fig. 19. The "wrapping around" property of the phase component. θ denotes the phase angle.

The other difficulty in modeling involves representation of phase information. It is a common practice in statistical modeling experiments to use some representative measure as the suitable quantity to model, for example, the mean or the principal components. This not only simplifies the model considerably but also reduces model dimensionality to an extent. However, this can only be applied in case a representative quantity exists that is able to capture all the relevant variation present in the data at hand. To study to what extent the image identifiability is captured by the mean log-magnitude and mean phase or the corresponding principal components, we performed some empirical analyses from which we concluded that although both mean and principal components are capable of representing magnitude information, this is not possible with phase. Figures 20 and 21 show respectively reconstructed images of a person with the mean Fourier components and using the projections of the first five principal components.

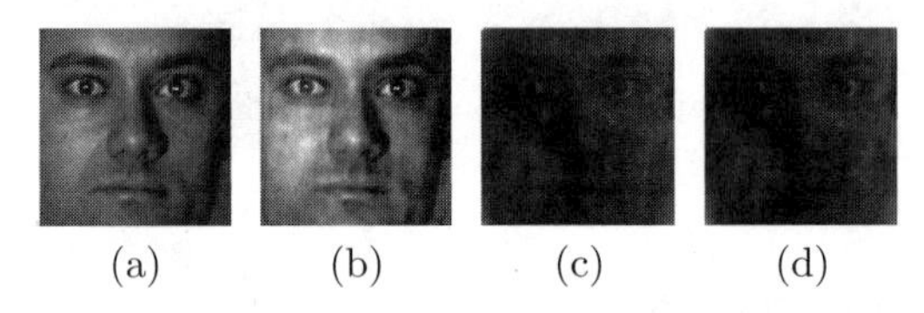

Fig. 20. (a) Original image, (b) reconstructed image using mean log-magnitude, (c) reconstructed image using mean phase, (d) reconstructed image using both mean log-magnitude and mean phase.

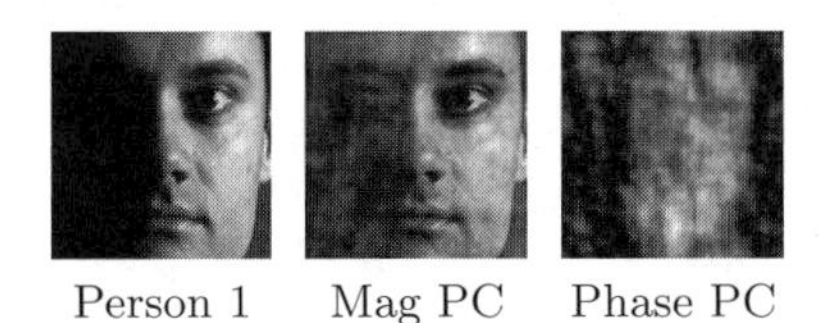

Fig. 21. Reconstructed images of two people using the projections of the top 5 principal components of log-magnitude and phase onto the original components.

Our exploratory analyses have thus shown that it is difficult to capture phase information effectively with the help of a single quantity. Moreover, phase is affected by variations in the images such as illumination and expression changes, and it is extremely difficult to understand how relevant information is distributed among the phase components at the different frequencies. Application of smoothing techniques and transformations also lose crucial information and do not prove helpful.

5.4 An Initial Simple Model

Given that phase changes considerably with illumination variations, we decided to take this into account by dividing the entire set of images for a person into a number of subsets depending on the nature and amount of illumination variations in them and building separate models for them. These subsets and the constituent images are tabulated in Table 7. Note that this division is subjective and done by eyeballing the images in Fig. 14.

We first look at the six images with completely balanced lighting. Figure 22 shows an original image and an image reconstructed using the mean phase and the mean log-magnitude. The reconstructed image is much more identifiable than that obtained earlier using the mean across all the 21 images of a person shown in Fig. 20.

Table 7. The division of the 21 images of a person into subsets of varying illumination. The numbers denote the image positions in Fig. 14 (rowwise)

Subset	Images
Balanced	7,8,10,11,19,20
Right shadows	12,15,16,21
Left shadows	1,2,4,9,17,18
Overall dark	3,5,6,13,14

Fig. 22. Reconstructed images of a person using mean phase and mean log-magnitude over the 6 images showing balanced lighting.

Encouraged by this, we proceed to fit the actual models. Our initial models are independent bivariate Gaussians for log-magnitude and phase at each frequency, one for each type of illumination condition for each person. Formally, we can define these models as:

$$\begin{aligned} L_{ij}^{(m,n)} &= \mu_{ij}^{(m,n)} + \varepsilon_{ij}^{(m,n)}, \\ P_{ij}^{(m,n)} &= \nu_{ij}^{(m,n)} + \eta_{ij}^{(m,n)}, \\ & i,j = 1,\ldots,K,\ m = 1,\ldots,4,\ n = 1,\ldots,N, \end{aligned} \tag{12}$$

where K denotes the number of frequencies that we wish to model, N the total number of individuals in the dataset, and m a specific illumination subset. L_{ij} and P_{ij} are respectively the log-magnitude and phase for frequency (i,j) for the mth illumination subset and the nth person, μ_{ij} and ν_{ij} are the corresponding means, and ε_{ij} and η_{ij} are the respective error terms. We build separate models for log-magnitude and phase since there does not seem to exist any significant cross-correlation among these two components, and we do this separately for each individual in the database. The errors are distributed as:

$$\varepsilon_{ij}^{(m,n)} \sim N(0, \sigma_{1,ij}^{2(m,n)}), \quad \eta_{ij}^{(m,n)} \sim N(0, \sigma_{2,ij}^{2(m,n)}), \tag{13}$$

and they are independent of each other. The frequencywise mean and variance parameters are estimated from all the sample images in each subset for each person.

The reconstructed images using the simulated models for the two Fourier components for the balanced lighting subset are shown in Fig. 23. They resemble the original image to an appreciable extent and moreover, we observe that it is sufficient to model only a few lower-frequency components around the origin (a 50×50 grid around the origin). Figure 24 shows the images using simulated models for only the low-frequency components based on the other three subsets of images. As expected, the images that are overall dark are the hardest to model.

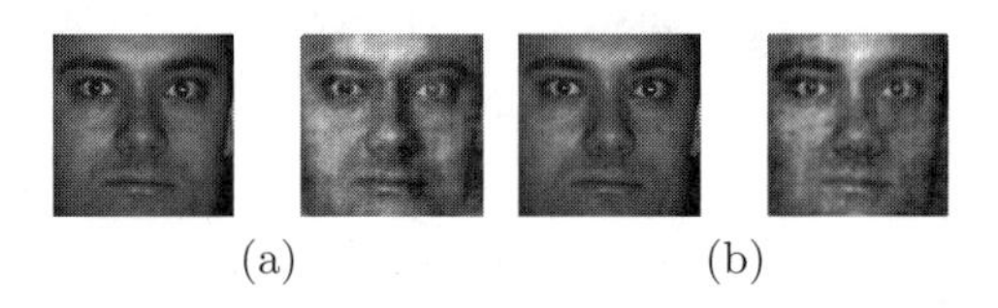

(a) (b)

Fig. 23. Reconstructed images with the simulated phase and log-magnitude components from Gaussian distributions using (a) all frequency components and (b) components within a 50×50 region around the origin and zeroing out higher frequencies.

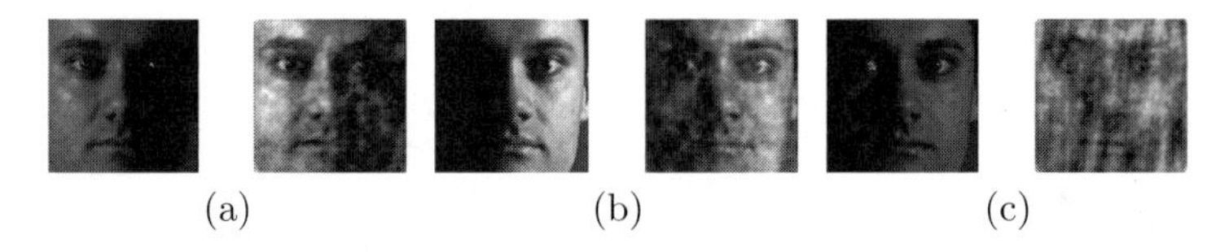

(a) (b) (c)

Fig. 24. Reconstructed images using simulated phase and log-magnitude components from Gaussian distributions within a 50×50 region around the origin using images in (a) right shadows, (b) left shadows, (c) overall darkness.

The simulation results are found to deteriorate much if we use the same standard deviation across all the frequencies in the 50×50 grid (assuming homoscedasticity), particularly for phase. This is clearly evident from the reconstructed images appearing in Fig. 25, which are much worse than those shown in Fig. 23. These correspond to the case with balanced lighting, but similar poor results are obtained for the other subsets, too.

This model can also help us study the illumination effects. Let us rewrite the model means as:

$$\begin{aligned} \mu_{ij}^{(m,n)} &= \mu_{ij}^{(n)} + \alpha_{ij}^{(m,n)}, \\ \nu_{ij}^{(m,n)} &= \nu_{ij}^{(n)} + \beta_{ij}^{(m,n)}, \; i,j = 1,\ldots,K, \; m = 1,\ldots,4, \; n = 1,\ldots,N, \end{aligned} \tag{14}$$

so that α_{ij} and β_{ij} respectively denote the effects of the four subsets representing illumination variations. $\mu_{ij}^{(n)}$ is then the common mean effect (over

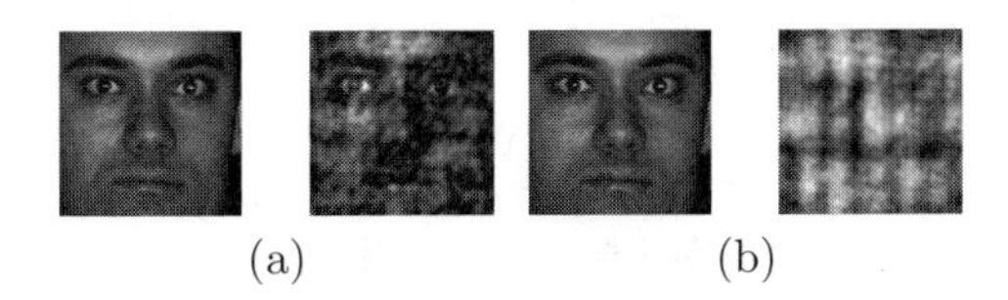

Fig. 25. Simulated images from independent Gaussian distribution using same standard deviation across frequencies within a 50×50 grid around origin for (a) log-magnitude, (b) log-magnitude and phase.

all the 21 images of each person) and the illumination effects can then be estimated by $\widehat{\mu_{ij}}^{(n)} - \widehat{\mu_{ij}}^{(m,n)}$. By isolating these effects, we can study how illumination changes affect the Fourier domain components and also study how these changes occur over the entire spectral plane since lights do not affect all regions uniformly.

Classification Results

Given that we obtain satisfactory model fit, the next step is to use these models for classification. If $f(L_{ij}^{(m,n)})$ and $g(P_{ij}^{(m,n)})$ respectively denote the probability densities of the above Gaussian distributions, the conditional likelihoods of log-magnitude and phase for each person and each subset are:

$$\begin{aligned} f(L|n,m) &= \Pi_{i=1}^{K} \Pi_{j=1}^{K} f(L_{ij}^{(m,n)}), \\ f(P|n,m) &= \Pi_{i=1}^{K} \Pi_{j=1}^{K} g(P_{ij}^{(m,n)}), \quad m = 1, \ldots, 4, \quad n = 1, \ldots, 65. \end{aligned} \tag{15}$$

In a Bayesian framework, the conditional likelihood or posterior probability of a test image belonging to a specific person under a given illumination condition is:

$$\begin{aligned} f(n,m|L) &\propto f(L|n,m)p(n,m), \\ f(n,m|P) &\propto f(P|n,m)p(n,m), \end{aligned} \tag{16}$$

where $p(n,m)$ denotes the prior joint probability for each subset for each person, assumed to be uniform for the time being. The normalizing constants here are $f(L) = \sum_{m,n} f(L|n,m)p(n,m)$ and $f(P) = \sum_{m,n} f(P|n,m)p(n,m)$ respectively for magnitude and phase. Since these are the same across all m, n, we can safely ignore them for the purpose of classification. Then according to Bayes' rule, a particular image with predetermined illumination condition m_0 is then assigned to class C if:

$$C = \arg\max_{n} \left\{ f(n, m_0|L) \times f(n, m_0|P) \right\}. \tag{17}$$

We obtain perfect classification results using this simple likelihood-based classification scheme, yet this model is restrictive given that it requires the illumination condition of a test image to be determined manually prior to training and classification. Given the vast amount of data usually available, it is imperative that such a process be automated for the method to be useful in practice. This leads us to consider a more flexible modeling approach using *mixture models*, which we present in the next section. Nevertheless, the model here acts as a useful baseline for our future modeling endeavors.

5.5 Gaussian Mixture Models

As any continuous distribution can be approximated arbitrarily well by a finite mixture of Gaussian densities, mixture models provide a convenient semiparametric framework in which to model unknown distributional shapes [MP00]. It can handle situations where a single parametric family is unable to provide a satisfactory model for local variations in the observed data. The model framework is briefly described below.

Let $(\mathbf{Y_1}, \ldots, \mathbf{Y_n})$ be a random sample of size n where $\mathbf{Y_j}$ is a p-dimensional random vector with probability distribution $f(\mathbf{y_j})$ on $\mathbb{R}^p$, and let $\boldsymbol{\theta}$ denote a vector of the model parameters to be estimated. A g-component mixture model can be written in parametric form as:

$$f(\mathbf{y_j}; \boldsymbol{\Psi}) = \sum_{i=1}^{g} \pi_i f_i(\mathbf{y_j}, \boldsymbol{\theta_i}), \tag{18}$$

where $\boldsymbol{\Psi} = (\pi_1, \ldots, \pi_{g-1}, \boldsymbol{\xi}^T)^T$ contains the unknown parameters and $\boldsymbol{\xi}$ is the vector of the parameters $\boldsymbol{\theta_1}, \ldots, \boldsymbol{\theta_g}$ known *a priori* to be distinct. Here, $\boldsymbol{\theta_i}$ represents the model parameters for the ith mixture component, and $\boldsymbol{\pi} = (\pi_1, \ldots, \pi_g)^T$ is the vector of the mixing proportions with $\sum_{i=1}^{g} \pi_i = 1$.

In case of Gaussian mixture models, the mixture components are multivariate Gaussian given by:

$$\begin{aligned} f(\mathbf{y_j}; \boldsymbol{\theta_i}) &= \phi(\mathbf{y_j}; \boldsymbol{\mu_i}, \Sigma_i) \\ &= (2\pi)^{-1} |\Sigma_i|^{-\frac{1}{2}} \exp\Big\{ -\frac{1}{2} (\mathbf{y_j} - \boldsymbol{\mu_i})^T \Sigma_i^{-1} (\mathbf{y_j} - \boldsymbol{\mu_i}) \Big\}, \end{aligned} \tag{19}$$

so that the parameters in $\boldsymbol{\Psi}$ are the component means, variances, and covariances, and the mixture model has the form:

$$f(\mathbf{y_j}; \boldsymbol{\Psi}) = \sum_{i=1}^{g} \pi_i \phi(\mathbf{y_j}; \boldsymbol{\mu_i}, \Sigma_i). \tag{20}$$

Of the several methods used to estimate mixture distributions, we use the MCMC-based Bayesian estimation method via posterior simulation (Gibbs sampler), which is now feasible and popular owing to the advent of computational power. According to Gelfand et al. [GHR90], the Gibbs sampler provides more refined numerical approximation for performing inference than expectation maximization (EM). It yields a Markov chain $\{\Psi^{(k)},\ k = 1, 2, \ldots\}$ whose distribution converges to the true posterior distribution of the parameters. For our parameter estimates, we use the posterior mean, which could be estimated by the average of the first N values of the Markov chain after discarding a sufficient burn-in, of say N_1. Thus our parameter estimates are:

$$E\{\widehat{\mathbf{\Psi}}|\mathbf{y}\} = \sum_{k=N_1+1}^{N} \frac{\mathbf{\Psi}^{(k)}}{(N - N_1)}. \tag{21}$$

The Phase Model

Owing to the difficulties associated with direct phase modeling (outlined in Sect. 5.3), we use an alternative representation of phase for modeling purposes derived as follows. First, we construct the "phase-only" images by removing the magnitude component from the frequency spectrum of the images. Since magnitude does not play as active a role in face identification, this is expected not to affect the system significantly. We then use the real and imaginary parts of these phase-only frequencies for modeling purposes. This is a simple and effective way of modeling phase, and at the same time does not suffer from the difficulties associated with direct phase modeling.

Let $R^{k,j}_{s,t}$ and $I^{k,j}_{s,t}$ respectively denote the real and the imaginary part at the (s,t)th frequency of the phase spectrum of the jth image from the kth person, $s, t = 1, 2, \ldots$, $k = 1, \ldots, 65$, $j = 1, \ldots, 21$. We model $(R^{k,j}_{s,t}, I^{k,j}_{s,t})$, $j = 1, \ldots, 21$ as a mixture of bivariate Gaussians whose density is given by (20), for each frequency (s,t) and each person k. We model only a few low frequencies within a 50×50 grid around the origin of the spectral plane since they capture all the image identifiability [Lim90], thus achieving considerable dimension reduction.

Classification Scheme

Classification of a new test image is done with the help of a MAP (maximum *a posteriori*) estimate based on the posterior likelihood of the data. For a new observation $Y = (R^j, I^j)$ extracted from the phase spectrum of a new image, if $f_k(\mathbf{y_j}; \mathbf{\Psi})$ denotes the Gaussian mixture models (GMM) for person k, we can compute the likelihood under the model for person k as:

$$g(Y|k) = \Pi_{\text{all freq.}} f_k(\mathbf{y_j}; \mathbf{\Psi}), \quad k = 1, \ldots, 65, \tag{22}$$

assuming independence among the frequencies. The convention is to use log-likelihoods for computational convenience to avoid numerical overflows and underflows in the evaluation of (22). The posterior likelihood of the observed data belonging to a specific person is given by:

$$f(k|Y) \propto g(Y|k)p(k), \tag{23}$$

where $p(k)$ denotes the prior probability for each person, which can be safely assumed to be uniform over all the possible people in the database. A particular image will then be assigned to class C if:

$$C = \arg\max_k f(k|Y). \tag{24}$$

Classification and Verification Results

We use $g = 2$ components to represent the illumination variations in the images of a person. A key step in the Bayesian estimation method consists of the specification of suitable priors for the unknown parameters in $\boldsymbol{\Psi}$. We choose suitable conjugate priors to ensure proper posteriors and simplified computations. We choose a burn-in of 2000 out of a total of $N = 5000$ iterations, by visual inspection of trace plots.

Table 8 shows the classification results for our database using a different number of training images. The training set in each case is randomly selected and the rest used for testing. This selection of the training set is repeated 20 times (in order to remove selection bias) and the final errors are obtained by averaging over those from the 20 iterations. The results are fairly good, which

Table 8. Error rates for GMM. The standard deviations are computed over the 20 repetitions in each case

# of Training Images	# of Test Images	Error Rate	Standard Deviation
15	6	1.25%	0.69%
10	11	2.25%	1.12%
6	15	9.67%	2.89%

demonstrates that the GMM is able to capture the illumination variation suitably. However, we notice that an adequate number of training images is required for the efficient estimation of the parameters; in our case, 10 is the optimal number of training images required. The associated standard errors in each case also proves the consistency of the results. Increasing the number of mixture components ($g = 3$ and $g = 4$) did not improve results significantly; hence a two-component GMM represents the best parsimonious model in this case.

Verification is performed by imposing a threshold on the posterior likelihood of the test images, so that a person is deemed authentic if the likelihood is greater than that threshold. Figure 26 shows the ROC curve obtained by plotting the FAR and FRR with varying thresholds on the posterior likelihood (for the optimal GMM with $g = 2$ and 10 training images). Satisfactory results are achieved with an EER of approximately 0.3% at a threshold log-likelihood value of -1700.

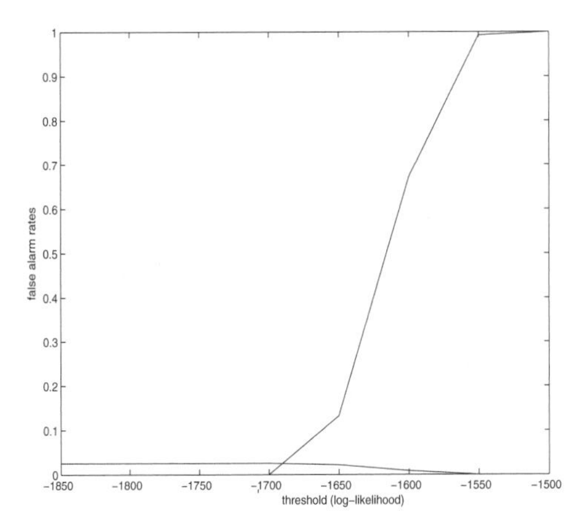

Fig. 26. ROC curve for authentication based on the phase model. The lower curve is the FAR.

Comparison of Model-Based Method with MACE

Our authentication experiments indicate that the mixture models yield better results than the MACE system (EER = 0.9 for MACE as in Fig. 5(a) and EER = 0.3 for mixture models). In applications as sensitive as authentication, even this little improvement is of tremendous consequences, and this establishes that our approach is more efficient than the MACE system. Apart from the results themselves, our model-based method uses the posterior likelihood as the match score for the authentication procedure and is a deterministic statistical quantity having nice distributional properties (efficiency, consistency) for constructing probability intervals and hypothesis tests. This greatly helps in assessing the reliability of these results. The MACE score PSR has no clear statistical interpretation of its own and this significantly limits its utility in inference problems. Model-based methods also are better capable of accounting for the image variability and hence are more flexible than MACE. We applied it to images with illumination changes, but it is a fairly general framework and can be easily extended to model other distortions such as noise and expression changes by defining the mixture components to represent different levels of those. Such robustness is the primary advantage of model-based methods, which is generally lacking in a non-model-based framework. Thus they are free from all the heuristics present in the MACE process (Sect. 4.1).

One potential disadvantage of model-based techniques is that they usually require more training samples than non-model-based methods. While MACE can yield satisfactory results with only 3 training images, the mixture model requires at least 10 for effective parameter estimation. So in case a sufficient number of images is not present, our model will not perform adequately. The training process is also time-consuming and is linear in the number of mixture components. However, as we have seen, in many cases we can obtain a sufficiently robust representation using as few mixture components as two. On the other hand, the number of training images required by MACE in a given situation may vary from one dataset to another, and there is no concrete way to determine the optimal number other than brute-force experimentation. For example, for a database with extreme expression variations, many more training samples will be required to synthesize an effective filter than that required for images with illumination variations. For the model-based system, on the other hand, the number of training images required in a given scenario may not vary as much, and sufficiently robust models can be devised with say 10 images in most cases. The implementation of the mixture model, however, is sufficiently fast, and this is what is crucial for a practical application. For example, in an airport, there is already a stored database of the trained templates (the fitted models, in this case) and when a person comes in at an immigration checkpoint, his face image is captured and classified using the MAP estimate, which can be done in realtime with no difficulties. This establishes that model-based systems are also useful from a practical point of view, and coupled with the statistical rigor they possess, they prove to be much superior to the MACE filter method and as such to any other non-model-based method, which is very likely to suffer from similar drawbacks.

6 Discussion

Statistically based methods have the potential to be flexible and reliable, both for handling large diverse databases and for providing a firm basis for the results. Besides, statistical methods can help us understand the potential performance of an authentication system as the complexity of a system changes (i.e., as the number of users increases) and assess the scalability of the results obtained on small- to moderate-sized databases. Such systems are henceforth expected to be more attractive to users and have wider applicability than non-model-based and empirical approaches. Although such methods are widely used and often found to yield satisfactory results, they are highly sensitive to the nature of the images and deteriorate quickly. For instance, the MACE filter yields good results but from a statistician's perspective, it still lacks rigor and validity and does not work as well in all situations, as pointed out in this chapter. These drawbacks can be overcome with the help of simple statistical tools. This chapter has shown that even simple models can achieve much and lays the ground for the application of more refined modeling strategies. Certain other tools to assess accuracies of different methods in identification and verification include using appropriate variations on the ROC tools that have proved so successful in other areas, for example, in polygraph testing [Com03] and in evaluating medical diagnostic tests [Cam94].

In principle, many of the techniques discussed in this chapter are also applicable in the study of authentication methods using other popular biometrics such as fingerprints and multimodal systems. In particular, fingerprints are gradually gaining in importance as a reliable biometric due to their uniqueness and permanence properties. Many authentication techniques have been developed for fingerprints, but like face-based methods, they are mostly empirical and not flawless. Venkataramani and Vijaya Kumar [VV03] applied the MACE authentication method to the National Institute of Standards and Technology (NIST) fingerprint database [Wat98] and claim perfect verification rates. The adaptation of the modeling approaches outlined in this chapter could also be expected to lead to similar gains in understanding as well as refining authentication schemes, especially since very few statistical analyses have been performed on fingerprint data.

Acknowledgments

This is joint work with Dr. Anthony Brockwell, Dr. Stephen E. Fienberg of the Department of Statistics, and Dr. Marios Savvides of the Electrical and Computer Engineering Department at Carnegie Mellon University. The research is funded in part by the Army Research Office contract DAAD19-02-1-3-0389 to CyLab.

References

[BPR00] Bolle, R. M., S. Pankanti, and N. K. Ratha. 2000. "Evaluation techniques for biometrics-based authentication systems (FRR)." *Proceedings of International Conference on Pattern Recognition.* 2831–2837.

[BRP99] Bolle, R. M., N. K. Ratha, and S. Pankanti. 1999. "Evaluating authentication systems using bootstrap confidence intervals." *Proceedings of AutoID, Summit, NJ.* 9–13.

[Cam94] Campbell, G. 1994. "Advances in statistical methodology for the evaluation of diagnostic and laboratory tests." *Statistics in Medicine* 13:499–508.

[Com03] Committee to Review the Scientific Evidence on the Polygraph. 2003. *The polygraph and lie detection.* Washington, DC: National Academies Press.

[Ega75] Egan, J. P. 1975. *Signal detection theory and ROC analysis.* New York: Academic Press.

[GHR90] Gelfand, A. E., S. E. Hills, A. Racine-Poon, and A. F. M. Smith. 1990. "Illustration of Bayesian inference in normal data models using Gibbs sampling." *Journal of the American Statistical Association* 85 (412): 972–985.

[GHW79] Golub, G. H., M. Heath, and G. Wahba. 1979. "Generalized cross-validation as a method for choosing a good ridge parameter." *Technometrics* 21 (2): 215–223.

[HM82] Hanley, A., and B. J. McNeil. 1982. "The meaning and use of the area under a receiver operating characteristic (ROC) curve." *Radiology* 143:29–36.

[HHC02] Havran, C., L. Hupet, J. Czyz, J. Lee, L. Vandendorpe, and M. Verleysen. 2002. "Independent component analysis for face authentication." *Proceedings of Knowledge-Based Intelligent Information and Engineering Systems (KES), Crema, Italy.*

[Hay82] Hayes, M. H. 1982. "The reconstruction of a multidimensional sequence from the phase or magnitude of its Fourier transform." *IEEE Transaction on Acoustics, Speech and Signal Processing* 30 (2): 140–154.

[IG00] Ishwaran, H., and C. Gatsonis. 2000. "A general class of hierarchical ordinal regression models with applications to correlated ROC analysis." *The Canadian Journal of Statistics* 28:731–750.

[JKL99] Jonsson, K., J. Kittler, Y. P. Li, and J. Matas. 1999. "Support vector machines for face authentication." *Proceedings of British Machine Vision Conference (BMVC).*

[KCT00] Kanade, T., J. F. Cohn, and Y. L. Tian. 2000. "Comprehensive database for facial expression analysis." *4th IEEE International Conference on Automatic Face and Gesture Recognition 2000.* 46–53.

[LKM99] Li, Y., J. Kittler, and J. Matas. 1999. "Effective implementation of linear discriminant analysis for face recognition and verification." *Proceedings of 8th International Conference on Computer Analysis and Patterns, Ljubljana, Slovenia.*

[Lim90] Lim, J. S. 1990. *Two-dimensional signal and image processing.* Indianapolis, IN: Prentice-Hall.

[LZS01] Liu, C., S. C. Zhu, and H. Y. Shum. 2001. "Learning inhomogeneous Gibbs model of faces by minimax entropy." *Proceedings of the IEEE International Conference on Computer Vision (ICCV).* 281–287.

[LCV02] Liu, X., T. Chen, and B. V. K. Vijaya Kumar. 2002. "On modeling variations for face authentication." *Proceedings of the Fifth IEEE International Conference on Automatic Face and Gesture Recognition.* Washington, DC: IEEE Computer Society, 384.

[MVC87] Mahalanobis, A., B. V. K. Vijaya Kumar, and D. Casasent. 1987. "Minimum average correlation energy filters." *Applied Optics* 26:3623–3630.

[MP00] McLachlan, G., and D. Peel. 2000. *Finite mixture models.* New York: John Wiley & Sons.

[PVY03] Palanivel, S., B. S. Venkatesh, and B. Yegnanarayana. 2003. "Real time face authentication system using autoassociative neutral network models." *Proceedings of IEEE International Conference on Multimedia and Expo, Baltimore, MD.*

[SV03] Savvides, M., and B. V. K. Vijaya Kumar. 2003. "Efficient design of advanced correlation filters for robust distortion-tolerant face identification." *Proceedings of IEEE Conference on Advanced Video and Signal Based Surveillance.* 45–52.

[SVK02] Savvides, M., B. V. K. Vijaya Kumar, and P. Khosla. 2002. "Face verification using correlation filters." *3rd IEEE Automatic Identification Advanced Technologies, Tarrytown, NY.* 56–61.

[Sch03] Schuckers, M. E. 2003. "Using the beta-binomial distribution to assess performance of a biometric identification device." *International Journal of Image Graphics* 3 (3): 523–529.

[SSK97] Shen, W., M. Surette, and R. Khanna. 1997. "Evaluation of automated biometrics-based identification and verification systems." *Proceedings of IEEE* 85 (9): 1464–1478.

[SBB02] Sim, T., S. Baker, and M. Bsat. 2002. "The CMU pose, illumination, and expression "PIE" database." *Proceedings of the 5th International Conference on Automatic Face and Gesture Recognition.*

[TP91] Turk, M., and A. Pentland. 1991. "Eigenfaces for recognition." *Cognitive Neuroscience* 3 (1): 71–96.

[VV03] Venkataramani, K., and B. V. K. Vijaya Kumar. 2003. "Fingerprint verification using correlation filters." In *Proceedings of Audio Visual Biometrics Based Person Authentication (AVBPA)*, edited by J. Kittler and M. S. Nixon, Volume 2688 of *Lecture Notes in Computer Science* 886–894. New York: Springer.

[Vij92] Vijaya Kumar, B. V. K. 1992. "Tutorial survey of composite filter designs for optical correlators." *Applied Optics* 31 (23): 4773–4801.

[VSV02] Vijaya Kumar, B. V. K., M. Savvides, K. Venkataramani, and C. Xie. 2002. "Spatial frequency domain image processing for biometric recognition." *Proceedings of International Conference on Image Processing ICIP, Rochester, NY.*

[Wat98] Watson, C. I. 1998. NIST special database 24.

Recognition Problem of Biometrics: Nonparametric Dependence Measures and Aggregated Algorithms

Andrew L. Rukhin

Statistical Engineering Division, National Institute of Standards and Technology, and Department of Mathematics and Statistics, University of Maryland, Baltimore County, andrew.rukhin@nist.gov

1 Introduction

This chapter explores the possibility of using nonparametric dependence characteristics to evaluate biometric systems or algorithms that play an important role in homeland security for the purpose of law enforcement, sensitive areas access, borders and airport control, etc. These systems, which are designed to either detect or verify a person's identity, are based on the fact that all members of the population possess unique characteristics (biometric signatures) such as facial features, eye irises, fingerprints, and gait, which cannot be easily stolen or forgotten. A variety of commercially available biometric systems are now in existence; however, in many instances there is no universally accepted optimal algorithm. For this reason it is of interest to investigate possible aggregations of two or more different algorithms. Kittler et al. [KHD98] and Jain et al. [JDM00, Sect. 6] review different schemes for combining multiple matchers.

We discuss here the mathematical aspects of a fusion for algorithms in the *recognition* or identification problem, where a biometric signature of an unknown person, also known as *probe*, is presented to a system. This probe is compared with a database of, say, N signatures of known individuals called the *gallery*. On the basis of this comparison, an algorithm produces the similarity scores of the probe to the signatures in the gallery, whose elements are then ranked accordingly. The top matches with the highest similarity scores are expected to contain the true identity.

A common feature of many recognition algorithms is representation of a biometric signature as a point in a multidimensional vector space. The similarity scores are based on the distance between the gallery and the query (probe) signatures in that space (or their projections onto a subspace of a smaller dimension). Because of inherent commonality of the systems, the similarity scores and their resulting orderings of the gallery can be dependent for two different algorithms. For this reason, traditional methods of combining different procedures, like classifiers in pattern recognition, are not appropriate. Another reason for failures of popular methods like bagging and boosting [Bre04, SFB98] is that the gallery size is much larger than the number of algorithms involved. Indeed the majority voting methods used by these techniques (as well as in analysis of multicandidate elections and social choice theory [Ste93]) are based on aggregated combined ranking of a fairly small number of candidates obtained from a large number of voters, judges, or classifiers. The axiomatic approach [Mar93] to this fusion leads to the combinations of classical weighted means (or random dictatorship).

As the exact nature of the similarity scores derivation is typically unknown, the use of nonparametric measures of association is appropriate. The utility of rank correlation statistics, like Spearman's rho or Kendall's tau, for measuring the relationship between different face recognition algorithms, was investigated by Rukhin and Osmoukhina [RO05]. In Sect. 2 the natural extensions of two classical rank correlation coefficients solely based on a given number of top matches are presented. We demonstrate difficulties with using these correlation coefficients for estimation of the correlation over the whole gallery. A version of a scan statistic, which measures co-occurrence of rankings for two arbitrary algorithms across the gallery, is employed as an alternative characteristic. The exact covariance structure of this statistic is found for a pair of independent algorithms; its asymptotic normality is derived in the general case.

An important methodological tool in nonparametric dependence characteristics studies is provided by the concept of copula [Joe90]. Special tail-dependence properties of copulas arising in the biometric algorithms analysis are established in Sect. 3. For common image recognition algorithms, the strongest (positive) correlation between algorithms similarity scores is shown to hold for both large and small rankings. Thus, in all observed cases the algorithms behave somewhat similarly, not only by assigning the closest images in the gallery but also by deciding which gallery objects are most dissimilar to the given image. This finding is useful for the construction of new procedures designed to combine several algorithms and also underlines the difficulty with a direct application of boosting techniques.

As different recognition algorithms generally fail on different subjects, two or more methods could be fused to get improved performance. Several such methods for aggregating algorithms are discussed in Sect. 4. These methods are based on different metrics on the permutation group and include a simple version of linear fusion suggested by Rukhin and Malioutov [RM05].

Notice that the methods of averaging or combining ranks can be applied to several biometric algorithms, one of which, say, is a face recognition algorithm, and another is a fingerprint (or gait, or ear) recognition device. Jain et al. [JBP99] discuss experimental studies of multimodal biometrics, in particular, fusion techniques for face and fingerprint classifiers. Methods discussed in Sect. 4 can be useful in a *verification* problem when a person presents a set of biometric signatures and claims that a particular identity belongs to the provided signatures.

The continued example considered in this chapter comes from the FERET (face recognition technology) program [PMR00] in which four recognition algorithms each produced rankings from galleries in three FERET datasets of facial images. It is discussed in detail in Sect. 5.

2 Correlation Coefficients, Partially Ranked Data, and the Scan Statistic

One of the main performance characteristics of a biometric algorithm is the percentage of queries in which the correct answer can be found in the top few, say K, matches. To start quantifying dependence between two algorithms for a large gallery size N, it seems sensible to focus only at the images in the gallery receiving the best K ranks. The corresponding metrics for the so-called partial rankings were suggested by Diaconis [Dia88] and studied by Critchlow [Cri85]. A survey of these methods is given in Marden [Mar95, Chap. 11].

Let X_i and Y_i, $i = 1, \ldots, N$, be similarity scores given to the gallery elements by two distinct algorithms on the basis of a given probe. We assume that the similarity scores can be thought of as continuous random variables, so that the probabilities of ties within the original scores are negligible.

In image analysis it is common to write similarity scores of each algorithm in decreasing order, $X_{(1)} \geq \cdots \geq X_{(N)}$, $Y_{(1)} \geq \cdots \geq Y_{(N)}$, and rank them, so that $X_i = X_{(R(i))}$, and $Y_i = Y_{(S(i))}$. Thus, $X_{(1)}$ is the largest and $X_{(N)}$ is the smallest similarity score while the rank of the largest similarity score is 1, and that of the smallest score is N. We use the notation $\mathbf{R}$ and $\mathbf{S}$ for the vectors of ranks $\mathbf{R} = (R(1), \ldots, R(N))$ and $\mathbf{S} = (S(1), \ldots, S(N))$, which can be interpreted as elements of the permutation group $\mathcal{S}_N$. Given a ranking $\mathbf{R}$, introduce the new ranking $\tilde{\mathbf{R}}$ by giving the rank $(N + K + 1)/2$ to all images not belonging to the subset of the best K images (which maintain their ranks). More specifically, new ranks $\tilde{R}_i$ are obtained from the formula

$$\tilde{R}(i) = \begin{cases} R(i) & \text{if } R(i) \leq K \\ \frac{N+K+1}{2} & \text{otherwise.} \end{cases}$$

This assignment preserves the average of the largest $N - K$ ranks, so that $\sum_{i=1}^{N} \tilde{R}(i) = \sum_{i=1}^{N} R(i) = N(N+1)/2$. Define the analogue of the Spearman rho coefficient for partial rankings of two algorithms producing rankings $\mathbf{R}$ and $\mathbf{S}$ as the classical rho coefficient for the rankings $\tilde{R}(i)$ and $\tilde{S}(i)$,

$$\tilde{\varrho}_S = \frac{\sum_{i=1}^{N} \left(\tilde{R}(i) - \frac{N+1}{2}\right)\left(\tilde{S}(i) - \frac{N+1}{2}\right)}{\sqrt{\sum_{i=1}^{N} \left(\tilde{R}(i) - \frac{N+1}{2}\right)^2 \sum_{i=1}^{N} \left(\tilde{S}(i) - \frac{N+1}{2}\right)^2}}.$$

The advantage of this definition is that by using the central limit theorem for linear rank statistics one can establish, for example, asymptotic normality of the Spearman coefficient when $N \to \infty$. A general result (Theorem 2) is formulated later.

The analogue of the Kendall tau coefficient for partial rankings is similarly defined. Namely, for the rankings $\tilde{R}(i)$ and $\tilde{S}(i)$

$$\tilde{\varrho}_K = \frac{\sum_{i,j} \operatorname{sign}\left((\tilde{R}(i) - \tilde{R}(j))(\tilde{S}(i) - \tilde{S}(j))\right)}{K(2N - K - 1)}.$$

The denominator, $K(2N - K - 1) = K(N - 1) + (N - K)K$, can be interpreted as the total number of different pairs formed by the ranks $\tilde{R}(i)$ and $\tilde{S}(i)$. Unfortunately, both of these partial correlation coefficients exhibit the same problem of drastically underestimating the true correlation for small and moderate K.

In accordance with the FERET protocol, four algorithms (I:MIT, March 96; II:USC, March 97; III:MIT, Sept 96; IV:UMD, March 97) have produced similarity scores of items from a gallery consisting of $N = 1196$ images with 234 probe images. The rank correlation matrix based on Spearman rho coefficients is

$$S = \begin{pmatrix} 1 & 0.189 & 0.592 & 0.340 \\ & 1 & 0.205 & 0.324 \\ & & 1 & 0.314 \\ & & & 1 \end{pmatrix}.$$

Disappointingly, both coefficients $\tilde{\varrho}_S$ and $\tilde{\varrho}_K$ have very small values for small and moderate K (see Fig. 1). Although they have the tendency to increase as K increases, the largest value of $\tilde{\varrho}_S$ (for two most correlated MIT methods I and III) was only 0.29 for $K = 50$.

Another definition of the correlation coefficient for partially ranked data can be obtained from a distance $d(\mathbf{R}, \mathbf{S})$ on the coset space $\mathcal{S}_N/\mathcal{S}_{N-K}$ of partial rankings. The list of the most popular metrics [Dia88] includes Hamming's metric d_H, Spearman's L_2, Footrule L_1, Kendall's distance, Ulam's distance, and Cayley's distance. With $\bar{d} = \max_{\mathbf{R},\mathbf{S}} d(\mathbf{R}, \mathbf{S})$, let

$$\varrho_d = 1 - 2\frac{d(\mathbf{R}, \mathbf{S})}{\bar{d}}$$

be such a correlation coefficient. One can show that, as $N \to \infty$, $\varrho_d \to -1$ even for independent $\mathbf{R}$, $\mathbf{S}$, when d is the Kendall metric or the Spearman metric (including L_1 Footrule). For moderate N, $d(\mathbf{R}, \mathbf{S})$ has the expected value too close to $\bar{d}$ for ϱ_d to be of practical use. Indeed small variability of ϱ_d makes it similar in this regard to the coefficient based on Cayley's distance [DG97].

To understand the reasons for failure of partial rank correlation characteristics the following *scan* (or co-occurrence) statistic was employed. For two algorithms producing similarity scores X_i and Y_i with rankings $\mathbf{R}$ and $\mathbf{S}$, put for a fixed M and $u = 1, \ldots, N - M + 1$,

$$T(u) = \text{card}\,\{i : u \leq R(i),\ S(i) \leq u + M - 1\}. \tag{1}$$

For independent X_i and Y_i both $\mathbf{R}$ and $\mathbf{S}$ are uniformly distributed over the permutation group $\mathcal{S}_N$. In this case one only needs to consider $W_r = S\left(R^{-1}(r)\right)$. Let $Y_{[i]}$ be the similarity score of the second algorithm corresponding to $X_{(i)}$. These statistics are called *concomitants* of order statistics $X_{(1)}, \ldots, X_{(n)}$. Thus, W_r is the rank of $X_{(i)}$, whose concomitant $Y_{[i]}$ has the rank r, and

$$T(u) = \sum_{u \leq r, s \leq u+M-1} I(W_r = s),$$

where $I(\cdot)$ is the indicator function. The random variable $T(u)$ counts the common ranks between u and $u + M - 1$. Therefore, in the uniform case it follows a hypergeometric distribution with parameters (N, M, M),

$$P(T = t) = \frac{\binom{M}{t}\binom{N-M}{M-t}}{\binom{N}{M}}, \qquad t = 0, 1, \ldots, M.$$

The behavior of the scan statistic for biometric data is very different from that for independent $\mathbf{R}$ and $\mathbf{S}$. Indeed, for all datasets in FERET, the scan statistic exhibits a "bathtub" effect, i.e., its typical plot looks bowl-shaped

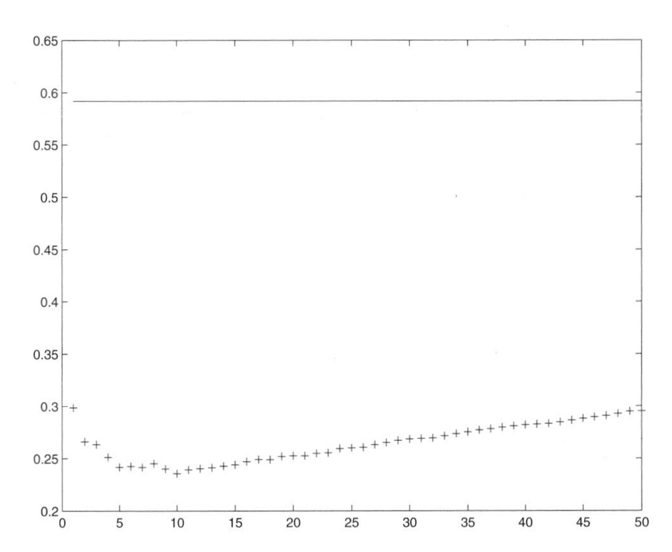

Fig. 1. The plot of the partial Spearman rho coefficient for algorithms I and III as a function of K. The solid line represents the limiting value 0.592.

(see Fig. 2). The readings of the scan statistic $T(u)$ for the correlated scores are much larger than the corresponding values based on independent scores for both small and large u. These values for independent scores would oscillate around the mean $\mathrm{E}(T) = M^2/N$. As the variables $T(u)$ must be positively correlated, the covariance function is of interest.

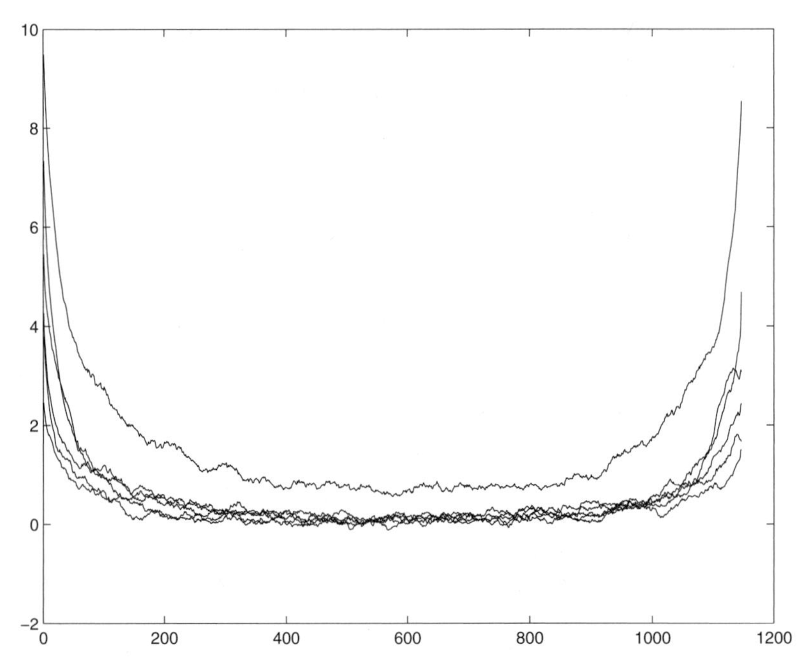

Fig. 2. The plots of the scan statistic for algorithms in the FERET study.

Theorem 1. *If the random scores* X_i *and* $Y_i, i = 1, \dots, N$, *are independent, then the covariance function of* $T(u)$, *for* $0 \le h \le N-1$, $1 \le u \le N-M-h+1$, *has the form*

$$\mathrm{Cov}(T(u), T(u+h)) = \begin{cases} \dfrac{[(M-h)N - M^2]^2}{N^2(N-1)}, & h < M, \\ \dfrac{M^4}{N^2(N-1)}, & h \ge M. \end{cases}$$

For independent scores neither the covariance between $T(u)$ and $T(u+h)$, nor the mean of $T(u)$ depends on u; $T(u)$ is then a stationary process, and the bathtub effect cannot take place.

3 Copulas and Asymptotic Normality

To study the structure of dependence of a pair of algorithms, one can employ the concept of *copula* defined for two random variables X and Y with cumulative distributions functions F_X and F_Y, respectively. In our context X and Y represent random similarity scores of the algorithms. Copula is a function $C(u,v), 0 < u, v < 1$, such that

$$P(F_X(X) \leq u,\ F_Y(Y) \leq v) = C_{X,Y}(u,v) = C(u,v).$$

Copulas are invariant under monotone transformations, i.e., if α and β are strictly increasing, then $C_{\alpha(X),\beta(Y)}(u,v) = C_{X,Y}(u,v)$. In this sense, copulas describe the structure of dependence. Each copula induces a probability distribution with uniform marginals on the unit square. Nelsen [Nel99] discusses further properties of copulas and methods for their construction.

We assume that the joint distribution of $F_X(X)$ and $F_Y(Y)$ is absolutely continuous, and refer to its density, $c(u,v)$, as copula density. On the basis of a sample, $(X_1,Y_1),\ldots,(X_N,Y_N)$, this function can be estimated by the empirical copula density,

$$c_N\left(\frac{i}{N},\frac{j}{N}\right) = \begin{cases} 1/N, & \text{if } \left(X_{(i)}, Y_{(j)}\right) \text{ is in the sample,} \\ 0, & \text{otherwise.} \end{cases}$$

Note that c_N is a probability mass function assigning the weight $1/N$ to the point $(R(i)/N, S(i)/N)$, where both $R(i)$ and $S(i)$ are the ranks of X_i and Y_i, respectively. The empirical copula is defined as

$$C_N\left(\frac{i}{N},\frac{j}{N}\right) = \sum_{p=1}^{i}\sum_{q=1}^{j} c_N\left(\frac{p}{N},\frac{q}{N}\right).$$

As the exact distribution of the scan statistic (1) for general (dependent) scores appears to be intractable, we give the limiting distribution of $T(u)$ when $N \to \infty$,

$$\frac{u}{N} \to \lambda, \quad \frac{M}{N} \to a, \quad \text{with} \quad 0 < \lambda < 1 - a, \quad 0 < a < 1. \tag{2}$$

With $C(u,v)$ denoting the copula for (X,Y),

$$N^{-1} \sum_{r=u}^{u+M-1} P(W_i = r) \to P(\lambda \leq F_X(X) \leq \lambda + a,\ \lambda \leq F_Y(Y) \leq \lambda + a)$$

$$= C(\lambda+a,\lambda+a) + C(\lambda,\lambda) - C(\lambda+a,\lambda) - C(\lambda,\lambda+a),$$

which gives the asymptotic behavior of the mean of the scan statistic.

Theorem 2. *Under regularity conditions R1–R5 in [RO05] when* $N \to \infty$*, the distribution of* $\sqrt{N}\left(\frac{T(u)}{N} - \int_\lambda^{\lambda+a}\int_\lambda^{\lambda+a} c(s,t)\,\mathrm{d}s\,\mathrm{d}t\right)$ *converges to the normal distribution with zero mean and with variance*

$$\sigma^2 = \mathrm{Var}\Big(I(\lambda \le U \le \lambda + a)I(\lambda \le V \le \lambda + a) + I(U \le \lambda)\int_\lambda^{\lambda+a} c(\lambda, v)\mathrm{d}v$$

$$-I(U \le \lambda + a)\int_\lambda^{\lambda+a} c(\lambda + a, v)\mathrm{d}v + I(V \le \lambda)\int_\lambda^{\lambda+a} c(u, \lambda)\mathrm{d}u$$

$$-I(V \le \lambda + a)\int_\lambda^{\lambda+a} c(u, \lambda + a)\mathrm{d}u\Big).$$

Here λ *and* a *are defined in (2), and* U *and* V *are random variables with the joint distribution function* $C(u,v)$ *and the joint density* $c(u,v)$.

Theorem 2 suggests that the observed bathtub behavior of the scan statistics reflects the form of the underlying copula for the scores. The copulas with a bowl-shaped function of u, $\int_{u/N}^{(u+M)/N}\int_{u/N}^{(u+M)/N} c(s,t)\,\mathrm{d}s\,\mathrm{d}t$, appear in all FERET algorithm pairs. These copulas correspond to mixtures of two unimodal copulas: one with the bulk of the mass concentrated at the origin $(0,0)$ (small ranks), and the second one concentrated around $(1,1)$ (large ranks). In other terms, the density $c(u,v)$ is bimodal: one peak is at $(0,0)$ and another at $(1,1)$. The set $\{(u,v) : c(u,v) \ge c\}$ is a union of two sets: C_0, which is star-shape about $(0,0)$, and C_1, which is star-shape about $(1,1)$.

In particular, the distribution having such a copula satisfies the definition of *left (right) tail monotonicity* of one random variable U in another random variable V [Nel99]. Namely, $P(U \le u|V \le v)$ is a nondecreasing function of v for any fixed u. Also $P(U > u|V > v)$ is a nondecreasing function of v for any fixed u. Each of these conditions implies positive quadrant dependence: $P(U \le u, V \le v) \ge P(U \le u)P(V \le v)$, (i.e., $C(u,v) \ge uv$) and under these monotonicity conditions, Spearman's rho is larger than Kendall's tau, which must be positive. All these properties have been observed in all FERET datasets.

In practical terms, tail monotonicity properties mean that the strongest correlations between algorithms' similarity scores happen for *both large and small rankings.* Thus, in all observed cases the algorithms behave somewhat similarly not only by assigning the closest images in the gallery, but also by deciding which gallery object is most dissimilar to the given image. The explored algorithm pairs behave more or less independently one from another only in the middle range of the rankings. In the FERET experiment only algorithms I and III (both MIT algorithms, MIT, March 96, and MIT, Sept 96) showed fairly high correlation even for the medium ranks. This finding leads to the conclusion that the partial correlation coefficients, which are based only on small ranks, in principle, cannot capture the full dependence between algorithms.

Verification of the suppressed regularity conditions in Theorem 2 for specific families of copulas is usually straightforward. For example, for $\alpha > 0, \beta \geq 1$,

$$C(u,v) = C_{\alpha,\beta}(u,v) = \left\{ \left[\left(u^{-\alpha} - 1\right)^{\beta} + \left(v^{-\alpha} - 1\right)^{\beta} \right]^{1/\beta} + 1 \right\}^{-1/\alpha} \tag{3}$$

satisfies these regularity conditions ensuring the asymptotic normality of the statistic $T(u)$. This family, for an appropriate choice of α and β, fits the observed similarity scores fairly well.

The next result concerns the asymptotic behavior of the partial correlation coefficient.

Theorem 3. *The asymptotic distribution of $\sqrt{N}(\tilde{\varrho}_S - \mu_\varrho)$ is normal with zero mean and variance σ_ϱ^2. Here $a = \lim_{N\to\infty} K/N$,*

$$\mu_\varrho = \left(\frac{a^3}{12} - \frac{a^2}{4} + \frac{a}{4} \right)^{-1}$$

$$\times \left[\int_0^a \int_0^a uv\, c(u,v)\, du\, dv - \frac{1}{2} \int_0^a \int_0^a (u+v)\, c(u,v)\, du\, dv + \frac{a}{2} \int_0^a \int_a^1 u\, c(u,v)\, du\, dv \right.$$

$$\left. + \frac{a}{2} \int_a^1 \int_0^a v\, c(u,v)\, dv\, du + \frac{1}{4} C(a,a)(a+1)^2 - \frac{a^2}{4}(2a+1) \right],$$

$$\sigma_\varrho^2 = \left(\frac{a^3}{12} - \frac{a^2}{4} + \frac{a}{4} \right)^{-2}$$

$$\times \mathrm{Var}\left(\left[\left(U - \frac{1}{2}\right) I(U \leq a) + \frac{a}{2} I(U > a) \right] \left[\left(V - \frac{1}{2}\right) I(V \leq a) + \frac{a}{2} I(V > a) \right] \right.$$

$$\left. + \int_U^a \int_0^a v\, c(u,v)\, dv\, du + \frac{1}{2}(a+1)(C(U,a) + C(a,V)) + \int_V^a \int_0^a u\, c(u,v)\, du\, dv \right).$$

U and V are random variables with joint distribution function $C(u,v)$, and the joint density $c(u,v)$.

Genest et al. [GGR95] discuss pseudo-likelihood estimation of copula parameters. The pseudo-loglikelihood is $l(\alpha, \beta, u, v) = \log c_{\alpha,\beta}(u,v)$. To estimate the parameters α and β, one has to maximize

$$\sum_{i=1}^{N} l\left(\alpha, \beta, \frac{S_i}{N+1}, \frac{R_i}{N+1}\right),$$

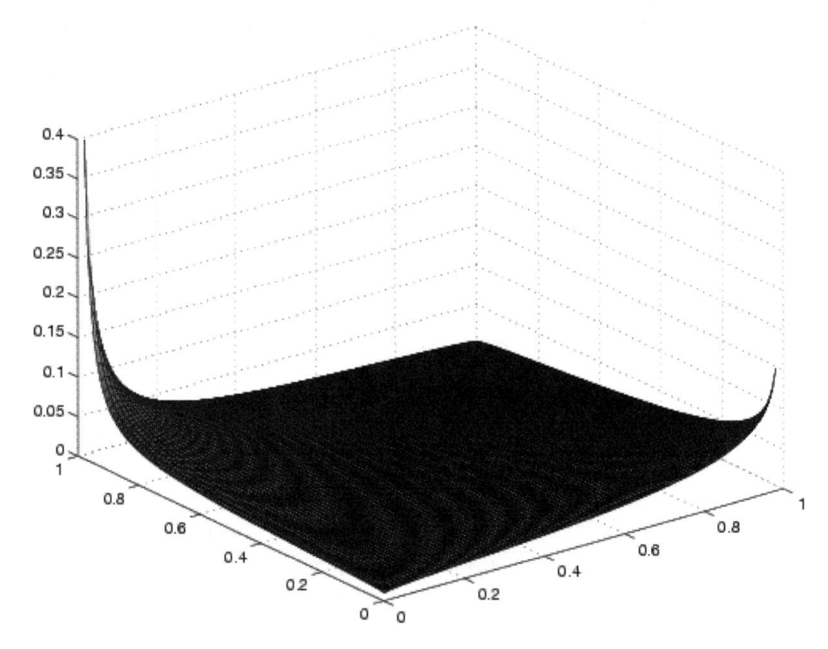

Fig. 3. The plot of the estimated theoretical copula for algorithms II and IV with $\alpha = 0.084$, $\beta = 1.227$.

which leads to the likelihood-type equations for $\hat{\alpha}$ and $\hat{\beta}$. The complicated form of these equations prevents an explicit form of the estimator. However, the numerical computation is quite feasible. The resulting estimators are asymptotically normal, if in (3) $\alpha < \frac{1}{2}$, $\alpha\beta < \frac{1}{2}$, and $\beta < 2$.

Figures 3 and 4 portray the empirical and theoretical copulas for (3) with pseudo-likelihood estimated α and β.

4 Averaging of Ranks via Minimum Distance and Linear Aggregation

A possible model for the combination of, say J, dependent algorithms representable by their random similarity scores $X_1, \ldots, X_J$, involves their joint copula $C_{X_1,\ldots,X_J}(u_1, \ldots, u_J)$, such that

$$C_{X_1,\ldots,X_J}(u_1, \ldots, u_J) = H(F_1^{-1}(u_1), \ldots F_J^{-1}(u_J)),$$

where $F_1, \ldots F_J$ are marginal distribution functions, and H is the joint distribution function of $X_1, \ldots, X_J$.

If $(X_1^j, \ldots, X_N^j)$ are similarity scores produced by the jth algorithm, the similarity scores of the aggregated algorithm are defined by a convex combination of N-dimensional random vectors $F_j = (F_j^{-1}(X_1^j), \ldots, F_j^{-1}(X_N^j))$, i.e., the score given to the kth element of the gallery is $\sum_{j=1}^{J} w_j F_j^{-1}(X_k^j), k = 1, 2, \ldots, N$. To find nonnegative weights (probabilities) $w_1, \ldots, w_J$, such that $w_1 + \cdots + w_J = 1$, we take

$$\operatorname{tr}\left(\operatorname{Var}(\sum_{j=1}^{J} w_j F_j)\right) = \sum_{j,\ell} w_j w_\ell \operatorname{tr}\left(\operatorname{Cov}(F_j, F_\ell)\right)$$

as the objective function to be minimized. With the vectors $w = (w_1, \ldots, w_J)^T$, $e = (1, \ldots, 1)^T$, and the matrix S formed by elements $\operatorname{tr}\left(\operatorname{Cov}(F_j, F_\ell)\right)$, the optimization problem reduces to the minimization of $w^T S w$ under condition $w^T e = 1$ with the solution

$$w^0 = \frac{S^{-1} e}{e^T S^{-1} e} \tag{4}$$

(assuming that S is nonsingular).

The matrix S can be estimated from archive data, for example, as the rank correlation matrix based on Spearman rho coefficients in Sect. 2. Another possibility is to use the pseudo-likelihood estimators of copula parameters (say, α and β in (3)) as discussed in the previous section by plugging them into the formula for $\operatorname{Cov}(F_j, F_\ell)$. This typically involves additional numerical integration.

A different (but related) approach is to think of the action of an algorithm (its ranking) as an element of the permutation group $\mathcal{S}_N$. Since the goal is to

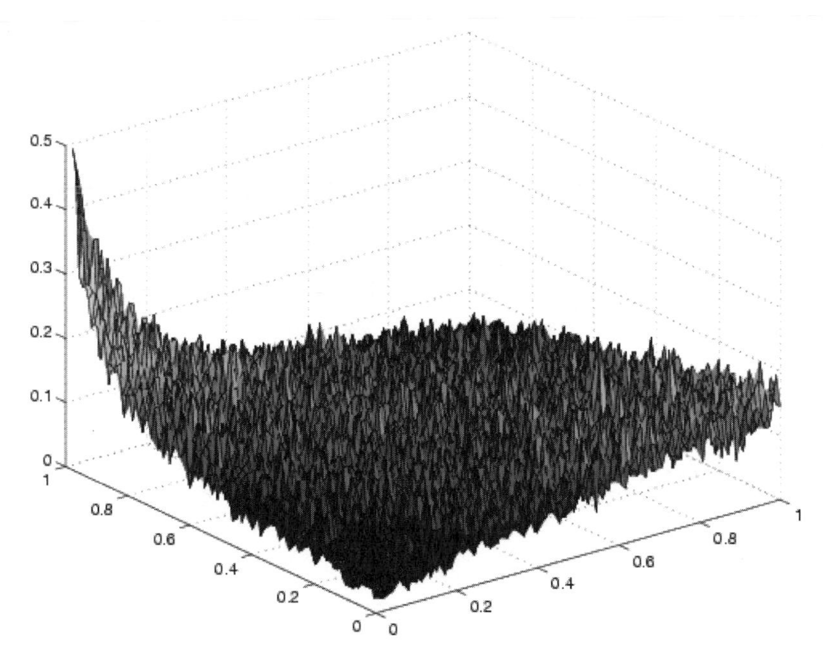

Fig. 4. The plot of the empirical copulas for algorithms II and IV.

combine J algorithms whose actions π_j can be considered as permutations of a gallery of size N, the "average permutation," $\hat{\pi}$, of $\pi_1, \ldots, \pi_J$ can be defined by the analogy with classical means. Namely, if $d(\pi, \sigma)$ is a distance between two permutations π and σ, then $\hat{\pi}$ is the minimizer (in π) of $\sum_{j=1}^{J} d(\pi_j, \pi)$. However, this approach does not take into account different precisions of different algorithms. Indeed, equal weights are implicitly given to all π_i, and the dependence structure of algorithms, which are to be combined, is neglected.

To form a fusion of dependent algorithms, a distance $d((\pi_1, \ldots, \pi_J), (\sigma_1, \ldots, \sigma_J))$, on the direct product $\mathcal{S}_N \bigotimes \cdots \bigotimes \mathcal{S}_N$ of J copies of the permutation group can be used. Then the combined (average) ranking $\hat{\pi}$ of observed rankings $\pi_1, \ldots, \pi_J$ is the minimizer of $d((\pi_1, \ldots, \pi_J), (\pi, \ldots, \pi))$. The simplest metric is the sum $\sum_{j=1}^{J} d(\pi_j, \pi)$, as above.

To define a more appropriate distance, we associate with a permutation π the $N \times N$ permutation matrix P with elements $p_{i\ell} = 1$, if $\ell = \pi(i); = 0$, otherwise. A distance between two permutations π and σ can be defined as the matrix norm of the difference between the corresponding permutation matrices. For a matrix P, one of the most useful matrix norms is $||P||^2 = \mathrm{tr}(PP^T) = \sum_{i,\ell} p_{i\ell}^2$. For two permutation matrices P and S corresponding to permutations π and σ, the resulting distance $d(\pi, \sigma) = ||P - S||$ essentially coincides with Hamming's metric,

$$d_H(\pi, \sigma) = N - \mathrm{card}\,\{i : \pi(i) = \sigma(i)\}.$$

For a positive definite symmetric matrix C, a convenient distance $d((\pi_1, \ldots, \pi_J), (\sigma_1, \ldots, \sigma_J))$ is defined as

$$d_C((\pi_1, \ldots, \pi_J), (\sigma_1, \ldots, \sigma_J)) = \mathrm{tr}((\Psi - \Sigma)C(\Psi - \Sigma)^T),$$

with $\Psi = P_1 \oplus \cdots \oplus P_J$ denoting the direct sum of permutation matrices corresponding to $\pi_1, \ldots, \pi_J$, and Σ having a similar meaning for $\sigma_1, \ldots, \sigma_J$.

The optimization problem, which one has to solve for this metric, consists of finding the permutation matrix Π minimizing the trace of the block matrix formed by submatrices $(P_j - \Pi)C_{jk}(P_k - \Pi)^T$, with $C_{jk}, j, k = 1, \ldots, J$ denoting $N \times N$ submatrices of the partitioned matrix C. In other terms, one has to minimize

$$\sum_{j=1}^{J} \mathrm{tr}((P_j - \Pi)C_{jj}(P_j - \Pi)^T)$$

$$= \mathrm{tr}\left(\Pi \sum_j C_{jj} \Pi^T\right) - 2\mathrm{tr}\left(\Pi \sum_j C_{jj} P_j^T\right) + \mathrm{tr}\left(\sum_j P_j C_{jj} P_j^T\right).$$

Matrix differentiation shows that the minimum is attained at the matrix

$$\Pi_0 = \left[\sum_j P_j C_{jj}\right] \left[\sum_j C_{jj}\right]^{-1}.$$

The matrix Π_0^T is stochastic, i.e., $e^T \Pi_0 = e^T$, but typically it is not a permutation matrix, and the problem of finding the closest permutation matrix, determined by a permutation π, remains. In this problem with $\Pi_0 = \{\hat{p}_{i\ell}\}$ we seek the permutation $\hat{\pi}$, which maximizes $\sum_i \hat{p}_{i\pi(i)}$,

$$\hat{\pi} = \arg\max_{\pi} \sum_i \hat{p}_{i\pi(i)}.$$

An efficient solution to this problem can be obtained from the Hungarian method for the assignment problem of linear programming (see Bazaraa et al. [BJS90, Sect. 10.7] for details).

In this setting one has to use an appropriate matrix C, which must be estimated on the basis of the training data; C^{-1} is the covariance matrix of all permutations $\pi_1, \ldots, \pi_J$ in the training sample.

A simpler aggregated algorithm suggested by Rukhin and Malioutov [RM05] can be defined by the matrix P, which is a convex combination of the permutation matrices $P_1, \ldots, P_J$, $P = \sum_{j=1}^J w_j P_j$. Again the problem is that of assigning nonnegative weights $w_1, \ldots, w_J$, such that $w_1 + \cdots + w_J = 1$, to matrices $P_1, \ldots, P_J$. The fairness of all (dependent) algorithms can be interpreted as $\mathrm{E}(P_i) = \mu$ with the same "central" matrix μ (in average, for a given probe, all algorithms measure the same quantity), the main difference between them is the accuracy. The optimal weights $w_1^0, \ldots, w_J^0$ minimize $E||\sum_j w_j(P_j - \mu)||^2$. Let Σ denote the positive definite matrix formed by the elements $E\mathrm{tr}((P_k - \mu)(P_j - \mu)^T)$, $k, j = 1, \ldots, J$. The optimization problem still consists in minimization of $w^T \Sigma w$ under condition $w^T e = 1$. The solution has the form

$$w^0 = \frac{\Sigma^{-1} e}{e^T \Sigma^{-1} e},$$

provided that Σ is nonsingular.

The "covariance matrix" Σ can be estimated by, say, $\hat{\Sigma}$, from the available training data. Note that for all k,

$$\mathrm{E}(\mathrm{tr}(P_k P_k^T)) = \mathrm{E}\left(\sum_{r,q} \delta_{r\pi(q)} \right) = N,$$

and for $k \neq j$,

$$\mathrm{E}(\mathrm{tr}(P_j P_k^T)) = \mathrm{E}(\mathrm{card}\, \{\ell : \pi_k(\ell) = \pi_j(\ell)\}).$$

Also the training data can be used to estimate μ by the sample mean $\hat{\mu}$ of all matrices in the training set.

Thus, to implement this linear fusion, these estimates are employed to get the estimated optimal weights,

$$\hat{w} = \frac{\hat{\Sigma}^{-1} e}{e^T \hat{\Sigma}^{-1} e}. \tag{5}$$

After these weights have been determined from the available data and found to be nonnegative, define a new combined ranking $\hat{\pi}_0$ on the basis of newly observed rankings $\pi_1, \ldots, \pi_J$ as follows. Let the N-dimensional vector $Z = (Z_1, \ldots, Z_N)$ be formed by coordinates $Z_i = \sum_{j=1}^J \hat{w}_j \pi_j(i)$, representing a combined score of element i. Put $\pi_0(i) = \ell$ if and only if Z_i is the ℓth smallest of $Z_1, \ldots, Z_N$. In other terms, π_0 is merely the rank corresponding to Z. In particular, according to π_0, the closest image in the gallery is k_0 such that

$$\sum_{j=1}^J \hat{w}_j \pi_j(k_0) = \min_k \sum_{j=1}^J \hat{w}_j \pi_j(k).$$

This ranking π_0 is characterized by the property

$$\sum_{i=1}^N \left(\sum_{j=1}^J \hat{w}_j \pi_j(i) - \pi_0(i) \right)^2 = \min_\pi \sum_{i=1}^N \left(\sum_{j=1}^J \hat{w}_j \pi_j(i) - \pi(i) \right)^2,$$

i.e., π_0 is the permutation that is the closest in the L_2 norm to $\sum_{j=1}^J \hat{w}_j \pi_j$ (see Theorem 2.2, p. 29 in Marden [Mar95]).

If some of the weights $\hat{w}$ are negative, they must be replaced by 0, and the remaining positive weights are to be renormalized by dividing by their sum. This method can be easily extended to the situation discussed in Sect. 2 when only partial rankings are available.

A more general approach is to look for matrix-valued weights W_i. These matrices must be nonnegative definite and sum up to identity matrix, $W_1 + \cdots + W_k = I$. The optimization problem remains as above.

The solution has the following, a bit more complicated, form. Let R be the $kN \times kN$ matrix formed by $N \times N$ blocks of the form $\mathrm{E}(P_i P_j^T), i, j = 1, \ldots, k$. Partition the inverse matrix $Q = R^{-1}$ in a similar way into submatrices $Q_{ij}, i, j = 1, \ldots, k$. Then the optimal solution is

$$W_i^0 = \sum_j Q_{ij} \left[\sum_{\ell,j} Q_{ij} \right]^{-1}.$$

After the matrix $\hat{P} = \sum_i W_i^0 P_i$ has been found, the combined algorithm ranks the gallery elements as follows:

$$\hat{p}(i) = \arg\max_j p_{ij}.$$

This solution is more computationally intensive as the dimension kN is large, and the matrix R can be ill-conditioned.

5 Example: FERET Data

To evaluate the proposed fusion methods, four face-recognition algorithms (I–IV), introduced earlier, were run on three 1996 FERET datasets of facial images, dupI (D1), dupII training (D2), and dupII testing (D3) (see Table 1) yielding similarity scores between gallery and probe images. The set D1 was discussed already in Sect. 2; the gallery consists of $N = 1196$ images, and 234 probe images were taken between 540 and 1031 days after its gallery match. For the sets D2 and D3 the probe image was taken before 1031 days. The similarity scores were used for training and evaluating the new classifiers; all methods were trained and tested on different datasets.

The primary measures of performance used for evaluation were the recognition rate, or the percentage of probe images classified at rank 1 by a method, and the mean rank assigned to the true images. Moreover, the relative recognition abilities were differentiated by the cumulative match characteristic (CMC) curve, which is a plot of the rank against the cumulative match score (the percentage of images identified below the rank). Finally, the receiver operating characteristic (ROC) curves were used for measuring the discriminating power of classifiers by plotting the true positive rate against the false positive rate for varying thresholds. The area under the ROC curve can be used as another quantitative measure of performance.

Table 1. Size of FERET datasets

	D1	D2	D3
Gallery size	1196	552	644
Probe size	234	323	399

Both methods of weighted averaging (4) and (5) produced similar weights. For example, the weights obtained from the correlation matrix S based on Spearman rho coefficients for the training set D1 are $w = (0.22, 0.32, 0.22, 0.24)$, while the weights via (4) are $w = (0.24, 0.27, 0.24, 0.25)$.

These two methods outperformed all but the best of constituent algorithms, II. On different pairs of training and testing datasets, the overall recognition rate of these methods fell short of this algorithm by 15% in the worst case and surpassed it by 2% in the best case (Table 2). The mean ranks of the two algorithms were generally within 5 ranks of each other.

Table 2. Percentage of images at rank 1

Dataset	Weights	(5)	I	II	III	IV
D2	D3	48.6	26.0	59.8	47.1	37.1
D3	D2	67.2	48.4	65.7	72.4	61.4
D1	D3	36.3	17.1	52.1	26.1	20.9

In terms of CMC curves, the methods of weighted averaging of ranks (4) or (5) improved on all but the best of constituent algorithms, II, which was better in the range of ranks from 1 to 30. It looks like this phenomenon is general for linear weighting, namely, for small ranks the best algorithm outperforms (4) and (5) for all weights giving this particular algorithm a weight smaller than 1. However, the weighted averaging method (4) was better than all of the four algorithms in the interval of ranks larger than 30 in the D2 dataset (Fig. 4). For each of these methods there was about an 85% chance of the true image being ranked 50 or below, which significantly narrowed down the number of possible candidates from more than 1000 images to only 50.

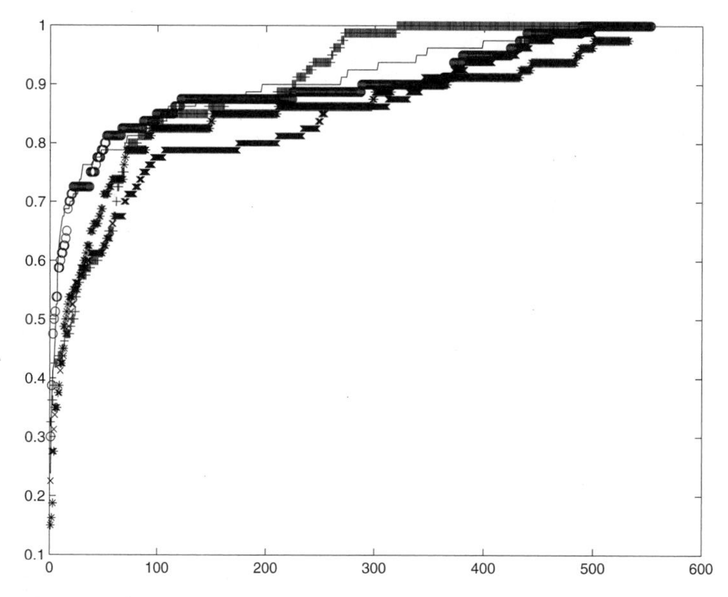

Fig. 5. Graphs of the cumulative match curves for algorithms I–IV (marked by *, +,o, x) and the linear aggregation (5) (marked by −).

The experiment showed that the weights derived from training for the different algorithms were all very close, which suggested that equal weights might be given to the different rankings. Although a simple averaging of ranks is a viable alternative to weighted averaging in terms of its computational efficiency, in our examples it was consistently inferior to method (4) or (5), and the benefit of training seems apparent.

References

[BJS90] Bazaraa, M. S., J. J. Jarvis, and H. D. Sherali. 1990. *Linear programming and network flows.* New York: Wiley.

[Bre04] Breiman, L. 2004. "Population theory for boosting ensembles." *Annals of Statistics* 32:1–11.

[Cri85] Critchlow, D. E. 1985. *Metric methods for analyzing partially ranked data.* New York: Springer.

[Dia88] Diaconis, P. 1988. *Group representations in probability and statistics.* Hayward, CA: Institute of Mathematical Statistics.

[DG97] Diaconis, P., and R. L. Graham. 1997. "Spearman's footrule as a measure of disarray." *Journal of the Royal Statistical Society Series B* 39:262–268.

[GGR95] Genest, C., K. Ghoudy, and L. P. Rivest. 1995. "A semiparametric estimation procedure of dependence parameters in multivariate families of distributions." *Biometrika* 82:543–552.

[JBP99] Jain, A. K., R. Bolle, and S. Pankanti. 1999. *Personal identification in networked society.* Dordrecht: Kluwer.

[JDM00] Jain, A. K., R. P. W. Duin, and J. Mao. 2000. "Statistical pattern recognition: A review." *IEEE Transactions on Pattern Analysis and Machine Intelligence* 22:4–37.

[Joe90] Joe, H. 1990. *Multivariate models and dependence.* London: Chapman & Hall.

[KHD98] Kittler, J., M. Hatef, R. P. W. Duin, and J. Matas. 1998. "On combining classifiers." *IEEE Transactions on Pattern Analysis and Machine Intelligence* 20:66–75.

[Mar95] Marden, J. I. 1995. *Analyzing and modeling rank data.* London: Chapman & Hall.

[Mar93] Marley, A. A. M. 1993. "Aggregation theorems and the combination of probabilistic rank orders." In *Probability models and statistical analyses for ranking data*, edited by M. A. Fligner and J. S. Verducci, Volume 80 of *Lecture Notes in Statistics*, 245–272. New York: Springer.

[Nel99] Nelsen, R. 1999. *An introduction to copulas.* Volume 139 of *Lecture Notes in Statistics.* New York: Springer.

[PMR00] Phillips, P. J., H. Moon, S. A. Rizvi, and P. J. Rauss. 2000. "The FERET evaluation methodology for face-recognition algorithms." *IEEE Transactions on Pattern Analysis and Machine Intelligence* 22:1090–1104.

[RM05] Rukhin, A. L., and I. Malioutov. 2005. "Fusion of algorithms via weighted averaging of ranks." *Pattern Recognition Letters* 26:679–684.

[RO05] Rukhin, A. L., and A. Osmoukhina. 2005. "Nonparametric measures of dependence for biometric data studies." *Journal of Statistical Planning and Inference* 131:1–18.

[SFB98] Schapire, R. E., Y. Freund, P. Bartlett, and W. S. Lee. 1998. "Boosting the margin: a new explanation for the effectiveness of voting methods." *Annals of Statistics* 26:1651–1686.

[Ste93] Stern, H. 1993. "Probability models on rankings and the electoral process." In *Probability models and statistical analyses for ranking data*, edited by M. A. Fligner and J. S. Verducci, Volume 80 of *Lecture Notes in Statistics.* New York: Springer.

Part III

Syndromic Surveillance

Data Analysis Research Issues and Emerging Public Health Biosurveillance Directions

Henry Rolka

National Center for Public Health Informatics, Centers for Disease Control and Prevention, HRolka@cdc.gov

Statistical and data analysis issues for new surveillance paradigms are quickly emerging in public health. Among the key factors motivating their evolution and development are

1. New requirements and resources to address a perceived bioterrorism threat as well as emerging diseases.
2. Information system technology growth in general.
3. Recognition of surveillance integration as a priority.
4. Widely available data with unrealized potential for useful information.

The term *syndromic surveillance* is used here somewhat as a catch-all for referring to new surveillance system paradigms and should be interpreted broadly [MH03, Hen04]. Biosurveillance has a much longer history for naturally occurring morbidity and mortality (i.e., infectious diseases, birth defects, injuries, immunization coverage, sexually transmitted diseases, HIV, medical product adverse events, etc.) than for deliberately malicious exposures. The professional relationships and established roles among public health levels (local, state, and federal) must be considered carefully as the context in which public health surveillance activity and system maturity take place. To ignore this extant infrastructure in advancing surveillance methodology involves the risk of developing irrelevant ideas because they may not be feasible to implement. However, if we do not extend beyond our applied creativity, we risk stagnation and incompetence. The balance is to identify the right size research, development, and implementation steps that will enable palatable progress and then take these steps quickly, frequently, and repeatedly in the same direction. Implementation of national scope public health information system change is extremely complex. This is a prologue designed to introduce and sensitize the reader to factors that may serve as enablers in that complexity for taking advantage of how to best consider the concepts asserted in the rest of the articles dealing with biosurveillance.

1 Evaluation

There is a growing body of literature on evaluation of "syndromic surveillance" that ranges broadly. Topics include

1. Advice on what to consider as a framework [CDC04]
2. Assessment of specific data source validity [FSS04]
3. Algorithm performance [MRC04, SD03, BBC05]
4. General policy discourse [Rei03, SSM04]
5. Activity overview, etc. [BBC05]

The breadth of this subtopic attests to the interest of evaluation for a relatively immature area in public health surveillance system development. It does not seem reasonable to expect meaningful evaluative conclusions about surveillance systems (e.g., cost/benefit utility) without a means to rigorously evaluate system *components* individually.

Consider that for modern biosurveillance systems, there are

1. Information technology process segments for recording electronic transactions and moving data.
2. Data preprocessing functions that include structuring an accessible analytic database architecture and ensuring data quality.
3. Data analysis components to apply methods for inference as well as deduction.
4. Support tools that operate in a decision theoretic framework for combining evidence, other information, and communication to facilitate action in near realtime.

To acquire useful evaluation measures for a surveillance system, subcategories are required so that specific enough objectives could be established. By this approach, evaluation for the provincial notion of "whether or not" to do surveillance gets replaced with the more practical notion of "how to do it better." Also, system complexity is reduced by decomposition. A risk here is to over-segregate interdependent activities and create operational stovepipes among professional skill sets. Good management and leadership must be alert and proactive to prevent maladaptive marginalization of system development, data management, statistical subject matter or end-user professionals in evaluation research and subsequent development activities for surveillance. This is essential in order to "conquer" after dividing; or more specifically in this context, to make sense out of algorithm performance characteristics after considering them separately from other operational surveillance components.

Since much of the data used for public health surveillance are not collected specifically for that purpose and/or are spontaneously generated, (1) they are referred to as "secondary" or "opportunistic," (2) the data require substantial preprocessing for analytic use, (3) a sample-to-population mapping is not probabilistically defined, and therefore, (4) the analytic signal detection methodologies are empirical in nature and do not lend themselves to conclusions bearing well-defined inferential quantities such as confidence intervals or p-values. There is generally no sampling design to define the probabilistic relationship between the data and a specified population of interest, and a design-based guide for an analytic strategy in a traditional sense is absent. Therefore effective use of these systems is primarily empirical.

Detection algorithm performance evaluation in an empirical setting is problematic when events for detection are rare. In the absence of recorded events of importance to train upon, thoughtful and informed simulation is much needed to accelerate learning. Ideally, a realistically described scenario can be translated into representation in data as a response to people's behavior. Characteristics of the scenario that would affect representation in the data could be modified with a consequential data representation. Monte Carlo iterations of the simulated signal structured over real data absent of events of interest could then be cycled with detection activities recorded. Thus, the usual means to evaluate a statistical detection approach for its operating characteristics under varying conditions could be established. What frequently takes place is that the people or groups who develop and promote a detection approach are the same ones who establish the simulation and the evaluation criteria and interpret the outcome. This is certainly a reasonable first step but this process leaves too much opportunity for scientific confounding — designing the evaluation criteria to fit the object of evaluation. A more objective approach would serve to advance the field more effectively.

In addition to (1) well-defined signals of importance, (2) the use of simulation, and (3) increased objectivity, the results of evaluation studies for surveillance system performance are of much greater practical value if they consider the realistic operational conditions under which data analysts must make decisions. Three considerably influential factors are data "lag time," "time alignment," and the "unlinked multiple data source" problem. Two ways that data lag time can be considered are (1) the average time between a population event (e.g., patient encounter or some other health-seeking behavioral event) and the event's data representation in an analytic system interface or (2) the proportion of data available at the time a decision is needed (versus at some later time). "Time alignment" refers to the differential health-seeking behavior times relevant for various data sources that may be available in one analytic system. For example, if one were able to view time series signals in response to a population exposure that caused illness, it may appear earlier for sales data than for emergency department (ED) data. The reasoning is that people may generally purchase products for self-treatment before their symptoms would be severe enough to warrant a trip to the ED. The "unlinked data source" problem is an issue for the secondary use of data sources when record linkage is either not possible or avoided for other reasons. Given that much of the data used in automated surveillance is gathered for some other purpose (treating patients, billing, market analysis, inventory, etc.) and that protecting individual confidentiality is a motive, broad linkage of records is not generally feasible. Therefore, the extent of information overlap is unknown across data streams. For example, if a system uses over-the-counter sales, ED, and laboratory test order data, it is not known to what extent the same people and their reactions to illness are manifest in the different sources. Without consideration of these operational realities, simulations for determining operating characteristics of new surveillance paradigms are incomplete at best and of marginal practical value.

2 Coordination for Information Exchange among Jurisdictions: BioSense

This is an aspect of analyzing and using information that easily goes unnoticed or is not well understood by the technical data analysis professionals who develop the analytic methodologies of surveillance systems. In public health as well as many public service industries, local jurisdictions are the primary users of information systems relating to situational awareness and their potential need to respond in their communities. When situations cross jurisdictional borders, coordinating response becomes a shared challenge. When public health threats cross state borders, the federal government becomes responsible for coordinating information. The time and efficiency of meeting this challenge are facilitated greatly through the use of technology standards [Bra05]. Conversely, multiple and diverse system outputs are difficult to exchange and consequently interpret. Thus, considering the potential public health threat that bioterrorism poses, there is a critical need for standards in data coding and preprocessing, data management procedures, analytic algorithms, data monitor operating procedures, and documentation of anomaly investigations. Further, since it could be any part of the nation that is at risk, these standards need to be national in scope. The Centers for Disease Control and Prevention has launched an initiative called BioSense to serve as a platform for standards development as part of the Public Health Information Network [Loo04]. BioSense is intended to provide a national safety net ensuring that early detection is enabled in all major metropolitan areas and works to support and integrate with existing regional surveillance systems. Requirements, data characteristics, threshold tolerances for response potential, etc. will likely continue to be different among local areas but it is certainly in our interests to enable rapid exchange of analytic results across jurisdictions. The goal is to have standard statistical and other data analytic conceptual approaches that can be tailored to local needs using various user-defined settings, results from which can be described coherently in a way that provides interoperable information for national situation awareness.

3 An Open Issue: Null Hypothesis Dilemma

An open question that is worth consideration in both advancing probabilistic methods for surveillance data analysis relates to the type I and type II error concepts. If we consider the null condition to be the assumption of "no event of importance in progress" and the alternative to be supported when there is sufficient data to conclude that a countermeasure response is needed, then the type I error is defined to be falsely concluding that a response is needed when in fact it is not necessary. This seems like the less important "mistake" in that if something were occurring that warranted a reaction and we did not respond, lives would be lost and precious time would have passed in stopping an event of importance. Thus, our general approach to controlling the type I error using "alpha" for threshold setting is questionable in this setting. On the other hand, being overly conservative at the expense of allowing too many false alerts may fatigue readiness resulting in an inability to respond when truly needed. The goal is to strike an informed balance between sensitivity maintenance and false alert toleration. Currently implemented surveillance systems in public health are based on inferential concepts that use p-values for thresholds under the null assumption that the situation is expected with relation to the temporal and/or geographical context. Given the situational consequences of failing to alert to true events and too frequently alerting to unimportant events, more refined bases for conclusions must be established as standard operating procedures using decision theoretic approaches and specifying risk and utility functions.

4 Summary and Directions

What has been commonly referred to as "syndromic surveillance" is not well-defined and is quickly growing out of its previous characterization. The implementation of new operational models for early event detection and subsequent situational awareness is creating opportunities for statistical and other data analytic applications in public health. Challenges include the following:

- There is little collective working experience with secondary data use among analysts.
- Data systems are new relative to the statistical methodologies employed.
- Data management tasks are large and the human resource skill sets for accomplishing those tasks are rare and underrated.
- Successful information system operations require close communications among staff of several interdependent disciplines.
- Analysis of these data requires inductive and deductive reasoning in combination (results may be difficult to communicate concisely).
- Multiple data streams:
    1. How can we best approach analysis: multiunivariate or multivariate?

2. There is a knowledge gap for population behavioral response patterns (the time alignment question).

The practice of binning population events into categories of likely association with syndromes relating to known serious biological agents, counting, comparing, and looking for patterns is currently the basis for most of the work in this area. This seems a logical first iteration of maturity for a surveillance system to enable earlier detection than would be possible otherwise. There is a need to apply decision science concepts to support end-user's threshold determination. The use of prior knowledge in a Bayesian framework and more refined pattern recognition seems like a promising direction for detection refinement, especially as more detailed data can be consolidated and means to process it are built. As more diverse data sources are integrated (human health, animal health, plant health, water quality, Internet traffic, utilities, intelligence, etc.), analytic approaches and applied methodologies for combining evidence from multiple and often conflicting sources will become even more important [SF02]. In the meantime, simulation appears to be the most promising method for accelerating available working knowledge of empirical surveillance.

In the chapters of Part III that follow, Shmueli and Fienberg provide an informed listing and brief conceptual characterization for a spectrum of detection approaches that either have already been implemented or hold promise for utility in surveillance. Their attention is primarily on the statistical methodologies and use of data from multiple sources, a logical focus given the current state of systems in application. Stoto et al. continue in this topic by creatively comparing the empirical detection performance of algorithms using simulated changes in patterns embedded in real health care data from Washington, DC. Finally, Forsberg et al. develop in an elegant historical context, the elucidation of how to take advantage of the space and time dimensions simultaneously in identifying clusters of events.

References

[Bra05] Bradwell, W. R. 2005. Enterprise architecture for health departments. Public Health Informatics Institute, http://www.phii.org/pages/Bradwell.html.

[BBC05] Buckeridge, D. L., H. Burkom, M. Campbell, W. R. Hogan, and A. W. Moore. 2005. "Algorithms for rapid outbreak detection: A research synthesis." *Journal of Biomedical Informatics* 38:99–113.

[CDC04] Centers for Disease Control and Prevention. 2004. "Framework for evaluating public health surveillance systems for early detection of outbreaks; recommendations from the CDC Working Group." *Morbidity and Mortality Weekly Report* 53 (RR-5): 1–13.

[FSS04] Fleischauer, A. T., B. J. Silk, M. Schumacher, K. Komatsu, S. Santana, V. Vaz, M. Wolfe, L. Hutwagner, J. Cono, R. Berkelman, and T. Treadwell. 2004. "The validity of chief complaint and discharge diagnosis in emergency department-based syndromic surveillance." *Academic Emergency Medicine Journal* 11 (12): 1262–1267 (December).

[Hen04] Henning, K. J. 2004. "What is syndromic surveillance?" *Morbidity and Mortality Weekly Report* 53 (Supplement): 7–11. Syndromic Surveillance: Reports from a National Conference, 2003.

[Loo04] Loonsk, J. W. 2004. "BioSense — A national initiative for early detection and quantification of public health emergencies." *Morbidity and Mortality Weekly Report* 53 (Supplement): 53–55. Syndromic Surveillance: Reports from a National Conference, 2003.

[MRC04] Mandl, K. D., B. Reis, and C. Cassa. 2004. "Measuring outbreak-detection performance by using controlled feature set simulations." *Morbidity and Mortality Weekly Report* 53 (Supplement): 130–136. Syndromic Surveillance: Reports from a National Conference, 2003.

[MH03] Mostashari, F., and J. Hartman. 2003. "Syndromic surveillance: A local perspective." *Journal of Urban Health* 80 (Suppl. 1): i1–i7.

[Rei03] Reingold, A. 2003. "If syndromic surveillance is the answer, what is the question?" *Biosecurity and Bioterrorism: Biodefense Strategy, Practice, and Science* 1 (2): 1–5.

[SF02] Sentz, K., and S. Ferson. 2002. "Combination of evidence in Dempster-Shafer theory." Sandia report SAND2002-0835, Sandia National Laboratories, Albuquerque, NM.

[SD03] Sosin, D. M., and J. DeThomasis. 2004. "Evaluation challenges for syndromic surveillance — making incremental progress." *Morbidity and Mortality Weekly Report* 53 (Supplement): 125–129. Syndromic Surveillance: Reports from a National Conference, 2003.

[SSM04] Stoto, M. A., M. Schonlau, and L. T. Mariano. 2004. "Syndromic surveillance: Is it worth the effort?" *Chance* 17:19–24.

Current and Potential Statistical Methods for Monitoring Multiple Data Streams for Biosurveillance

Galit Shmueli[1] and Stephen E. Fienberg[2]

[1] Decision and Information Technologies, Robert H. Smith School of Business, University of Maryland, College Park, gshmueli@rhsmith.umd.edu
[2] Department of Statistics, The Center for Automated Learning and Discovery, and Cylab, Carnegie Mellon University, fienberg@stat.cmu.edu

1 Introduction

A recent review of the literature on surveillance systems revealed an enormous number of research-related articles, a host of websites, and a relatively small (but rapidly increasing) number of actual surveillance systems, especially for the early detection of a bioterrorist attack [BMS04]. Modern bioterrorism surveillance systems such as those deployed in New York City, western Pennsylvania, Atlanta, and Washington, DC, routinely collect data from multiple sources, both traditional and nontraditional, with the dual goal of the rapid detection of localized bioterrorist attacks and related infectious diseases. There is an intuitive notion underlying such detection systems, namely, that detecting an outbreak early enough would enable public health and medical systems to react in a timely fashion and thus save many lives. Demonstrating the real efficacy of such systems, however, remains a challenge that has yet to be met, and several authors and analysts have questioned their value (e.g., see Reingold [Rei03] and Stoto et al. (2004) [SSM04, SFJ06]). This article explores the potential and initial evidence adduced in support of such systems and describes some of what seems to be emerging as relevant statistical methodology to be employed in them.

Public health and medical data sources include mortality rates, lab results, emergency room (ER) visits, school absences, veterinary reports, and 911 calls. Such data are directly related to the treatment and diagnosis that would follow a bioterrorist attack. They might not, however, detect the outbreak sufficiently fast. Several recent national efforts have been focused on monitoring "earlier" data sources for the detection of bioterrorist attacks or other outbreaks, such as over-the-counter (OTC) medication sales, nurse hotlines, or even searches on medical websites (e.g., WebMD). This assumes that people who are not aware of the outbreak and are feeling sick, would generally seek self-treatment before approaching the medical system and that an outbreak signature will manifest itself earlier in such data. According to Wagner et al. [WRT03], preliminary studies suggest that sales of OTC health care products can be used for the early detection of outbreaks, but research progress has been slow due to the difficulty that investigators have in acquiring suitable data to test this hypothesis for sizable outbreaks. Some data of this sort are already being collected (e.g., pharmacy and grocery sales). Other potential nontraditional data sources that are currently not collected (e.g., browsing in medical websites, automatic body sensor devices) could contain even earlier signatures of an outbreak.[3]

To achieve rapid detection there are several requirements that a surveillance system must satisfy: frequent data collection, fast data transfer (electronic reporting), real-time analysis of incoming data, and immediate reporting. Since the goal is to detect a large, localized bioterrorist attack, the collected information must be local, but sufficiently large to contain a detectable signal. Of course, the different sources must carry an early signal of the attack. There are, however, trade-offs between these features; although we require frequent data for rapid detection, too frequent data might be too noisy to the degree that the signal is too weak for detection. A typical solution for too frequent data is temporal aggregation. Two examples where aggregation is used for biosurveillance are aggregating OTC medication sales from hourly to daily counts [GSC02] and aggregating daily hospital visits into multiday counts [RPM03]. A similar trade-off occurs between the level of localization of the data and their amount. If the data are too localized, there might be insufficient data for detection, whereas spatial aggregation might dampen the signal.

Another important set of considerations that limit the frequency and locality of collected data relate to confidentiality and data disclosure issues (concerns over ownership, agreements with retailers, personal and organizational privacy, etc.). Finding a level of aggregation that contains a strong enough signal, that is readily available for collection without confronting legal obstacles, and yet is sufficiently rapid and localized for rapid detection, is clearly a challenge. We describe some of the confidentiality and privacy issues briefly here.

There are many additional challenges associated with the phases of data collection, storage, and transfer. These include standardization, quality control, confidentiality, etc. [FS05]. In this paper we focus on the statistical challenges associated with the data monitoring phase, and in particular, data in the form of multiple time series. We start by describing data sources that are

[3] While our focus in this article is on passive data collection systems for syndromic surveillance, there are other active approaches that have been suggested (e.g., screening of blood donors [KPF03]), as well as more technological fixes, such as biosensors [Sul03] and "Zebra" chips for clinical medical diagnostic recording, data analysis, and transmission [Cas04].

collected by some major surveillance systems and their characteristics. We then examine various traditional monitoring tools and approaches that have been in use in statistics in general, and in biosurveillance in particular. We discuss their assumptions and evaluate their strengths and weaknesses in the context of biosurveillance. The evaluation criteria are based on the requirements of an automated, nearly real-time surveillance system that performs online (or prospective) monitoring of incoming data. These are clearly different than for retrospective analysis [SB03] and include computational complexity, ease of interpretation, roll-forward features, and flexibility for different types of data and outbreaks.

Currently, the most advanced surveillance systems routinely collect data from multiple sources on multiple data streams. Most of the actual statistical monitoring, however, is typically done at the univariate time series level, using a wide array of statistical prediction methodologies. Ideally, multivariate methods should be used so that the data can be treated in an integrated way, accounting for the relationships between the data sources. We describe the traditional statistical methods for multivariate monitoring and their shortcomings in the context of biosurveillance. Finally, we describe monitoring methods, in both the univariate and multivariate sections, that have evolved in other fields and appear potentially useful for biosurveillance of traditional and nontraditional temporal data. We describe the methods and describe their strengths and weaknesses for modern biosurveillance.

2 Types of Data Collected in Surveillance Systems

Several surveillance systems aimed at rapid detection of disease outbreaks and bioterror attacks have been deployed across the United States in the last few years, including the *Realtime Outbreak and Disease Surveillance* system (RODS) and *National Retail Data Monitor* (NRDM) in western Pennsylvania, the *Early Notification of Community-Based Epidemics system* (ESSENCE) in the Washington, DC, area (which also monitors many Army, Navy, Air Force, and Coast Guard data worldwide), the *New York City Department of Health and Mental Hygiene* (NYC-DOHMH) system, and recently the *BioSense* system by the Centers for Disease Control and Prevention. Each system collects information on multiple data sources with the intent of increasing the certainty of a true alarm by verifying anomalies found in various data sources [PMK03]. All of these systems collect data from medical facilities, usually at a daily frequency. These include emergency rooms admissions (RODS, NYC-DOHMH), visits to military treatment facilities (ESSENCE), and 911 calls (NYC-DOHMH). Nontraditional data include OTC medication and healthcare product sales at grocery stores and pharmacies (NRDM, NYC-DOHMH, ESSENCE), prescription medication sales (ESSENCE), HMO billing data (ESSENCE), and school/work absenteeism records (ESSENCE). We can think of the data in a hierarchical structure; the first level consists of the data source (e.g., ER or pharmacy), and then within each data source there might be multiple time series, as illustrated in Fig. 1.

This structure suggests that series that come from the same source should be more similar to each other than to series from different sources. This can influence the type of monitoring methods used within a source as opposed to methods for monitoring the entire system. For instance, within-source series will tend to share variation sources such as holidays, closing dates, and seasonal effects. Pharmacy holiday closing hours will influence all medication categories equally but not school absences. From a modeling point of view this structure raises the question whether a hierarchical model is needed or else all series can be monitored using a flat multivariate model. In practice, most traditional multivariate monitoring schemes and a wide range of applications consider similar data streams. Very flexible methods are needed to integrate all the data within a system that is automatic, computationally efficient, timely, and with low false alarms. In the following sections we describe univariate and multivariate methods that are currently used or can potentially be used for monitoring the various multiple data streams. We organize and group the different methods by their original or main field of application and discuss their assumptions, strengths, and limitations in the context of biosurveillance data.

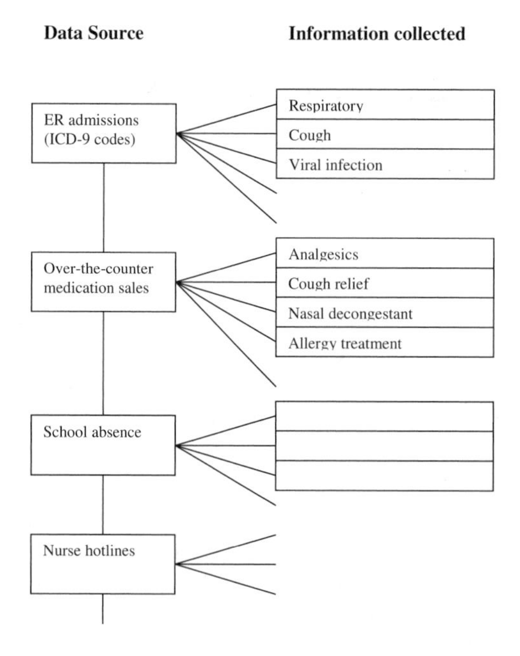

Fig. 1. Sketch of data hierarchy; each data source can contain multiple time series.

3 Monitoring Univariate Data Streams

The methods used in biosurveillance borrow from several fields, with statistical process control being the most influential. Methods from other fields have also been used occasionally, with most relying on traditional statistical methods such as regression and time series models. Although different methods might be more suitable for different data streams or sources, there are advantages to using a small set of methods for all data streams within a single surveillance system. This simplifies automation, interpretability, and coherence, and the ability to integrate results from multiple univariate outputs. The principle of parsimony, which balances performance and simplicity, should be the guideline.

We start by evaluating some of the commonly used monitoring methods and then describe other methods that have evolved or have been applied in other fields, which are potentially useful for biosurveillance.

3.1 Current Common Approaches

Statistical Process Control

Monitoring is central to the field of statistical process control. Deming, Shewhart, and others revolutionized the field by introducing the need and tools for monitoring a process to detect abnormalities at the early stages of production. Since the 1920s the use of control charts has permeated into many other fields including the service industry. One of the central tools for process control is the control chart, which is used for monitoring a parameter of a distribution. In its simplest form the chart consists of a centerline, which reflects the target of the monitored parameter and control limits. A statistic is computed from an *iid* sample every time point, and its value is plotted on the chart. If it exceeds the control limits, the chart flags an alarm, indicating a change in the monitored parameter. Statistical methods for monitoring univariate and multivariate time series tend to be model-based. The most widely used control charts are Shewhart charts, moving average (MA) charts, and cumulative sum (CuSum) charts. Each of these methods specializes in detecting a particular type of change in the monitored parameter [BL97].

We now briefly describe the different charts. Let $\mathbf{y}_t$ be a random sample of measurements taken at time t ($t = 1, 2, 3, \ldots$). In a Shewhart chart the monitoring statistic at time t, denoted by S_t, is a function of $\mathbf{y}_t$:

$$S_t = f(\mathbf{y}_t). \tag{1}$$

The statistic of choice depends on the parameter that is monitored. For instance, if the process mean is monitored, then the sample mean ($f(\mathbf{y}_t) = \overline{\mathbf{y}}_t$) is used. If the process variation is monitored, a popular choice is the sample standard deviation ($f(\mathbf{y}_t) = s_t$). The monitoring statistic is drawn on a time plot, with lower and upper control limits. When a new sample is taken, the point is plotted on the chart, and if it exceeds the control limits, it raises an alarm. The assumption behind the classic Shewhart chart is that the monitoring statistic follows a normal distribution. This is reasonable when the sample size is large enough relative to the skewness of the distribution of $\mathbf{y}_t$. Based on this assumption, the control limits are commonly selected as ± 3 standard deviations of the monitoring statistic (e.g., if the sample mean is the monitoring statistic, then the control limits are $\pm 3\sigma/\sqrt{n}$) to achieve a low, false-alarm rate of $2\phi(-3) = 0.0027$. Of course, the control limits can be chosen differently to achieve a different false-alarm rate. If the sample size at each time point is $n = 1$, then we must assume that $\mathbf{y}_t$ are normally distributed for the chart to yield valid results. Alternatively, if the distribution of $f(\mathbf{y}_t)$ (or $\mathbf{y}_t$) is known, then a valid Shewhart chart can be constructed by choosing the appropriate percentiles of that distribution for the control limits as discussed in [SFJ06].

Shewhart charts are very popular because of their simplicity. They are very efficient at detecting moderate-to-large, spike-type changes in the monitored parameter. Since they do not have a "memory," a large spike is immediately detected by exceeding the control limits. However, Shewhart charts are not useful for detecting small spikes or longer-term changes. In those instances we need to retain a longer "memory." One solution is to use the "Western Electric" rules. These rules raise an alarm when a few points in a row are too close to a control limit, even if they do not exceed it. Although such rules are popular and are imbedded in many software programs, their addition improves detection of real aberrations at the cost of increased false alarms. The trade-off turns out to be between the expected time-to-signal and its variability [SC03].

An alternative is to use statistics that have longer memories. Three such statistics are the MA, the exponentially weighted moving average (EWMA), and the CuSum. MA charts use a statistic that relies on a constant-size window of the k last observations:

$$\mathrm{MA}_t = \sum_{j=1}^{k} f(\mathbf{y}_{t-j+1})/k. \tag{2}$$

The most popular statistic is a grand mean $(\sum_{j=1}^{k} \overline{y}_{t-j+1}/k)$. These charts are most efficient for detecting a step increase/decrease that lasts k time points.

The original CuSum statistic defined by $\frac{1}{\sigma}\sum_{i=1}^{t}(y_t - \mu)$ keeps track of all the data until time t [HO98]. However, charts based on this statistic are awkward graphically. A widely used adaptation is the tabular CuSum, which restarts the statistic whenever it exceeds zero. The one-sided tabular CuSum for detecting an increase is defined as

$$\mathrm{CuSum}_t = \max\{0, (y_t^* - k) + \mathrm{CuSum}_{t-1}\}, \tag{3}$$

where $y_t^* = (y_t - \mu)/\sigma$ are the standardized observations, and k is proportional to the size of the abnormality that we want to detect. This is the most efficient statistic for detecting a step change in the monitored parameter. However, it is less useful for detecting a spike since it would be masked by the long memory. In general, time series methods that place heavier weight on recent values are more suitable for short-term forecasts [Arm01].

The EWMA statistic is similar to the CuSum, except that it weights the observations as a function of their recency, with recent observations taking the highest weight:

$$\text{EWMA}_t = \alpha y_t + (1-\alpha)\text{EWMA}_{t-1} = \alpha \sum_{j=0}^{t-1} (1-\alpha)^j f(y_{t-j}) + (1-\alpha)\text{EWMA}_0, \quad (4)$$

where $0 < \alpha \leq 1$ is the smoothing constant [Mon01]. This statistic is best at detecting an exponential increase in the monitored parameter. It is also directly related to exponential smoothing methods (see below). For further details on these methods, see Montgomery [Mon01]. In biosurveillance, the EWMA chart was used for monitoring weekly sales of OTC electrolytes to detect pediatric respiratory and diarrheal outbreaks [HTI03] and is used in ESSENCE II to monitor ER admissions in small geographic regions [Bur03a].

Since the statistic in these last three cases is a weighted average/sum over time, the normality assumption of y_t is less crucial for adequate performance due to the central limit theorem, especially in the case of the EWMA [RS04b, ACV04]. The main disadvantage of all these monitoring tools is that they assume statistical independence of the observations. Their original and most popular use is in industrial process control where samples are taken from the production line at regular intervals, and a statistic based on these assumably independent samples is computed and drawn on the control chart. The *iid* assumption is made in most industrial applications, whether correct or not. Sometimes the time between samples is increased to minimize correlation between close samples. In comparison, the types of data collected for biosurveillance are usually time series that are collected on a frequent basis to achieve timeliness of detection, and therefore autocorrelation is inherent. For such dependent data streams the use of distribution-based or distribution-free control charts can be misleading in the direction of increased false-alarm rates [Mon01, p. 375].

A common approach to dealing with autocorrelated measurements is to approximate them using a time series model and monitor the residual error using a control chart [GDV04]. The assumption is that the model accounts for the dependence and therefore the residuals should be nearly independent. Such residuals will almost never be completely independent, however, and the use of control charts to monitor them should be done cautiously. This is where time series analysis emerges in anomaly detection applications in general, and in biosurveillance in particular. Moreover, because the forecast at every time point is used to "test" for anomalies, we need to deal with the multiple testing problem for dependent tests and possibly use variations on the new literature on false discovery rates (FDR) to control familywise type I errors [BH95, EST01].

Time Series Methods

The most well-known class of time series models used by statisticians is autoregressive moving average (ARMA) models. Conceptually they are similar to regressing the current observations on a window of previous observations while assuming a particular autocovariance structure. An ARMA(p,q) model is defined as

$$y_t = \mu + \sum_{i=1}^{p} \alpha_i y_{t-i} + \varepsilon_t - \sum_{j=1}^{q} \theta_j \varepsilon_{t-j}, \tag{5}$$

where α_i and θ_j are parameters and $\varepsilon_{t-q} \ldots \varepsilon_t$ are white noise (having mean 0 and standard deviation σ_ε). To fit an ARMA model, the modeler must determine the order of the autoregression p and the order of the MA component, q. This task is not straightforward and requires experience and expertise (for example, not every selection of p and q yields a causal model). After p and q are determined, there are $p+q+1$ parameters to estimate, usually through nonlinear least squares (LS) and conditional maximum likelihood. The process of selecting p and q and estimating the parameters is cyclical [BJR94] and typically takes several cycles until a satisfactory model is found. Some software packages do exist that have automated procedures for determining p and q and estimating those parameters.

ARMA models can combine external information by adding predictors in the model. This allows to control for particular time points that are known to have a different mean by adding indicators with those time points. Such modifications are especially useful in the biosurveillance context, since effects such as weekend/weekday and holidays are normally present in medical and non-traditional data. ESSENCE II, for instance, uses an autoregressive model that controls for weekends, holidays, and postholidays through predictors [Bur03a].

ARMA models assume that the series is stationary over time (i.e., the mean, variance, and autocovariance of the series remain constant throughout the period of interest). In practice, fitting of an ARMA model to data usually requires an initial preprocessing step where the data are transformed in one or more ways until a stationary or approximately stationary series emerges. The most popular generalization of ARMA models for handling seasonality and trends is to add a differencing transformation, thereby yielding an autoregressive integrated moving average (ARIMA) model of the form

$$(1-\alpha_1 B-\alpha_2 B^2-\ldots-\alpha_p B^p)[(1-B)^d(1-Bs)^D y_t-\mu] = (1-\theta_1 B-\ldots-\theta_q B^q)\varepsilon_t, \tag{6}$$

where B is the back-shift operator ($By_t = y_{t-1}$), $d > 0$ is the degree of nonseasonal differencing, $D > 0$ is the degree of seasonal differencing, and s is the length of a seasonal cycle. Determining the level of differencing is not trivial, and over- and underdifferencing can lead to problems in modeling and forecasting [CR96]. Although this model allows flexibility, in practice the model identification step is complicated and highly data specific, and requires expertise of the modeler. Another disadvantage of ARIMA models is their computational complexity. With thousands of observations, the method requires considerable computer time and memory [SAS04b].

To summarize, the common statistical approach towards monitoring has been mostly distribution based. Recent advances in data availability and collection in the process industry have led authors such as Willemain and Runger [WR96] to emphasize the importance of model-free methods. It appears, though, that such methods have already evolved and have been used in other fields! Next, we describe a few such methods that are distribution-free.

3.2 Monitoring Approaches in Other Fields

Monitoring methods have been developed and used in other fields such as machine learning, computer science, geophysics, and chemical engineering. Also, forecasting, which is related to monitoring, has had advances in fields such as finance and economics. In these fields there exist a wealth of very frequent autocorrelated data; the goal is the rapid detection of abnormalities ("anomaly detection") or forecasting, and the developed algorithms are flexible and computationally efficient. We describe a few of the methods used in these fields and evaluate their usefulness for biosurveillance.

Anomaly detection in machine learning emphasizes automated and usually model-free algorithms that are designed to detect local abnormalities. Even within the class of model-free algorithms, there is a continuum between those that are intended to be completely "user-independent" and those that require expert knowledge integration by the user. An example for the former is the symbolic aggregate approximation (SAX), which is a symbolic representation for time series that allows for dimensionality reduction and indexing [LKL03]. According to its creators, "anomaly detection algorithms should have as few parameters as possible, ideally none. A parameter free algorithm would prevent us from imposing our prejudices, expectations, and presumptions on the problem at hand, and would let the data itself speak to us" [KLR04]. In biosurveillance there exists expert knowledge about the progress of a disease, its manifestation in medical and public health data, etc. An optimal method would then be distribution-free and parsimonious, but would allow the integration of expert knowledge in a simple way.

Exponential Smoothing

Exponential smoothing (ES) is a class of methods that is very widely used in practice (e.g., for production planning, inventory control, and marketing [PA89]) but not so in the biosurveillance field. ES has gained popularity mostly because of its usefulness as a short-term forecasting tool. Empirical research by

Makridakis et al. [MAC82] has shown simple exponential smoothing (SES) to be the best choice for one-step-ahead forecasting, from among 24 other time series methods and using a variety of accuracy measures. Although the goal in biosurveillance is not forecasting, ES methods are relevant because they can be formulated as models [CKO01]. Nontraditional biosurveillance data include economic series such as sales of medications, health-care products, and grocery items. Since trends, cycles, and seasonality are normally present in sales data, more advanced ES models have been developed to accommodate nonstationary time series with additive multiplicative seasonality/linear/exponential/dampened trend components. A general formulation of an ES model assumes that the series is comprised of a level, trend (the change in level from last period), seasonality (with M seasons), and error. To illustrate the model formulation, estimation, and forecasting processes, consider an additive model of the form

$$y_t = \text{local mean} + \text{seasonal factor} + \text{error}, \tag{7}$$

where the local mean is assumed to have an additive trend term and the error is assumed to have zero mean and constant variance. At each time t, the smoothing model estimates these time-varying components with level, trend, and seasonal smoothing states denoted by L_t, T_t, and $S_{t-i} (i = 0, \ldots, M-1)$, respectively.[4] The smoothing process starts with an initial estimate of the smoothing state, which is subsequently updated for each observation using the *updating equations*:

$$\begin{aligned} L_{t+1} &= \alpha(y_{t+1} - S_{t+1-M}) + (1-\alpha)(L_t + T_t), \\ T_{t+1} &= \beta(L_{t+1} - L_t) + (1-\beta)T_t, \\ S_{t+1} &= \gamma(y_{t+1} - L_{t+1}) + (1-\gamma)S_{t+1-M}, \end{aligned} \tag{8}$$

where α, β, and γ are the smoothing constants. The m-step-ahead forecast at time t is

$$\hat{y}_{t+m} = L_t + mT_t + S_{t+m-M}. \tag{9}$$

A multiplicative model of the form $Y_t = (L_{t-1} + tT_{t-1})S_{t-i}\,\varepsilon_t$ can be obtained by applying the updating equations in (8) to $\log(y_t)$. The initial values L_0, T_0, and the M seasonal components at time 0 can be estimated from the data using a centered MA (see Pfeffermann and Allon [PA89] and the NIST Handbook [NIS04] for details). The three smoothing constants are either determined by expert knowledge, or estimated from the data by maximizing a well-defined loss function (e.g., mean of squared one-step-ahead forecast errors).

From a modeling point of view, many ES methods have ARIMA, seasonal ARIMA (SARIMA), and structural models equivalents, and they even include a class of dynamic nonlinear state space models that allow for changing

[4] The smoothing state is normalized so that the seasonal factors S_{t-i} for $i = 0, 1, \ldots, M$ sum to zero for models that assume additive seasonality, and average to one for models that assume multiplicative seasonality [CY88].

variance [CKO01]. Table 1 summarizes some of these equivalences. It is noteworthy that some of the SARIMA equivalents are so complicated that they are most unlikely to be identified in practice [CKO01]. Furthermore, Chatfield et al. [CKO01] show that there are multiple generating processes for which a particular ES method is optimal in the sense of forecast accuracy, which explains their robust nature. The advantage of these models is their simplicity of implementation and interpretation, their flexibility for handling many types of series, and their suitability for automation [CY88] because of the small number of parameters involved and the low computational complexity. They are widely used and have proved empirically useful, and automated versions of them are available in major software packages such as the high-performance forecasting module by SAS®[SAS04a].

Table 1. The equivalence between some exponential smoothing and (seasonal)-ARIMA models. The notation $ARIMA(p,d,q)(P,D,Q)_s$ corresponds to an ARIMA(p,d,q) with seasonal cycle of length s, P-order autoregressive seasonality, seasonal differencing of order D, and seasonal MA of order Q

Exponential Smoothing Method	ARIMA/SARIMA Equivalent
Simple exponential smoothing	ARIMA(0,1,1)
Holt's (double) linear trend method	ARIMA(0,2,2)
Damped-trend linear method	ARIMA(1,1,2)
Additive Holt-Winters (triple) method	SARIMA$(0,1,p+1)(0,1,0)_p$
Multiplicative Holt-Winters (triple) method	[KSO01]'s dynamic nonlinear state-space models

Singular Spectral Analysis

The methods of singular spectral analysis (SSA) were developed in the geosciences as an alternative for Fourier/spectral analysis and have been used mostly for modeling climatic time series such as global surface temperature records [GV91], and the Southern Oscillation Index that is related to the recurring El Niño/Southern Oscillations conditions in the Tropical Pacific [PGV95, YSG00] . It is suitable for decomposing a short, noisy time series into a (variable) trend, periodic oscillations, other aperiodic components, and noise [PGV95].

SSA is based on an eigenvalue-eigenvector decomposition of the estimated M-lag autocorrelation matrix of a time series, using a Karhunen-Loeve decomposition. The eigenvectors, denoted by $\varrho_1, \ldots, \varrho_M$, are called empirical orthogonal functions (EOFs) and form an optimal basis that is orthonormal at lag zero. Usually a single EOF is sufficient to capture a nonlinear oscillation. Using statistical terminology, principal components analysis (PCA) is applied to the estimated autocorrelation matrix, so that projecting the EOFs on the time series gives the principal components $(\Lambda_1, \ldots, \Lambda_M)$:

$$\Lambda_k(t) = \sum_{i=1}^{M} y(t+i)\varrho_k(i), \tag{10}$$

and the eigenvalues reflect the variability associated with the principal components [GY96]. The next step in SSA is to reconstruct the time series using only a subset $\mathcal{K}$ of the EOFs:

$$y_{\mathcal{K}}(t) = \frac{1}{M_t} \sum_{k \in \mathcal{K}} \sum_{i=1}^{M} \Lambda_k(t-i)\varrho_k(i), \tag{11}$$

where M_t is a normalizing constant (for details, see [GV91]). Choosing $\mathcal{K}$ is done heuristically or by Monte Carlo simulations.

SSA is used mostly for revealing the underlying components of a time series and separating signal from noise. However, it can be used for forecasting by using low-order autoregressive models for the separate reconstructed series [PGV95]. This means that SSA can be used for biosurveillance and monitoring in general by computing one-step-ahead forecasts and comparing them to the actual data. If the distance between a forecast and an actual observation is too large, a signal is triggered.

Although SSA assumes stationarity (by decomposing the estimated autocorrelation matrix), according to Yiou et al. [YSG00], it appears less sensitive to nonstationarity than spectral analysis. However, Yiou et al. [YSG00] suggested a combination of SSA with wavelets to form multiscale SSA (MS-SSA). The idea is to use the EOFs in a data-adaptive fashion with a varying window width, which is set as a multiple of the order M of the autocorrelation matrix. After applying the method to synthetic and real data, they conclude that MS-SSA behaves similarly to wavelet analysis, but in some cases it provides clearer insights into the data structure. From a computational point of view, MS-SSA is very computationally intensive, and in practice a subset of window widths is selected rather than exhaustively computing over all window widths [YSG00].

Wavelet-Based Methods

An alternative to ARIMA models that has gained momentum in the last several years is wavelet decomposition. The idea is to decompose the time series $y(t)$ using wavelet functions:

$$y(t) = \sum_{k=1}^{N} a_k \phi(t-k) + \sum_{k=1}^{N} \sum_{j=1}^{m} d_{j,k} \psi(2^j t - k), \tag{12}$$

where a_k is the scaled signal at time k at the coarsest scale m, $d_{j,k}$ is the detail coefficient at time k at scale j, ψ is a scaling function (known as the "father wavelet"), and ϕ is the mother wavelet function.

This method is very useful in practice, since data from most processes are multiscale in nature due to "events occurring at different locations and with different localization in time and frequency, stochastic processes whose energy or power spectrum changes with time and/or frequency, and variables measured at different sampling rates" [Bak98]. In traditional process control, the solution is to use not a single control chart but to combine different control charts (such as Shewhart-CuSum [Luc82] and Shewhart-EWMA charts [LS90]) to detect shifts at different scales. This, of course, leads to increased alarm rates (false and true). The wavelet decomposition method offers a more elegant and suitable solution. Aradhye et al. [ABS03] used the term multiscale statistical process control (MSSPC) to describe these methods. Wavelet methods are also more suitable for autocorrelated data, since the wavelet decomposition can approximately decorrelate the measurements. A survey of wavelet-based process monitoring methods and their history is given in Ganesan et al. [GDV04]. Here we focus on their features that are relevant to biosurveillance.

The main application of wavelets has been for denoising, compressing, and analyzing image, audio, and video signals. Although wavelets have been used by statisticians for smoothing/denoising data (e.g., [DJ95], for density estimation [DJK96], nonparametric regression [OP96], and other goals [PW00]), they have only very recently been applied to statistical process monitoring. The most recent developments in wavelet-based monitoring methods have been published mainly within the area of chemical engineering [SCR97, HLM98, ABS03]. The main difference between chemical engineering processes and biosurveillance data (traditional and nontraditional) is that in the former the definitions of normal and abnormal are usually well-defined, whereas in the latter it is much harder to establish such clear definitions. In that sense wavelets are even more valuable in biosurveillance because of their nonspecific nature. Aradhye et al. [ABS03] have shown that using wavelets for process monitoring yields better average performance than single-scale methods if the shape and magnitude of the abnormality are unknown.

The typical wavelet monitoring scheme works in four main steps:

1. Decompose the series into coefficients at multiple scales using the discrete wavelet transform (DWT). The DWT algorithm is as follows:
   - Convolve the series with a low-pass filter to obtain the approximation coefficient vector $\mathbf{a}_1$ and with a high-pass filter to obtain the detail coefficient vector $\mathbf{d}_1$. If we denote the low-pass decomposition filter by $\mathbf{h} = [h_0, h_1, \ldots, h_n]$, then the ith component of the high-pass decomposition filter, $\mathbf{g}$, is given by $g_i = (-1)^i h_{n-i}$.
   - Downsample the coefficients. Half of the coefficients can be eliminated according to the Nyquist rule, since the series now has a highest frequency of $\pi/2$ radians instead of π radians. Discarding every other coefficient downsamples the series by two, and the series will then

have half the number of points. The scale of the series is now doubled [Pol].
- Reconstruct the approximation vector A_1 and detail vector D_1 by upsampling and applying "reconstruction" filters (Inverse-DWT). The set of low-pass and high-pass reconstruction filters are given as $h_n* = h_{-n}$ and $g_n* = g_{-n}$.

If an orthogonal wavelet is used, then the original signal can be completely reconstructed by simple addition: $Y = A_1 + D_1$. The second level of decomposition is obtained by applying this sequence of operations to the first level approximation A_1. The next levels of decomposition are similarly obtained from the previous level approximations.

2. Perform some operation on the detail coefficients $\mathbf{d}_k$ $(k = 1, \ldots, m)$. Various operations were suggested for monitoring purposes. Among them:
   - Thresholding the coefficients at each scale for the purpose of smoothing or data reduction [LLW02].
   - Forecasting each of the details and the mth approximation at time $t+1$. This is done by fitting a model such as an autoregressive model [GSC02] or neural networks [AM97] to each scale and using it to obtain one-step-ahead forecasts.
   - Monitoring A_m and $D_1, D_2, \ldots, D_m$ by creating control limits at each scale [ABS03].
3. Reconstruct the series from the manipulated coefficients. After m levels of decomposition, an orthogonal wavelet will allow us to reconstruct the original series by simple addition of the approximation and detail vectors: $Y = A_m + D_1 + D_2 + \ldots + D_m$. If thresholding was applied, the reconstructed series will differ from the original series, usually resulting in a smoother series. In the case of single-scale monitoring [ABS03] use the control limits as thresholds and reconstruct the series only from the coefficients that exceeded the thresholds. In the forecasting scheme, the reconstruction is done to obtain a forecast of the series at time $t+1$ by combining the forecasts at the different scales.
4. Perform some operation on the reconstructed series. Aradhye et al. [ABS03] monitor the reconstructed series using a control chart. In the forecasting scheme the reconstructed forecast is used to create an upper control limit for the incoming observation [GSC02].

Although DWT appears to be suitable for biosurveillance, it has several limitations that must be addressed. The first is that the downsampling causes a delay in detection and thus compromises timeliness. This occurs because the downsampling causes a lag in the computation of the wavelet coefficients, which increases geometrically as the scale increases. An alternative is to avoid the downsampling stage. This is called stationary- or redundant-DWT. Although it solves the delay problem, it introduces a different challenge; it does not allow the use of orthonormal wavelets, which approximately decorrelate the series. This means that we cannot treat the resulting coefficients at each scale as normally distributed, uncorrelated, and with equal variance. Aradhye et al. [ABS03] conclude that for detecting large shifts it is preferable to use stationary-DWT if the series is uncorrelated or moderately correlated, whereas for highly nonstationary or autocorrelated series the use of downsampling is preferable. Both models perform similarly in detecting small changes. For further discussion of this issue and empirical results see Aradhye et al. [ABS03].

The second issue is related to the boundaries of the series, and especially the last observation. Since DWT involves convolving the series with filters, the beginning and end of the series need to be extrapolated (except when using the Haar). One approach is to use boundary-corrected wavelets. These have been shown to be computationally impractical [GDV04]. Another approach is to use an extrapolation method such as zero padding, periodic extension, and smooth padding. In surveillance applications the end of the series and the type of boundary correction are extremely important. Extrapolation methods such as zero padding and periodic extension (where the beginning and end of the series are concatenated) are clearly not suitable, since it is most likely that the next values will not be zeros or those from the beginning of the series. More suitable methods are the class of smooth padding, which consist of extrapolating the series by either replicating the last observation or linearly extrapolating from the last two values. An alternative would be to use exponential smoothing, which is known to have good forecasting performance in practice.

Finally, although wavelet-based methods require very little user input for the analysis, there are two selections that need to be made manually, namely, the depth of decomposition and the wavelet function. Ganesan et al. [GDV04] offer the following guidelines based on empirical evidence: the depth of decomposition should be half the maximal possible length. Regarding choice of wavelets, the main considerations are good time-frequency localizations, number of continuous derivatives (determine degrees of smoothness), and a large number of vanishing moments. We add to that computational complexity and interpretability. The Haar, which is the simplest and earliest wavelet function, is best suited for detecting step changes or piecewise constant signals. For detecting smoother changes, a Daubechies filter of higher order is more suitable.

4 Monitoring Multiple Data Streams

Modern biosurveillance systems such as the ones described earlier routinely collect data from multiple sources. Even within a single source there are usually multiple data streams. For instance, pharmacy sales might include sales of flu, allergy, and pain-relief medications, whereas ER visits record the daily number of several chief complaints. The idea behind syndromic surveillance is to monitor a collection of symptoms to learn about possible disease outbreaks. Therefore we expect multivariate monitoring methods to be superior to univariate methods in actual detection, since the hypothesized signal can be formulated in a multivariate fashion. Optimally, multivariate models should detect changes not only in single data streams but also in the functional relationships between them.

4.1 Merging Data Sources: Why Use Aggregated Data?

One of the main reasons for treating biosurveillance data at the aggregated level is the issue of privacy associated with individuals whose data are being used. Medical and public health data systems of relevance for surveillance systems are typically subject to formal rules and/or legal restrictions regarding their use in identifiable form (e.g., as provided for by the Health Insurance Portability and Accountability Act of 1996, Public Law 104-191 (HIPAA) under its recently issued and implemented privacy and confidentiality rules), although there are typically research and other permitted uses of the data provided that they are de-identified. The HIPAA Safe Harbor de-identification, for instance, requires the de-identification of 18 identifiers including name, social security number, zip code, medical record number, age, etc. The removal of such information clearly restricts the possibility of record linkage across data sources, although it also limits the value of the data for statistical analysis and prediction, especially in connection with the use of spatial algorithms [Won04]. Similar legal restrictions apply to prescription information from pharmacies. Other public and semipublic data systems, such as school records, are typically subject to a different form of privacy restriction but with similar intent. Finally, grocery and OTC medication sales information is typically the property of the commercial interests that are wary of sharing data in individually identifiable form even if there are no legal strictures against such access. Solutions do exist that would potentially allow record linkage to at least some degree (e.g., by using a trusted broker and related data sharing agreements) (see the discussion in Gesteland et al. [GGT03]). While the practical solution of independently and simultaneously monitoring the separate sources, especially at the aggregate level, avoids the issue of record linkage and privacy concerns, it also leads to loss of power to detect the onset of a bioterrorist attack! Thus ultimately, if the syndromic surveillance methodology is to prove successful in early detection of a bioterrorist attack, the HIPAA and other privacy rules will need to be adjusted either to allow special exceptions for this type of data use, or to recognize explicitly that privacy rights may need to be compromised somewhat to better protect the public as a whole through the increased utility of the use of linked multiple data sources.

A separate reason for using aggregated data is the difficulty of record linkage from multiple sources: "identifiers" that are attached to records in different sources will usually differ at least somewhat. Linking data from two or more sources either requires unique identifiers that are used across systems or variables that can be used for record linkage. In the absence of unique identifiers, matching names and fields, especially in the presence of substantial recording error, poses substantial statistical challenges. For further discussion of these issues see Fienberg and Shmueli [FS05] and especially Bilenko et al. [BMC03].

4.2 Current Common Approaches

Monitoring multiple time series is central in the fields of quality/process control, intrusion detection [Ye02], and anomaly detection in general. When the goal is to detect abnormalities in independent series, then multiple univariate tools can be used, and then merged to form a single alarm mechanism. However, the underlying assumption behind the data collected for biosurveillance is that the different sources are related and are measuring the same phenomenon. In this case, multivariate methods are more suitable. The idea is to detect not only abnormalities in single series, but also abnormal relationships between the series (also termed "counterrelationships"). In the following we describe multivariate methods that have been used in different applications for the purpose of detecting anomalies.

Statistical Process Control

The quality control literature includes several multivariate monitoring methods. Some are extensions of univariate methods, such as the χ^2 and Hotelling T^2 control charts, the multivariate CuSum chart, and the multivariate EWMA chart [ASJ97]. The multivariate versions are aimed at detecting shifts in single series as well as counterrelationships between the series. As in the univariate case, they are all based on the assumptions of independent and normally distributed observations. Also, like their univariate counterparts they suffer from problems of underdetection. In practice they are sometimes combined with a Shewhart chart, but this solution comes at the cost of slowing down the detection of small shifts [ASJ97]. When the multiple series are independent of each other, they do not require a multivariate model to monitor counterrelationships. An example is monitoring multiple levels of activity in an information system to detect intrusions, where Ye [Ye02] found that the different activity measures were not related to each other, and therefore a simple χ^2 chart outperformed a Hotelling T^2 chart. A multivariate model is still needed here, however, instead of a set of univariate control charts. One reason is the inflated false-alarm rate that results from multiple testing. If each of p univariate charts has a false-alarm probability α, then the combined false-alarm probability is given by

$$1 - (1 - \alpha)^p. \tag{13}$$

One solution is to use a small enough α in each univariate chart; however, this approach becomes extremely conservative and is impractical for the moderate to high number of series collected by biosurveillance systems. This issue is also related to the problem of interpreting an alarm by the multivariate control chart. Although it may seem intuitive to determine which of the univariate measures is causing the alarm by examining the univariate charts, this is not a good approach not only because of the α-inflation but also because the alarm might be a result of changes in the covariance or correlation between the variables. Solutions for the α inflation based on Bonferroni-type adjustments have been shown to be conservative. A better approach is to decompose the monitoring statistic into components that reflect the contribution of each variable [Mon01]. For example, if the monitoring statistic is the Hotelling T^2, then for each variable i ($i = 1, \ldots, n$) we compute

$$d_i = T^2 - T^2_{(i)}, \tag{14}$$

where $T^2_{(i)}$ is the value of the statistic for all the $p-1$ variables except the ith variable. This is another place where the use of FDR methodology may be appropriate and of help. One also needs to consider monitoring the covariance in parallel.

Other methods within this approach have tried to resolve the shortcomings of these control charts. One example is using Shewhart and CuSum charts to monitor "regression-adjusted variables," which is the vector of scaled residuals from regressing each variable on the remaining variables [Haw91]. Another example is a Bayesian approach for monitoring a multivariate mean (with known covariance matrix), where a normal prior is imposed on the process mean. A quadratic form that multiplies the posterior mean vector and the posterior covariance matrix is then used as the monitoring statistic [Jai93].

The second statistical approach towards multivariate monitoring is based on reducing the dimension of the data and then using univariate charts to monitor the reduced series and the residuals. PCA and partial least squares (PLS) are the most popular dimension reduction techniques. In PCA, principal components are linear combinations of the standardized p variables. We denote them by $PC_1, \ldots, PC_p$. They have two advantages. First, unlike the original variables the principal components are approximately uncorrelated. Second, in many cases a small number of components captures the variability in the entire set of data [NIS04]. Kaiser's rule of thumb for determining the number of components that is needed to capture most of the variability is to retain only components associated with an eigenvalue larger than 1 [Kai60]. There are alternatives, such as the number of components that explain a sufficient level of variability. In quality control usually the first two components, PC_1, PC_2, are plotted using a Hotelling-T^2 chart, but the number of components (k) can be larger. A second plot monitors the "residuals" using

$$Q = \sum_{i=k+1}^{p} \frac{PC_i^2}{\lambda_i}, \tag{15}$$

where λ_i is the eigenvalue corresponding to the ith principal component (which is also equal to its variance). Bakshi [Bak98] points out that these charts suffer from the same problems of T^2 charts, as described above, in the sense of being insensitive to small changes in the process. Solutions are to monitor these statistics using a CuSum or EWMA scheme. The main shortcoming of these charts is their reliance on the assumption that the observations follow a multivariate normal distribution. In practice, multivariate normality is usually difficult to justify [CLL00]. This is especially true in biosurveillance where the different data sources come from very diverse environments.

Time Series Models

The multivariate form of ARMA models is called vector-ARMA models. The basic model is equivalent to (5), except that $\mathbf{y}_t$ $(t = 1, 2, \ldots)$ are now vectors. This structure allows for autocorrelation as well as cross-correlation between different series at different lags. In addition to the complications mentioned in relation to ordinary ARMA models, in vector-ARMA the number of α and θ parameters is larger ($(p + q + 1)$ multiplied by the number of series). The parameter covariance matrix to be estimated is therefore much larger. Since estimating the MA part adds a layer of complication, vector-AR models are more popular. In the context of biosurveillance, vector-AR models have advantages and disadvantages. Their strength lies in their ability to model lagged and counterrelationships between different series. This is especially useful for learning the pattern of delay between, for instance, medication sales and ER visits. However, vector-AR models have several weaknesses that are especially relevant in our context. First, their underlying assumption regarding the stationarity of the data is almost never true in data streams such as sales and hospital visits. This nonstationarity becomes even more acute as the frequency of the data increases. Second, although in some cases a set of transformations can be used to obtain stationarity, this preprocessing stage is highly series-specific and requires experience and special attention from the modeler. Furthermore, the application of different transformations can cause the series that were originally aligned to lose this feature. For example, by differentiating one series once and another series three times, the resulting series are of different length. Finally, any series that cannot be transformed properly into a stationary series must be dropped from the analysis. The third weakness of vector-AR models is that they are hard to automate. The model identification, estimation, and validation process is computationally heavy and relies on user expertise. Automated procedures do exist in software such as SAS (the VARMAX procedure [SAS00]). For determining the order of the model they use numerical measures such as Akaike information criterion (AIC), criterion final prediction error (FPE), and Bayesian information criterion (BIC). However, it is not guaranteed that the chosen model is indeed useful in a practical sense, and experienced statisticians would insist on examining other graphical measures such as auto- and cross-correlation plots to decide on the order of the model.

Estimation of the vector-AR parameters can be done using maximum likelihood. More often, for computational reasons, it is framed as an ordinary regression model and estimated using LS. Casting an AR model in the form of a regression model is an approximation in that in a regression model the predictors are assumed to be constant, whereas in an AR process they are a realization of a stochastic process. The parameter estimates are still consistent and asymptotically unbiased estimates for the AR model [NS01]. Thus, this estimation method is suitable for sufficiently long series, as is the case in typical biosurveillance data. However, collinearity and overparametrization are typical problems. One solution is to use a Bayesian approach and to impose priors on the AR coefficients [Ham94]. An alternative used by Bay et al. [BSU04] is to use ridge regression. The basic idea is to zero estimates that are below a certain threshold. Ridge regression yields biased estimates, but their variance is much smaller than their LS counterparts [MS75]. The estimated parameters are those that solve the equation

$$\boldsymbol{\beta} = (X'X + \lambda I)^{-1} X'\mathbf{y}, \tag{16}$$

where $\lambda \geq 0$ is the ridge parameter and I is the identity matrix. In the context of a vector-AR model we set $\mathbf{y} = \mathbf{y}_t$ (the multiple measurements at time t) and X is the matrix of lagged measurements at lags $1, \ldots, p$.

As with univariate ARIMA models, the stationarity assumption, the need in expert knowledge in model identification and estimation, the computational complexity, and overparametrization limit the usefulness of multivariate ARIMA models for integration into automated biosurveillance systems.

4.3 Alternative Approaches

Multichannel Singular Spectral Analysis

A generalization of SSA to multivariate time series, called multichannel-SSA (M-SSA), was described by Ghil and Yiou [GY96] and applied to several climate series. The idea is similar to the univariate SSA, except that now the lag-covariance matrix is a block-Toeplitz matrix T, where T_{ij} is an $M \times M$ lag-covariance matrix between series i and series j.

From a practical point of view, as the space increases in the number of series (L) and/or window width (M), the diagonalization of T, which is a $(T \times M) \times (T \times M)$ matrix, becomes cumbersome. Solutions include projecting the data onto a reduced subspace using PCA, undersampling the data to reduce M, and using expert knowledge to reduce the frequencies of interest. To give a feeling of the dimensions that can be handled, Plaut and Vautard [PV94] applied M-SSA to $L = 13$ series of 5-day observations, with $M = 40$ (equivalent to a maximum lag of 200 days).

There are several reasons why M-SSA should be investigated for biosurveillance. First, climatic data and syndromic data share components such as weekly, seasonal, and annual patterns. Second, its relative insensitivity to the stationarity assumption makes it attractive for biosurveillance data. Finally, the ability to generalize it to the analysis of multiple time series (although computationally challenging) is useful not only for monitoring purposes but also for shedding light on the cross-relationship between different biosurveillance series, both within a data source and across data sources. The SSA-MTM toolkit is a software package for applying M-SSA (and other techniques), and is freely available at `http://www.atmos.ucla.edu/tcd/ssa/`.

Multivariate Wavelet Method

DWT has proven to be a powerful tool for monitoring nonstationary univariate time series for abnormalities of an unknown nature. Several authors created generalizations of the univariate method to a multivariate monitoring setting mostly by combining it with PCA. The most recent method, by Bakshi [Bak98], uses a combination of DWT and PCA to create a multiscale principal components analysis (MSPCA) for online monitoring of multivariate observations. The idea is to combine the ability of PCA to extract the cross-correlation between the different series with the wavelets' ability to decorrelate the autocorrelation within each series. As with control chart methodology, there are two phases: In phase I it is assumed that there are no abnormalities, and the control limits for the charts are computed. In phase II new data are monitored using these limits. The process used in MSPCA consists of

1. Decomposing each univariate series using DWT (the same orthonormal wavelet is used for all series).
2. Applying PCA to the vectors of coefficients in the same scale, independently of other scales.
3. Using T^2- and Q-charts to monitor the principal components at each scale. During phase I the control limits for these charts are computed.
4. Combining the scales that have coefficients exceeding the control limits to form a "reconstructed multivariate signal" and monitoring it using T^2- and Q-charts. During phase I the control limits for these two charts are computed. In phase II the reconstructed series is obtained by combining the scales that indicate an abnormality at the most recent time point.

The idea is that a change in one or more of the series will create a large coefficient first at the finest scale, and as it persists, it will appear at coarser scales (similar to the delay in detecting spike changes with CuSum and EWMA charts). This might cause a delay in detection, and therefore the reconstructed data are monitored in parallel. The overall control chart is used for raising an alarm, while the scale-specific charts can assist in extracting the features representing abnormal operation.

As in the univariate case, the downsampling operation causes delays in detection. Bakshi [Bak98] therefore suggests using a stationary-wavelet transform, which requires the adjustment of the control limits to account for the coefficient autocorrelation that is now present and its effect on the global false-alarm rate. An enhancement to the Bonferroni-type adjustment suggested by Bakshi [Bak98] would be to use the more powerful FDR approach, which controls the expected proportion of false positives.

Multivariate Exponential Smoothing

Although research and application of univariate exponential smoothing is widespread there is a surprising scant number of papers on multivariate exponential smoothing, as a generalization of the univariate exponential smoothing methods. Two papers that have addressed this topic are Pfeffermann and Allon [PA89] and Harvey [Har86]. Since then, it appears, there has been little new on the topic.

The generalized exponential smoothing model suggested by Harvey [Har86] includes linear and polynomial trends and seasonal factors and can be estimated using algorithms designed for the univariate case. Pfeffermann and Allon [PA89] suggest a generalization of the Holt-Winters additive exponential smoothing, simply by expressing the decomposition and updating equations in matrix form. The only additional assumption is that the error term $\boldsymbol{\varepsilon}_t$ is assumed to have $\mathrm{E}(\boldsymbol{\varepsilon}_t) = \mathbf{0}$, $\mathrm{Var}(\boldsymbol{\varepsilon}_t) = \Sigma$, and $\mathrm{E}(\boldsymbol{\varepsilon}_t \boldsymbol{\varepsilon}'_{t-i}) = \mathbf{0}$ for $i > 0$. The set of updating equations are given by

$$
\begin{aligned}
\mathbf{L}_{t+1} &= A(\mathbf{Y}_{t+1} - \mathbf{S}_{t+1-M}) + (I - A)(\mathbf{L}_t + \mathbf{T}_t), \\
\mathbf{T}_{t+1} &= B(\mathbf{L}_{t+1} - \mathbf{L}_t) + (I - B)\mathbf{T}_t, \\
\mathbf{S}_{t+1} &= C(\mathbf{Y}_{t+1} - \mathbf{L}_{t+1}) + (I - C)\mathbf{S}_{t+1-M},
\end{aligned} \tag{17}
$$

where A, B, and C are three convergent matrices of smoothing constants. The m-step-ahead forecast at time t is

$$
\hat{\mathbf{Y}}_{t+m} = \mathbf{L}_t + m\mathbf{T}_t + \mathbf{S}_{t+m-M}. \tag{18}
$$

These are similar to the univariate smoothing updating and prediction equations. In fact, the updating equations can be written as weighted averages of estimates derived by the univariate components and correction factors based on information from the other series (the off-diagonal elements of the matrices A, B, and C). Pfeffermann and Allon [PA89] show that the forecasts from this model are optimal under particular state space models. They also illustrate and evaluate the method by applying it to two bivariate time series: one related to tourism in Israel, and the other on retail sales and private consumption in Israel. They conclude that the multivariate exponential smoothing (MES) forecasts and seasonal estimates are superior to univariate exponential smoothing and comparable to ARIMA models for short-term forecasts. Although the model formulation is distribution free, to forecast all series the specification of the smoothing matrices and initial values for the different components requires a distributional assumption or prior subjective judgments (which are much harder in a multivariate setting). This is the most challenging part of the method. However, once specified, this process need not be repeated. Also, once specified, the estimated smoothing matrices can shed light on the cross-relationships between the different time series in terms of seasonal, trend, and level components.

Data Depth

The pattern recognition literature discusses nonparametric multivariate models such as those associated with data depth methodology. This approach was developed through techniques at the interface between computational geometry and statistics and is suitable for nonelliptically structured multivariate data [Liu03]. A data depth is a measure of how deep or how central a given point is with respect to a multivariate distribution. The data depth concept leads to new nonparametric, distribution-free multivariate statistical analyses [RS04a], and in particular, it has been used to create multivariate monitoring charts [Liu03, LS02]. These charts allow the detection of both a location change and a scale increase in the process simultaneously. In practice, they have been shown to be more sensitive to abnormalities relative to a Hotelling-T^2 chart in monitoring aviation safety, where the data are not multivariate normal [CLL00]. There are several control charts that are based on data depth measures, the simplest being the r chart. In this time-preserving chart the monitoring statistic is the rank of the data depth measure, denoted by r. Liu and Singh [LS93] proved that r converges in distribution to a U(0,1) distribution. Therefore, the lower control limit on the r-chart equals the α of choice, and if the statistic exceeds this limit, it means that the multivariate observation is very far from the distribution center, and a flag is raised. The computation of the data depth measures becomes prohibitively intensive as the dimension of the space increases. Solutions have been to use probabilistic algorithms [CC03].

4.4 Spatial Approaches to Biosurveillance

A different approach to monitoring multiple data sources has been to focus on the spatial information and look for geographical areas with abnormal counts. Two major approaches have been used for monitoring biosurveillance data using a spatial approach. Both operate on discrete, multidimensional temporal datasets. The first method uses the algorithm What's Strange About Recent Events (WSARE), which is applied in RODS and uses a rule-based technique that compares recent emergency department admission data against data from a baseline distribution and finds subgroups of patients whose proportions have changed the most in the recent data [WMC03]. In particular, recent data are defined as all patient records from the past 24 hours. The definition of the baseline was originally the events occurring exactly five, six, seven, and eight weeks prior to the day under consideration (WSARE version 2.0) [WMC02]. Such a comparison eliminates short-term effects such as day-of-week, and longer-term seasonal effects (by ignoring weeks that are farther in the past). The baseline was then modified to include all historic days with similar attributes (WSARE version 2.5), and in the current version (WSARE 3.0) a Bayesian Network represents the joint distribution of the baseline [WMC03]. One limitation of WSARE is that it is practically limited to treating a maximum of two rules (i.e., variables), due to computational complexity [WMC02, WMC03]. Another computational limitation is the randomization tests used to account for the multiple testing, which are also computationally intensive. Finally, WSARE can use only discrete data as input, so that continuous information such as age must be categorized into groups. This, of course, requires expert knowledge and might be specific to the type of data monitored and/or the outbreak of interest.

A different method, implemented in ESSENCE II and in the NYC-DOHMH system, is the spatial-temporal scan statistic [Kul01], which compares the counts of occurrences at time t in a certain geographical location with its neighboring locations and past times, and flags when the actual counts differ consistently from the expected number under a nonhomogeneous Poisson model. The purely spatial approach is based on representing a geographical map by a uniform two-dimensional grid and aggregating the records within families of circles of varying radii centered at different grid points. The underlying assumption is that the number of records in a circle come from a nonhomogeneous Poisson process with mean qp_{ij} where q is the underlying disease rate and p_{ij} is the baseline rate for that circle. The purely spatial scan statistic is the maximum likelihood ratio over all possible circles, thereby identifying the circle that constitutes the most likely cluster. This requires the estimation of the expected number of cases within each circle and outside of it given that there is no outbreak. The p-value for the statistic is obtained through Monte Carlo hypothesis testing [Kul01]. The spatial-temporal scan statistic adds time as another dimension, thereby forming cylinders instead of circles. The varying heights of the cylinders represent different windows in time. The multiple testing is then accounted for both in space and in time domains. Lawson [Law01] mentions two main challenges of the spatial-temporal scan statistic, which are relevant to biosurveillance. First, the use of circular forms limits the types of clusters that can be detected efficiently. Second, the timeliness of detection and false-alarm rates need further improvement. In an application of the scan statistic to multiple data sources in ESSENCE II, Burkom [Bur03b] describes a modified scan-statistic methodology where the counts from various series are aggregated and used as the monitored data, and these are assumed to follow an ordinary Poisson model. A few modifications were needed to address features of biosurveillance data. The uniform spatial incidence is usually inappropriate and requires the estimation of expected counts for each of the data sources (which is challenging in and of itself); the aggregation of counts from disparate sources with different scales was adjusted by using a "stratified" version of the scan statistic. It appears that such data-specific and time-varying tailoring is necessary and therefore challenges the automation of this method for biosurveillance.

Both methods are flexible in the sense that they can be applied to different levels of geographical and temporal aggregation and for different types of diseases. With respect to automation and user input the two methods slightly differ. In the scan-statistic methods the user is required to specify the maximal spatial cluster size (represented by the circle radius) and the maximal temporal cluster length (represented by the cylinder height). In addition, since neither the Poisson nor the Bernoulli model is likely to be a good approximation of the baseline counts in each area, a nonhomogeneous Poisson will most likely be needed. This requires the specification of the relevant variables and the estimation of the corresponding expected counts inside and outside each cylinder. For WSARE the user need only specify the time window that is used for updating the Bayesian network. Finally, the major challenge in these two spatial methods as well as other methods (e.g., the modified Poisson CuSum method by Rogerson [Rog01]) is their limitation to monitoring only count data and the use of just categorical covariates.

5 Concluding Remarks

The collection of data streams that are now routinely collected by biosurveillance systems is diverse in its sources and structure. Since some data sources comprise multiple data streams (e.g., different medication sales or different chief complaints at ER admission), there are two types of multivariate relationships to consider: "within sets" and "across sets." Data streams within a single source tend to be qualitatively more similar to each other as they are measured, collected, and stored by the same system and share common influencing factors such as seasonal effects. Data streams across different sources are obviously less similar, even if the technical issues such as frequency of measurement and missing observations are ignored. The additional challenge is that the signature of a disease or bioterrorism-related outbreak is usually not specified and can only be hypothesized for some of the data sources (e.g., how does a large-scale anthrax attack manifest itself in grocery sales?). Stoto et al. [SFJ06] discuss the utility of univariate methods in biosurveillance by comparing univariate and multivariate Shewhart and CuSum chart performance. Their discussion and analyses are provocative, but there is need for a serious testbed of data to examine the utility of the different approaches.

The task of monitoring multivariate time series is complicated even if we consider a single data source. Traditional statistical approaches are based on a range of assumptions that are typically violated in syndromic data. These range from multivariate normal distribution, independence over time, to stationarity. Highly advanced methods that relax these assumptions tend to be very complicated, computationally intensive, and require expert knowledge to apply them to real data. On the other hand, advances in other fields where automation and computational complexity is important, and where nonstationary data are typical, have been in the direction of nonparametric model-free methods. They are also aimed at detecting an abnormality, without specifying its exact nature. Methods such as wavelets, data depth, and exponential smoothing have proven successful in anomaly detection in various applications and superior to traditional monitoring methods. We therefore believe that they should be investigated for use with biosurveillance data.

Exponential smoothing models appear to be a promising class of methods for biosurveillance because of their simplicity, interpretability, easy automation, inherent one-step-ahead forecast notion, adaptiveness, and parsimony. They can also handle missing values, which are probable in biosurveillance data. Although their generalization to multivariate monitoring has not been researched or applied to data as widely, it appears to be potentially useful as a basis for a monitoring tool that can be used for biosurveillance. Multiscale (wavelet) methods do not require preprocessing, perform better in detecting unspecified abnormality patterns, and are suitable for nonstationary data. A few challenges remain with these methods, such as boundary correction. Data depth does not assume a multivariate normal distribution of the observations although it still suffers from high computational complexity for high-dimensional data.

Another issue that is related to monitoring is the cross-correlation structure between different data sources. Although we know that an increase in deaths is preceded by an increase in ER visits, and both are preceded by a possible increase in medication sales, the exact structure of this relationship is unknown. The direct modeling of such relationships has been attempted by Najmi and Magruder [NM04, NM05], who use a filtering approach to model the time-dependent correlations between clinical data and OTC medication sales. They find that respiratory illness data can be predicted using OTC sales data. In general, the degree to which this structure is visible through multivariate time series methods differs; vector-AR models and M-SSA yield a cross-correlation matrix, which directly shows the relationship. Multivariate exponential smoothing gives estimates of the relationships between the level, trend, and seasonal components of the different series. In contrast, in MSPCA the relationship between the series is indirect and requires examining the loadings related to peaking coefficients. A pragmatic approach would use a collection of analysis and monitoring methods to examine the different types of information that they reveal.

There are many other data-analytical methods that can be used as a basis for monitoring, e.g., biologically inspired techniques widely used in the artificial intelligence area such as neural networks and genetic and evolutionary algorithms. Their application in real-time forecasting has proven empirically useful (e.g., Cortez et al. [CAR02]) and they tend to be computationally efficient, easily automated, and very flexible. But the proof of the pudding is in the eating, and any method needs to be subjected to serious reality checks. Unfortunately, in the area of biosurveillance these can usually only come from simulations. The methods that we describe here have the potential of being useful for biosurveillance from a theoretical point of view as well as from their usefulness in practical applications in other fields. However, it is difficult to assess their actual performance for biosurveillance in the absence of relevant databases involving bioterror attacks to test them on. This situation can change only through the collection of new and elaborate data from surveillance systems where we are observing no attacks, and then overlaying attacks through some kind of simulation. In such simulations we can actually carry out tests of methodology. For example, Goldenberg et al. [GSC02] seeded their data with the "footprint" of the onset of an anthrax attack in OTC-medication sales data and were able to achieve high rates of detection, although at the expense of a relatively high rate of false alarms. Stoto et al. [SSM04] demonstrated that many methods were able to detect a similar kind of "fast"-onset outbreak (except at peak flu season!), but that none was especially good at detecting "slow"-onset outbreak. Thus the real challenge is to find a set of flexible yet powerful monitoring tools that are useful in detecting multiple bioterrorism disease signatures with high accuracy and relatively low false-alarm rates. We hope that the methodology described here represents a first step in this direction.

Acknowledgments

The work of the second author was supported in part by National Science Foundation grant IIS–0131884 to the National Institute of Statistical Sciences and by Army contract DAAD19-02-1-3-0389 to CyLab at Carnegie Mellon University.

References

[ASJ97] Alt, F. B., N. D. Smith, and K. Jain. 1997. "Multivariate quality control." In *Handbook of statistical methods for engineers and scientists*, edited by H. M. Wadsworth, 2nd ed., 21.1–21.15. New York: McGraw-Hill.

[ABS03] Aradhye, H. B., B. R. Bakshi, R. A. Strauss, and J. F. Davis. 2003. "Multiscale statistical process control using wavelets — theoretical analysis and properties." *AIChE Journal* 49 (4): 939–958.

[Arm01] Armstrong, J. S. 2001. *Principles of forecasting: A handbook for researchers and practitioners.* Boston, MA: Kluwer Academic Publishers.

[ACV04] Arts, G. R. J., F. P. A. Coolen, and P. van der Laan. 2004. "Nonparametric predictive inference in statistical process control." *Quality Technology & Quantitative Management* 1 (2): 201–216.

[AM97] Aussem, A., and F. Murtagh. 1997. "Combining neural network forecasts on wavelet-transformed time series." *Connection Science* 9:113–121.

[Bak98] Bakshi, B. R. 1998. "Multiscale PCA with application to multivariate statistical process monitoring." *AIChE Journal* 44:1596–1610.

[BSU04] Bay, S., K. Saito, N. Ueda, and P. Langley. 2004. "A framework for discovering anomalous regimes in multivariate time-series data with local models." Technical Report, Institute for the Study of Learning and Expertise. http://www.isle.org.

[BH95] Benjamini, Y., and Y. Hochberg. 1995. "Controlling the false discovery rate: A practical and powerful approach to multiple testing." *Journal of the Royal Statistical Society Series B* 57:289–300.

[BMC03] Bilenko, M., R. Mooney, W. Cohen, P. Ravikumar, and S. Fienberg. 2003. "Adaptive name matching in information integration." *IEEE Intelligent Systems* 18 (5): 16–23.

[BL97] Box, G., and A. Luceño. 1997. *Statistical control: By monitoring and feedback adjustment.* Hoboken, NJ: Wiley-Interscience.

[BJR94] Box, G. E. P., G. M. Jenkins, and G. C. Reinsel. 1994. *Time series analysis, forecasting, and control*, 3rd ed. Englewood Cliffs, NJ: Prentice–Hall.

[BMS04] Bravata, D. M., K. M. McDonald, W. M. Smith, C. Rydzak, H. Szeto, D. L. Buckeridge, C. Haberland, and D. K. Owens. 2004. "Systematic review: Surveillance systems for early detection of bioterrorism-related diseases." *Annals of Internal Medicine* 140:910–922.

[Bur03a] Burkom, H. S. 2003. "Development, adaptation, and assessment of alerting algorithms for biosurveillance." *Johns Hopkins APL Technical Digest* 24 (4): 335–342.

[Bur03b] Burkom, H. S. 2003. "Biosurveillance applying scan statistics with multiple, disparate data sources." *Journal of Urban Health* 80:57–65.

[Cas04] Casman, E. A. 2004. "The potential of next-generation microbiological diagnostics to improve bioterrorism detection speed." *Risk Analysis* 24:521–535.

[CC03] Chakraborty, B., and P. Chaudhuri. 2003. "On the use of genetic algorithm with elitism in robust and nonparametric multivariate analysis." *Austrian Journal of Statistics* 32:13–27.

[CKO01] Chatfield, C., A. Koehler, J. Ord, and R. Snyder. 2001. "A new look at models for exponential smoothing." *The Statistician, Journal of the Royal Statistical Society Series D* 50 (2): 147–159.

[CY88] Chatfield, C., and M. Yar. 1988. "Holt-Winters forecasting: Some practical issues." *The Statistician, Journal of the Royal Statistical Society Series D* 37:129–140.

[CLL00] Cheng, A., R. Liu, and J. Luxhoj. 2000. "Monitoring multivariate aviation safety data by data depth: Control charts and threshold systems." *IIE Transactions on Operations Engineering* 32:861–872.

[CAR02] Cortez, P., F. S. Allegro, M. Rocha, and J. Neves. 2002. "Real-time forecasting by bio-inspired models." Edited by M. Hamza, *Proceedings of the Second IASTED International Conference on Artificial Intelligence and Applications, Malaga, Spain.* ACTA Press, 52–57.

[CR96] Crato, N., and B. K. Ray. 1996. "Some problems in the overspecification of ARMA and processes using ARFIMA models." *Proceedings of the 3rd Congress of the Portuguese Statistical Society.* Lisbon: Salamandra Publishing, 527–539.

[DJ95] Donoho, D. L., and I. M. Johnstone. 1995. "Adapting to unknown smoothness via wavelet shrinkage." *Journal of the American Statistical Association* 90 (432): 1200–1224.

[DJK96] Donoho, D. L., I. M. Johnstone, G. Kerkyacharian, and D. Picard. 1996. "Density estimation by wavelet thresholding." *The Annals of Statistics* 24 (2): 508–539.

[EST01] Efron, B., J. Storey, and R. Tibshirani. 2001. "Microarrays, empirical Bayes methods, and false discovery rates." *Journal of the American Statistical Association* 96:1051–1060.

[FS05] Fienberg, S. E., and G. Shmueli. 2005. "Statistical issues and challenges associated with rapid detection of bio-terrorist attacks." *Statistics in Medicine* 24 (4): 513–529.

[GDV04] Ganesan, R., T. K. Das, and V. Venkataraman. 2004. "Wavelet based multiscale statistical process monitoring — a literature review." *IIE Transactions on Quality and Reliability* 36 (9): 787–806.

[GGT03] Gesteland, P., R. Gardner, F. C. Tsui, J. Espino, R. Rolfs, B. James, W. Chapman, A. Moore, and M. Wagner. 2003. "Automated syndromic surveillance for the 2002 Winter Olympics." *Journal of the American Medical Informatics Association* 10:547–554.

[GV91] Ghil, M., and R. Vautard. 1991. "Interdecadal oscillations and the warming trend in global temperature time series." *Nature* 350 (6316): 324–327.

[GY96] Ghil, M., and P. Yiou. 1996. "Decadal climate variability: Dynamics and predictability." In *Spectral methods: What they can and cannot do for climatic time series*, 445–482. Amsterdam: Elsevier.

[GSC02] Goldenberg, A., G. Shmueli, R. A. Caruana, and S. E. Fienberg. 2002. "Early statistical detection of anthrax outbreaks by tracking over-the-counter medication sales." *Proceedings of the National Academy of Sciences* 99:5237–5240.

[Ham94] Hamilton, J. D. 1994. *Time series analysis.* Princeton, NJ: Princeton University Press.

[Har86] Harvey, A. C. 1986. "Analysis and generalization of a multivariate exponential smoothing model." *Management Science* 32:374–380.

[Haw91] Hawkins, D. M. 1991. "Multivariate quality control based on regression-adjusted variables." *Technometrics* 33:61–75.

[HO98] Hawkins, D. M., and D. H. Olwell. 1998. *Cumulative sum control charts and charting for quality improvement.* New York: Springer.

[HLM98] Himmelblau, D., R. Luo, M. Misra, S. J. Qin, and R. Barton. 1998. "Sensor fault detection via multi-scale analysis and nonparametric statistical inference." *Industrial and Engineering Chemistry Research* 37:1024–1032.

[HTI03] Hogan, W. R., F. C. Tsui, O. Ivanov, P. H. Gesteland, S. Grannis, J. M. Overhage, J. M. Robinson, and M. M. Wagner. 2003. "Detection of pediatric respiratory and diarrheal outbreaks from sales of over-the-counter electrolyte products." *Journal of the American Medical Informatics Association* 10 (6): 555–562.

[Jai93] Jain, K. 1993. "A Bayesian approach to multivariate quality control." Ph.D. diss., University of Maryland, College Park.

[Kai60] Kaiser, H. 1960. "The application of electronic computers to factor analysis." *Psychometrika* 23:187–200.

[KPF03] Kaplan, E., C. Patton, W. FitzGerald, and L. Wein. 2003. "Detecting bioterror attacks by screening blood donors: A best-case analysis." *Emerging Infectious Diseases* 9:909–914.

[KLR04] Keogh, E., S. Lonardi, and C. Ratanamahatana. 2004. "Towards parameter-free data mining." *Proceedings of the 10th ACM SIGKDD International Conference on Knowledge Discovery and Data Mining, Seattle, WA.*

[KSO01] Koehler, A., R. Snyder, and J. Ord. 2001. "Forecasting models and prediction intervals for the multiplicative Holt-Winters method." *International Journal of Forecasting* 17:269–286.

[Kul01] Kulldorff, M. 2001. "Prospective time-periodic geographical disease surveillance using a scan statistic." *Journal of the Royal Statistical Society Series A* 164 (1): 61–72.

[LLW02] Lada, E. K., J. C. Lu, and J. R. Wilson. 2002. "A wavelet-based procedure for process fault detection." *IEEE Transactions on Semiconductor Manufacturing* 15 (1): 79–90.

[Law01] Lawson, A. 2001. "Comments on the papers by Williams et al., Kulldorff, Knorrheld and Best, and Rogerson." *Journal of the Royal Statistical Society Series A* 164 (1): 97–99.

[LKL03] Lin, J., E. Keogh, S. Lonardi, and B. Chiu. 2003. "A symbolic representation of time series, with implications for streaming algorithms." *Proceedings of the 8th ACM SIGMOD Workshop on Research Issues in Data Mining and Knowledge Discovery.*

[LS93] Liu, R., and K. Singh. 1993. "A quality index based on data depth and multivariate rank tests." *Journal of the American Statistical Association* 88:252–260.

[LS02] Liu, R., and K. Singh. 2002. "DDMA-charts: nonparametric multivariate moving average control charts based on data depth." Technical Report, Rutgers University.

[Liu03] Liu, R. Y. 2003. "Data depth: center-outward ordering of multivariate data and nonparametric multivariate statistics." In *Recent advances and trends in nonparametric statistics*, edited by M. G. Akritas and D. N. Politis, 155–167. Amsterdam: Elsevier.

[Luc82] Lucas, J. M. 1982. "Combined Shewhart-CuSum quality control schemes." *Journal of Quality Technology* 14 (2): 51–59.

[LS90] Lucas, J. M., and M. S. Saccucci. 1990. "Exponentially weighted moving average control schemes: Properties and enhancements, with discussion." *Technometrics* 32:1–29.

[MAC82] Makridakis, S., A. Andersen, R. Carbone, R. Fildes, M. Hibon, R. Lewandowski, J. Newton, E. Parzen, and R. Winkler. 1982. "The accuracy of extrapolative time series methods: Results of a forecasting competition." *Journal of Forecasting* 1 (2): 111–153.

[MS75] Marquardt, D. W., and R. D. Snee. 1975. "Ridge regression in practice." *American Statistician* 29 (1): 3–20.

[Mon01] Montgomery, D. C. 2001. *Introduction to statistical quality control*, 3rd ed. Hoboken, NJ: John Wiley & Sons.

[NM04] Najmi, A. H., and S. F. Magruder. 2004. "Estimation of hospital emergency room data using OTC pharmaceutical sales and least mean square filters." *BioMed Central Medical Informatics and Decision Making* 4 (5): 1–5.

[NM05] Najmi, A. H., and S. F. Magruder. 2005. "An adaptive prediction and detection algorithm for multistream syndromic surveillance." *BioMed Central Medical Informatics and Decision Making* 5:33.

[NIS04] National Institute of Standards and Technology. 2004. *NIST/SEMATECH e-Handbook of statistical methods.*

[NS01] Neumaier, A., and T. Schneider. 2001. "Estimation of parameters and eigenmodes of multivariate autoregressive models." *ACM Transactions on Mathematical Software* 27:27–57.

[OP96] Ogden, R. T., and E. Parzen. 1996. "Data dependent wavelet thresholding in nonparametric regression with change point applications." *Computational Statistics and Data Analysis* 6:93–99.

[PMK03] Pavlin, J. A., F. Mostashari, M. G. Kortepeter, N. A. Hynes, R. A. Chotani, Y. B. Mikol, M. A. K. Ryan, J. S. Neville, D. T. Gantz, J. V. Writer, J. E. Florance, R. C. Culpepper, F. M. Henretig, and P. W. Kelley. 2003. "Innovative surveillance methods for rapid detection of disease outbreaks and bioterrorism: Results of an interagency workshop on health indicator surveillance." *American Journal of Public Health* 93 (8): 1230–1235.

[PW00] Percival, D., and A. Walden. 2000. *Wavelet methods for time series analysis.* Cambridge, UK: Cambridge University Press.

[PA89] Pfeffermann, D., and J. Allon. 1989. "Multivariate exponential smoothing: Method and practice." *International Journal of Forecasting* 5 (1): 83–98.

[PGV95] Plaut, G., M. Ghil, and R. Vautard. 1995. "Interannual and interdecadal variability in 335 years of central England temperatures." *Science* 268 (5211): 710–713.

[PV94] Plaut, G., and R. Vautard. 1994. "Spells of low-frequency oscillations and weather regimes in the northern hemisphere." *Journal of the Atmospheric Sciences* 51 (2): 210–236.

[Pol] Polikar, R. The engineer's ultimate guide to wavelet analysis: The wavelet tutorial. http://users.rowan.edu/polikar/WAVELETS/WTtutorial.html.

[RS04a] Rafalin, E., and D. Souvaine. 2004. "Theory and applications of recent robust methods." In *Computational geometry and statistical depth measures.* Basel: Birkhauser.

[Rei03] Reingold, A. 2003. "If syndromic surveillance is the answer, what is the question?" *Biosecurity and Bioterrorism: Biodefense Strategy, Practice, and Science* 1 (2): 1–5.

[RPM03] Reis, B. Y., M. Pagano, and K. D. Mandl. 2003. "Using temporal context to improve biosurveillance." *Proceedings of the National Academy of Sciences* 100:1961–1965.

[RS04b] Reynolds, M. R. J., and Z. G. Stoumbos. 2004. "Control charts and the efficient allocation of sampling resources." *Technometrics* 46 (2): 200–214.

[Rog01] Rogerson, P. 2001. "Monitoring point patterns for the development of spacetime clusters." *Journal of the Royal Statistical Society Series A* 164 (1): 87–96.

[SCR97] Safavi, A. A., J. Chen, and J. A. Romagnoli. 1997. "Wavelet-based density estimation and application to process monitoring." *AICHE Journal* 43 (5): 1227–1238.

[SAS00] SAS Institute, Inc. 2000. *SAS online documentation, The VARMAX procedure.* Cary, NC: SAS Institute, Inc.

[SAS04a] SAS Institute, Inc. 2004. *SAS high performance forecasting user's guide.* Version 9. Cary, NC: SAS Institute, Inc.

[SAS04b] SAS Institute, Inc. 2004. *SAS online documentation, The ARIMA procedure.* Version 9. Cary, NC: SAS Institute, Inc.

[SC03] Shmueli, G., and A. Cohen. 2003. "Run length distribution for control charts with runs rules." *Communications in Statistics — Theory & Methods* 32 (2): 475–495.

[SB03] Sonesson, C., and D. Bock. 2003. "A review and discussion of prospective statistical surveillance in public health." *Journal of the Royal Statistical Society Series A* 166 (1): 5–21.

[SFJ06] Stoto, M., R. D. Fricker, A. Jain, J. O. Davies-Cole, C. Glymph, G. Kidane, G. Lum, L. Jones, K. Dehan, and C. Yuan. 2006. "Evaluating statistical methods for syndromic surveillance." In *Statistical methods in counterterrorism: Game theory, modeling, syndromic surveillance, and biometric authentication,* edited by A. Wilson, G. Wilson, and D. Olwell. New York: Springer.

[SSM04] Stoto, M. A., M. Schonlau, and L. T. Mariano. 2004. "Syndromic surveillance: Is it worth the effort?" *Chance* 17:19–24.

[Sul03] Sullivan, B. M. 2003. "Bioterrorism detection: The smoke alarm and the canary." *Technology Review Journal* 11 (1): 135–140.

[WRT03] Wagner, M. W., J. M. Robinson, F. C. Tsui, J. U. Espino, and W. R. Hogan. 2003. "Design of a national retail data monitor for public health." *Journal of the American Medical Informatics Association* 10 (5): 409–418.

[WR96] Willemain, T. R., and G. C. Runger. 1996. "Designing control charts using an empirical reference distribution." *Journal of Quality Technology* 28 (1): 31–38.

[Won04] Wong, W. K. 2004. "Data mining for early disease outbreak detection." Ph.D. diss., Carnegie Mellon University.

[WMC02] Wong, W. K., A. Moore, G. Cooper, and M. Wagner. 2002. "Rule-based anomaly pattern detection for detecting disease outbreaks." *Proceedings of the 18th National Conference on Artificial Intelligence.* MIT Press.

[WMC03] Wong, W. K., A. Moore, G. Cooper, and M. Wagner. 2003. "Bayesian network anomaly pattern detection for disease outbreaks." Edited by T. Fawcett and N. Mishra, *Proceedings of the Twentieth International Conference on Machine Learning.* Menlo Park, CA: AAAI Press, 808–815.

[Ye02] Ye, N. 2002. "Multivariate statistical analysis of audit trails for host-based intrusion detection." *IEEE Transactions on Computers* 51 (7): 810–820.

[YSG00] Yiou, P., D. Sornette, and M. Ghil. 2000. "Data-adaptive wavelets and multi-scale SSA." *Physica D* 142:254–290.

Evaluating Statistical Methods for Syndromic Surveillance

Michael A. Stoto[1], Ronald D. Fricker, Jr.[2], Arvind Jain[3], Alexis Diamond[4], John O. Davies-Cole[5], Chevelle Glymph[6], Gebreyesus Kidane[7], Garrett Lum[8], LaVerne Jones[9], Kerda Dehan[10], and Christine Yuan[11]

[1] RAND Statistics Group, mstoto@rand.org
[2] Department of Operations Research, Naval Postgraduate School, rdfricker@nps.edu
[3] RAND Statistics Group, arvind_jain@rand.org
[4] The Institute for Quantitative Social Science, Harvard University, adiamond@fas.harvard.edu
[5] Bureau of Epidemiology and Health Risk Assessment, District of Columbia Department of Health, john.davies-cole@dc.gov
[6] Bureau of Epidemiology and Health Risk Assessment, District of Columbia Department of Health, chevelle.glymph@dc.gov
[7] Bureau of Epidemiology and Health Risk Assessment, District of Columbia Department of Health, gebreyesus.kidane@dc.gov
[8] Bureau of Epidemiology and Health Risk Assessment, District of Columbia Department of Health, garret.lum@dc.gov
[9] Bureau of Epidemiology and Health Risk Assessment, District of Columbia Department of Health, laverne.jones@dc.gov
[10] Bureau of Epidemiology and Health Risk Assessment, District of Columbia Department of Health, kerda.dehan@dc.gov
[11] Bureau of Epidemiology and Health Risk Assessment, District of Columbia Department of Health, christine.yuan@dc.gov

Since the terrorist attacks on September 11, 2001, many state and local health departments around the United States have started to develop *syndromic surveillance* systems. Syndromic surveillance — a new concept in epidemiology — is the statistical analyses of data on individuals seeking care in emergency rooms (ER) or other health care settings with preidentified sets of symptoms thought to be related to the precursors of diseases. Making use of existing health care or other data, often already in electronic form, these systems are intended to give early warnings of bioterrorist attacks or other emerging health conditions. By focusing on symptoms rather than confirmed diagnoses, syndromic surveillance aims to detect bioevents earlier than would be possible with traditional surveillance systems. Because potential bioterrorist agents such as anthrax, plague, brucellosis, tularemia, Q-fever, glanders, smallpox, and viral hemorrhagic fevers initially exhibit symptoms ("present" in medical terminology) of a flulike illness, data suggesting a sudden increase of individuals with fever, headache, muscle pain, and malaise might be the first indication of a bioterrorist attack or natural disease outbreak. Syndromic surveillance is also thought to be useful for early detection of natural disease outbreaks [Hen04].

Research groups based at universities, health departments, private firms, and other organizations have proposed and are developing and promoting a variety of surveillance systems purported to meet public health needs. These include methods for analysis of data from healthcare facilities, as well as reports to health departments of unusual cases. Many of these methods involve intensive, automated statistical analysis of large amounts of data and intensive use of informatics techniques to gather data for analysis and to communicate among physicians and public health officials [WTE01]. Some of these systems go beyond health care data to include nonhealth data such as over-the-counter (OTC) pharmaceutical sales and absenteeism that might indicate people with symptoms who have not sought health care [Hen04].

There are a number of technological, logistical, and legal constraints to obtaining appropriate data and effective operation of syndromic surveillance systems [Bue04]. However, even with access to the requisite data and perfect organizational coordination and cooperation, the statistical challenges in reliably and accurately detecting a bioevent are formidable. The object of these surveillance systems, of course, is to analyze a stream of data in realtime and determine whether there is an anomaly suggesting that an incident has occurred. All data streams, however, have some degree of natural variability. These include seasonal or weekly patterns, a flu season that appears at a different time each winter or perhaps not at all, differences in coding practices, sales promotions for OTC medications, and random fluctuations due to small numbers of individuals with particular symptoms. Furthermore, for some natural outbreaks or bioterrorist attacks the "signal" (the number of additional cases over baseline rates) may be small compared to the "noise" (the random or systematic variation in the data). As a result, even the most effective statistical detection algorithms face a trade-off among three factors: sensitivity, false positives, and timeliness.

The goals of this chapter are (1) to introduce the statistical issues in syndromic surveillance, (2) to describe and illustrate approaches to evaluating syndromic surveillance systems and characterizing their performance, and (3) to evaluate the performance of a couple of specific algorithms through both abstract simulations and simulations based on actual data. Section 1 of this chapter introduces and discusses the statistical concepts and issues in syndromic surveillance, illustrating them with data from an ER surveillance system from the District of Columbia. Section 2 presents methods from the statistical process control (SPC) literature, including variants on existing multivariate detection algorithms tailored to the syndromic surveillance problem, and compares and contrasts the performance of univariate and multivariate techniques via some abstract simulations. Section 3 then compares the new multivariate detection algorithms with commonly used approaches and illustrates the simulation approach to evaluation using simulations based on actual data from seven Washington, DC, hospital ERs. We conclude with a discussion about the implications for public health practice.

1 Background

Immediately following September 11, 2001, the District of Columbia Department of Health (DC DOH) began a surveillance program based on hospital ER visits. ER logs from nine hospitals are faxed on a daily basis to the health department, where health department staff code them on the basis of chief complaint, that is, the primary symptom or reason that the patient sought care, recording the number of patients in each of the following syndromic categories: death, sepsis, rash, respiratory complaints, gastrointestinal complaints, unspecified infection, neurological, or other complaints. These data are analyzed daily using a variety of statistical detection algorithms, and when a syndromic category shows an unusually high occurrence, a patient chart review is initiated to determine if the irregularity is a real threat.

Simply displaying the daily number of ER visits for any given symptom group results in a figure in which day-to-day stochastic variation dominates any subtle changes in numbers of cases over time. To address this problem, the DC DOH employs a number of statistical detection algorithms to analyze data on a daily basis and raise an "alarm" when the count is significantly greater than expected, which may suggest a possible outbreak or attack. This type of analysis can help to identify the onset of the annual influenza season. The data also reveal indications of the "worried well" who sought care during the 2001 anthrax attacks and a previously undetected series of gastrointestinal illness outbreaks that occurred over a four-month period in different hospitals. No single symptom group or detection algorithm consistently signaled each of the events [SSM04].

1.1 Characterizing the Performance of Statistical Detection Algorithms

Although it is possible to have different levels of certainty for an alarm, syndromic surveillance algorithms typically operate in a binary fashion; on any given day they either alarm or they do not. Operating in this way, the performance of a detection algorithm in the context of a particular dataset can be characterized according to its sensitivity, false-positive rate, and timeliness. *Sensitivity*, sometimes called the true positive rate and similar to the power of a statistical hypothesis test, is the probability that an outbreak will be detected in a given period when there in fact is an outbreak. Clearly, a surveillance system should have as much sensitivity as possible. Lowering the threshold at which an alarm is sounded can generally increase sensitivity, but only at the expense of false positives.

A *false positive* occurs when an algorithm alarms on a day when there is no actual outbreak. In medical or epidemiological terminology, *specificity* is 1 minus the probability of a false positive, or the probability that an alarm will not be raised on a day that there is no outbreak. Ideally, the probability of a false positive would be zero, but practically it is always positive. Intrinsic variability in the data means that every methodology can alarm when in fact there is no event.

It is usually possible to make the false-positive rate tolerably small. There are two difficulties, however. First, lowering the false-positive rate generally involves either decreasing sensitivity or lowering timeliness (or both). Second, even with a very low false-positive rate for a single algorithm or system, it is still possible — even likely — that in the aggregate the number of false positives may be unacceptably large.

For example, sometime in the near future it is possible that thousands of syndromic surveillance systems will be running simultaneously in towns, cities, counties, states, and other jurisdictions throughout the United States. Each of these jurisdictions might be looking at data in six to eight symptom categories, separately from every hospital in the area, and so on. Suppose every county in the United States had a detection algorithm in place that was used daily and that had a 0.1% false-positive rate. Because there are approximately 3,000 counties, nationwide three counties a day on average would have a false-positive alarm. While any particular county would only experience a false positive about once every three years, which may be an acceptable rate at the county level, is the nationwide false-positive rate acceptable? The impacts of excessive false alarms are both monetary, as resources must respond to phantom events, and operational, as too many false alarms desensitize responders to real events.

Because a rapid response to a bioterrorist attack or natural disease outbreak is essential to minimizing the health consequences, timeliness is an important characteristic of all surveillance systems. With its focus on symptoms that occur before formal diagnosis, syndromic surveillance is specifically designed to enhance timeliness. While timeliness does not have a well-established definition to parallel sensitivity and specificity, we think of it as the speed at which an algorithm alarms during an outbreak.

Stoto, Schonlau, and Mariano [SSM04] characterized the trade-off between sensitivity and timeliness in a simulation study. Using the daily number of admissions of patients with influenzalike illness (ILI) over a three-year period to the emergency department of a typical urban hospital, which averages three per day outside the winter flu season, they added a hypothetical number of extra cases spread over a number of days to mimic the pattern of a potential bioterror attack. A "fast" outbreak was defined as 18 additional cases over three days — 3 on the first day, 6 on the second, and 9 on the third. A simulated "slow" outbreak involved the same total number of cases, but they were distributed over nine days as follows: 1, 1, 1, 2, 2, 2, 3, 3, 3. Each of these simulated outbreaks was added on each day in the database outside the winter flu season. Four different detection algorithms were examined. The first used ER admissions from a single day; the others used data from multiple days using various CuSum (cumulative sums) methods (such as those to be defined in Sect. 2.3), with the algorithms varying in the weight they gave to more recent data.

The simulation results suggest the minimum size and speed of outbreaks that are detectable. Even with an excess of 9 cases over two days, which is three times the daily average, there was only a 50% chance that the alarm would go off on the second day of an outbreak. Figure 1 indicates how this probability — the sensitivity of the algorithm — varies by day. In the slow outbreak, when 18 cases were spread over nine days (see Fig. 2), chances were no better than 50–50 that the alarm would sound by the ninth day.

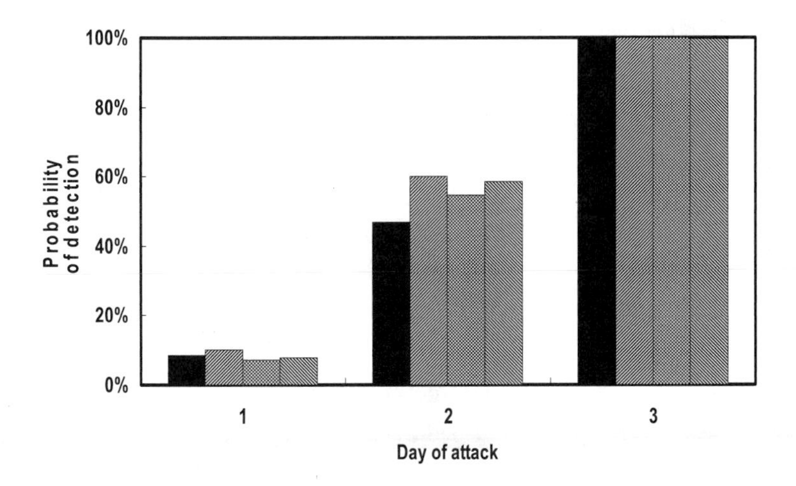

Fig. 1. Shaded bars correspond to four detection algorithms: the first using only one day's data, the other three combining data from multiple days. All four syndromic surveillance methods worked equally well for fast-spreading bioterrorist attacks, but had only about a 50–50 chance of detecting the outbreak by day two. See Stoto et al. [SSM04] for more information.

1.2 Evaluation of Syndromic Surveillance Systems

There are a number of ways to evaluate syndromic surveillance systems, formal and informal. For example, the Centers for Disease Control and Prevention's (CDC) "Framework for Evaluating Public Health Surveillance Systems for Early Detection of Outbreaks" [CDC04a] offers a useful framework to guide

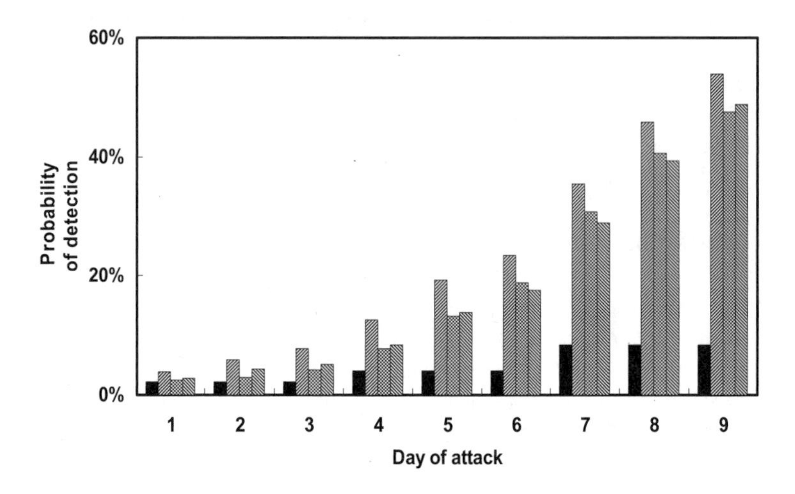

Fig. 2. Methods that combine data from multiple days (the hatched bars) were more effective at detecting slow-spreading attacks, but even the best method took until day nine to have a 50–50 chance of detecting a slow outbreak. See Stoto et al. [SSM04] for more information.

evaluation efforts. Other approaches focus on the completeness, timeliness, and quality of the data [BBM04], or on how syndromic surveillance efforts relate to public health practice [Rei03]. The annual national syndromic surveillance conference (see `http://www.syndromic.org`) offers many examples of such evaluations.

Formal approaches tend to focus on characterizing the statistical performance of detection algorithms applied to particular data streams. The Stoto, Schonlau, and Mariano [SSM04] analysis described above illustrates the simulation approach, and Sect. 3 of this paper presents a more detailed example. Both of these examples use real data as a baseline and add a simple simulated outbreak. As a perhaps more realistic alternative, Stacey [Sta04] has described an approach in which real data are used to model simulated outbreaks for testing purposes.

The retrospective analysis of known natural outbreaks is an alternative approach to evaluation. Siegrist and Pavlin [SP04], for instance, report on an exercise in which four leading biosurveillance research teams compared the sensitivity, specificity, and timeliness of their detection algorithms in two steps. First, an outbreak detection team identified actual natural disease outbreaks — eight involving respiratory illness and seven involving gastrointestinal illness — in data from five metropolitan areas over a 23-month period but did not reveal them to the research teams. Second, each research team applied its own detection algorithms to the same data, to determine whether and how quickly each event could be detected. When the false-alarm rate was set at one every 2 to 6 weeks, the best algorithms from each research team were able to detect all of the respiratory outbreaks; for two of the four teams detection typically occurred on the first day that the outbreak detection team determined as the start of the outbreak; for the other two teams, detection occurred approximately three days later. For gastrointestinal illness, the teams typically were able to detect six of seven outbreaks, one to three days after onset. (Of course, as previously discussed, such detection times are partially a function of the false-alarm rate — decreasing the false-alarm rate will increase the detection time.)

One can also look at the epidemiological characteristics of various pathogens to clarify the implications for syndromic surveillance [Bue04]. For instance, Fig. 3 gives two examples that differentiate between attacks in which many people are exposed at the same time, and those in which a contagious agent might cause large numbers of cases in multiple generations. Example A (the line with the triangles) illustrates what might be found if 90 people were exposed to a noncontagious agent (such as anthrax) and symptoms first appeared eight days on average after exposure. Example B (the line with the squares) illustrates the impact of a smaller number of people (24) exposed to a contagious agent (such as smallpox) with an average incubation period of 10 days. Two waves of cases appear, the second larger and 10 days after the first. Because the two epidemic curves are similar on days one through three, it is difficult to know what can be expected, but if the agent were contagious (Example B), early intervention could save some or all of the second generation of cases. In Example A, however, everyone would already have been exposed by the time that the outbreak was detected.

1.3 Improving the Performance of Syndromic Surveillance

Faced with results like those in Figs. 1 and 2, one naturally asks whether more effective systems can be developed. There are a number of alternatives that could be considered and actually are the subject of current research.

Most detection algorithms can be characterized in three respects: (1) what they assume as the background level and pattern of diseases or symptoms, (2) the type of departures from normal that they are tuned to detect (an exponential increase in the number of cases, a geographic cluster of cases, and so on), and (3) the statistical algorithm they use to determine when the data indicate a departure from normal (i.e., an "anomaly"). Each presents opportunities to improve the performance of detection algorithms. Ultimately, however, there really is no free lunch. As is the case in other areas of statistics, there is an inherent trade-off between sensitivity and specificity, and the special need for timeliness makes it even more difficult in this application. Every approach to increasing sensitivity to one type of attack is likely to cause a detection algorithm to be less sensitive to some other scenario. To circumvent this trade-off, we would have to have some knowledge about how a terrorist may attack.

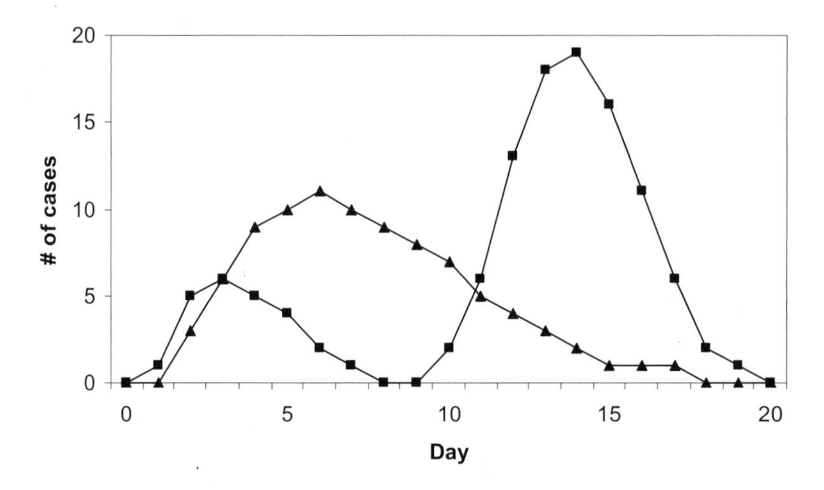

Fig. 3. Two epidemic curves that are similar on days one through three, but then diverge. The line with the triangles results from a gamma distribution with $\mu = 8$ and $\sigma = 4$. The line with the squares simulates an infectious agent with a mean latency of 10 days. It is made up of the sum of observations from two gamma distributions, one with $\mu = 4$ and $\sigma = 2$, and the second with $\mu = 14$ and $\sigma = 2$.

Siegrist's retrospective analysis [Sie04] summarizes the details of some of the leading syndromic surveillance systems, illustrating each of the approaches described below.

Background Level and Pattern

Models to adjust for background patterns can be simple or complex. At one extreme, a method may assume a constant mean number of cases and standard deviation over the entire year for each data series monitored. In other models, the expected number of cases varies seasonally, in a constant weekly pattern (reflecting availability of health services on weekends, for instance), or as represented in an autoregressive process.

Syndromic surveillance systems typically compare current cases to the number in the previous day or week, the number in the same day in the previous year, or some average of past values. More sophisticated approaches use statistical models to "filter" or reduce the noise in the background data to try to make the signal more obvious so that an outbreak would be easier to detect. For instance, if a hospital ER typically sees more ILI patients on weekend days (when other facilities are not open), a statistical model can be developed to account for this effect. With a long enough data series, annual effects can also be incorporated. Some patterns are not so easy to adjust for, however. Winter flu outbreaks, for instance, appear most years but vary in size and timing.

Departures from Routine Conditions

Better performance might also be obtained by carefully "tuning" the detection algorithm to detect specific types of outbreaks or perhaps one might choose to analyze a syndrome that is less common than ILI. Stoto, Schonlau, and Mariano [SSM04] used the same methods, for instance, to analyze the data on the number of patients with "Viral NOS" (NOS=not otherwise specified) symptoms, which averaged 1 per day. Outside of the flu season, they were able to detect a fast outbreak on day two 50% to 60% of the time, only a small improvement over ILI. With a slow outbreak, however, integrated methods had a 50% chance of detecting outbreaks on day 5 to 7, compared to day 9 for the same chance for ILI.

This improved performance, however, has a cost — it is only sensitive to symptoms that ER physicians would classify as Viral NOS. The combination of fever and rash is rare and suggests the early stages of smallpox. A syndromic surveillance system set up to look at this combination would likely be more effective than the results above suggest, but would only be sensitive to smallpox and not terrorist agents that have other symptoms.

Data also can be analyzed geographically, tuning detection algorithms to outbreaks that are focused in a small geographic area. For instance, if there were an extra 18 cases of ILI in a city, and all lived in the same neighborhood, that would surely be more informative than 18 cases scattered throughout the city — it would suggest a biological agent released at night in that area. This is only effective, however, for such a geographically focused attack. It would not work if terrorists chose to expose people in an office building during the workday or at an airport but the data were analyzed by home address.

Detection Algorithms

Finally, more sophisticated detection algorithms could lead to better performance. The simplest detection algorithms focus on the number of excess cases on a given day (the actual number minus some baseline value). If this is more than some number of standard deviations, an alarm is sounded.

Within this simple statement, however, are many choices, each of which affects the detection algorithm's sensitivity, false-positive rate, and timeliness. First, the normal background level and standard deviation must be determined. As indicated above, many choices — simple to complex — are possible for these variables. Second, the observation period must be chosen. Syndromic surveillance systems typically choose one day as the period for reasons of timeliness; any longer period would require waiting for data before the detection algorithm could be run. However, day-to-day variability in syndromic data due to small numbers sometimes means that adequate sensitivity can only be obtained at the cost of a high, false-positive rate. An alternative, therefore, would be to aggregate data over the period of one week, or to use a running average for the daily value. Both of these solutions are obviously less timely.

Current syndromic surveillance systems are typically set up to monitor eight or more separate sets of symptoms, perhaps in different geographical areas and from different hospital ERs. Doing so increases sensitivity simply because more conditions are monitored. If each set of symptoms has a 1% false-positive rate, however, increasing the number monitored will also increase the number of false positives.

One possibility is to pool data over multiple ERs, perhaps all hospitals in a metropolitan area or state, and indeed that is what cities such as Boston and New York are currently doing. If this results in both the signal and the background increasing proportionally, it will result in a more effective system. If, for instance, nine hospitals in the Washington area report daily, each with a daily average of 3 ILI cases, and outbreaks were nine times as large in the example above, the performance of detection algorithms would be substantially improved. If, however, there were 18 extra cases of ILI in the city and they all appeared in one hospital, this signal would be lost in the noise of the entire city's cases.

An alternative is to search for patterns in the set of symptoms; fever up but rash down, for instance, might lead to better performing detection algorithms. Statistical algorithms to determine whether a departure is sufficient to signal an alarm range from simple to sophisticated. The sophisticated What's Strange About Recent Events (WSARE) system developed at the Real Outbreak and Disease Surveillance (RODS) lab, for instance, is based on Bayesian belief networks [WMC03].

2 Statistical Process Control (SPC)

Quick detection of a change in a probability distribution is the fundamental problem of *statistical process control*. The problem arises in any monitoring situation, and lies at the foundation of the theory and practice of quality control. SPC methods use data to evaluate whether distributional parameters, such as the mean rate of a particular syndrome, have increased to an unacceptable level.

The simplest and best understood version of the problem specifies a one-parameter family of univariate distributions — the most studied family being the normal distribution with unknown mean — and aims to detect a change in the parameter from one value to another as quickly as possible after the change occurs. A number of popular and successful algorithms have been developed for this sort of problem, and a substantial body of theoretical and experimental research has accumulated.

Our interest here is in extending these methods to the problem of syndromic surveillance and, in particular, to the Washington, DC, ER data. That SPC is appropriate for syndromic surveillance is not immediately obvious, particularly since *a priori* one would expect a successful methodology would have to account for seasonal and perhaps other cycles in the data, and that methods specifically designed to detect monotonic changes in incident rates would outperform conventional SPC methods.

We address these and other issues below. In so doing, we introduce some modified multivariate algorithms that may be applied to health-related data for syndromic surveillance and then compare their performance to univariate SPC methods, both using simulated and actual syndromic surveillance data.

2.1 SPC Background and Literature

Walter A. Shewhart [She31] developed the concept of the *control chart*, a graphical statistical tool to help control the behavior of manufacturing processes, and in so doing became one of the founders of the quality control movement. Shewhart's methodology defined a scientific, statistical framework upon which to base decision-making and hence allow objective decisions to be made about how to manage systems. The field of SPC has since grown from Shewhart's seminal work. An excellent introductory text to quality control and SPC is *Introduction to Statistical Quality Control* by Montgomery [Mon85].

In addition to Shewhart's methodology, the classical approaches to SPC have generally been parametric and univariate. These include the CuSum ("cumulative sum") procedure of Page [Pag54] and Lorden [Lor71], the Bayesian procedure of Shiryayev [Shi63, Shi73] and Roberts [Rob66], and the EWMA ("exponentially weighted moving average") procedure of Roberts [Rob59].

The most basic SPC problem is that of monitoring a sequence of random variables over time with the goal of raising an alarm as soon as possible after the mean becomes too large. The CuSum has optimality properties if the mean experiences a one-time jump increase from one known level to another. However, syndromic surveillance is probably not realistically described by this type of change. Rather, a disease outbreak or bioterrorism attack is likely to be characterized by monotonically increasing numbers of people presenting to an ER as the pathogen spreads or the fraction of those who were exposed who develop symptoms increases (as illustrated in Fig. 3).

This difference would seem to cast doubt on the applicability of SPC to the problem of syndromic surveillance. However, Chang and Fricker [CF99] compared the performance of CuSum and EWMA versus a repeated generalized likelihood ratio (GLR) test designed specifically for the monotone problem. They found that the CuSum and EWMA, appropriately applied, performed surprisingly well in comparison to the GLR test, usually outperforming it, and concluded that the CuSum was probably the best overall choice. This result provides some evidence that the simple SPC methods may perform well in the syndromic surveillance problem.

Multivariate CuSum research has centered around detecting changes in either the normal mean vector or the covariance matrix. Seminal work was by

Hotelling [Hot47] in the manufacture of bomb sights in World War II who developed a Shewhart-like methodology for multivariate observations. More recent research includes Pignatiello and Runger [PR90] and Healy [Hea87]. Pignatiello and Runger [PR90] and Crosier [Cro88], as well as other researchers, have looked at the application of CuSum-like recursions to the product of the observation vector and an assumed known covariance matrix. Others have dealt with multivariate data by applying a number of individual univariate algorithms, one to each marginal distribution [WN85], for example. More detailed background information about multivariate SPC can be found in [Alt85].

2.2 Some Notation and Terminology

In the simple case of detecting a shift from one specific distribution to another, let f_0 denote the *in-control* distribution, which is the desired or preferred state of the system. For syndromic surveillance, for example, this could be the distribution of the daily counts of individuals diagnosed with a particular chief complaint at a specific hospital or within a particular geographic region under normal conditions. Let f_1 denote the *out-of-control* distribution where, under the standard SPC paradigm, this would be a particular distribution representing a condition or state that is important to detect. Within the syndromic surveillance problem, f_1 might be a specific, elevated mean daily count resulting from the release of a bioterrorism pathogen for example.

Let τ be the actual (unknown) time when the process shifts from f_0 to f_1 and let T be the length of time from τ to when an algorithm alarms (which we call the *delay*). We use the notation $\mathrm{E}_\tau(T|T \geq 0)$ to indicate the expected delay, which is the average time it takes an algorithm to alarm *once the shift has occurred.* We also use the notation $\mathrm{E}_\infty(T)$ to indicate the expected time to a false alarm, meaning that $\tau = \infty$ and the process never shifts to the out-of-control distribution.

In the SPC literature, algorithms are compared in terms of the expected time to alarm, where $\mathrm{E}_\infty(T)$ is first set equally for two algorithms and then the algorithm with the smallest $\mathrm{E}_\tau(T|T \geq 0)$, for a particular f_1, is deemed better. Often when conducting simulation comparisons, τ is set to be 0, so the conditioning in the expectation is automatic.

The term *average run length* (ARL) is frequently used for the expected time to alarm, where it is understood that when $\tau = \infty$ the ARL denotes the expected time to false alarm. Similarly, in simulation experiments, the performance of various algorithms is compared by setting the expected time to false alarms to be equal and then comparing ARLs when $\tau = 0$, where it is then understood that the ARL is the mean delay time. In general terms, an algorithm with a smaller ARL has a higher sensitivity for detecting anomalies, though this comes at the expense of an increased false-alarm rate.

For syndromic surveillance, the out-of-control situation can be more than a jump change from f_0 to f_1. For example, if μ_0 is the mean of f_0, then one possible out-of-control situation might be a monotonic increase in the mean so that for each time $i > \tau$, $\mu_1(i) = \mu_0 + (i - \tau)\delta$, for some positive δ. Yet, even for this type of out-of-control condition, algorithms can still be compared using $\mathrm{E}_\tau(T|T \geq 0)$.

Note that the specific value of τ is generally irrelevant to the analysis. What is important is how long an algorithm takes to alarm after time τ. However, setting $\tau = 0$ means that the algorithm is guaranteed to be in its initial condition when the shift to f_1 occurs (or starts to occur, in the case of something other than a jump change), which may be a help or hindrance to a particular algorithm.

Also, note that comparisons using the expected value are characterizing the distribution of the delay via a single number. This has the advantage of allowing many comparisons to be easily graphically summarized (as we will show), but comes with all the inherent limitations of such summaries. Hence, here we used both the ARL in our initial simulation investigations and then subsequently used the distribution of the delay in the final simulations with actual data.

2.3 Applying SPC to Syndromic Surveillance

This section presents two standard univariate algorithms (the Shewhart and the CuSum) and two multivariate extensions of these two algorithms (Hotelling's T^2 and one of Crosier's multivariate CuSums). Here we also discuss how to apply the univariate algorithms to multivariate syndromic surveillance data and describe how we modified the multivariate algorithms to best apply to the syndromic surveillance problem. We focus on the Shewhart and CuSum algorithms, and not the EWMA, because the EWMA can be made to perform very similarly to either of the Shewhart or CuSum through the appropriate selection of the EWMA's weighting parameter.

Furthermore, we chose to use Shewhart and CuSum SPC methods due to the nature of our data. Specifically, for these particular data:

- The mean rates for each of the syndromic groups were quite constant, and
- The logarithmically transformed counts (not shown here) were quite normally distributed.

It is important to note that most SPC procedures, including those described here, have been developed under the assumption that the observations are independent. In industrial applications, this can often be reasonably well achieved by taking observations sufficiently far apart in time. For syndromic surveillance data that exhibit characteristics such as seasonal cycles or other trends, which we were frankly surprised not to find in our data, other methods such as the EWMA or those proposed by Nomikos and MacGregor [NM95] might be more appropriate and effective.

Univariate Shewhart Algorithm

Shewhart's algorithm [She31] is probably the simplest and best known of all SPC methods and is widely applied in industry. The basic idea is to sequentially evaluate one observation (or period) at a time, alarming when an observation that is rare under f_0 occurs. The most common form of the algorithm, often known as the $\bar{X}$ *chart*, alarms when the absolute value of an observed sample mean exceeds a prespecified *threshold* h, often defined as the mean value plus some number of standard deviations of the mean. There are variants on the algorithm for monitoring the variability of processes and the algorithm can be defined to only alarm for deviations in one direction.

For application to the syndromic surveillance problem, we assume that only deviations in the positive direction that would indicate a potential outbreak are important to detect. For a univariate random variable X, and for some desired probability p, the threshold h is chosen to satisfy

$$\int_{\{x>h\}} f_0(x)\,\mathrm{d}x = p.$$

The algorithm proceeds by observing values of X_i; it stops and concludes $X_i \sim f_1$ at time $\hat{\tau} = \inf\{i : X_i > h\}$.

If the change to be detected is a one-time jump in the mean and the probability of an observation exceeding the threshold is known, then simulation is not required as the delay is geometrically distributed and exact calculations for the average run lengths can be directly calculated as $\mathrm{E}_\infty(T) = 1/p$ and

$$\mathrm{E}_\tau(T|T \geq 0) = \mathrm{E}_0(T) = \left[\int_{\{x>h\}} f_1(x)\,\mathrm{d}x\right]^{-1}.$$

Generally, however, it is quite simple to empirically estimate the ARLs via simulation. For a particular f_0, choose an h and run the algorithm m times, recording for each run the time t when the first $X_i > h$ (where each X_i is a random draw from f_0, of course). Estimate the in-control ARL as

$$\widehat{\mathrm{E}_\infty(T)} = \sum t/m,$$

adjusting h and rerunning as necessary to achieve the desired in-control ARL, where m is made large enough to make the standard error of $\widehat{\mathrm{E}_\infty(T)}$ acceptably small. Having established the threshold h for that f_0 with sufficient precision, then for each f_1 of interest rerun the algorithm n times (where n is often smaller than m), drawing the X_is from f_1 starting at time 1. As before, take the average of $t_1, \ldots, t_n$ to estimate the expected delay.

For the multivariate syndromic surveillance problem, multiple univariate algorithms are applied, one to each data stream. When comparing the performance of simultaneous univariate algorithms applied to multivariate data to a multivariate algorithm it is important to ensure that the expected times to false alarm are set equally. For multiple univariate algorithms running simultaneously, say j, one must choose how to set the j thresholds. If there is some reason to make the combined algorithms more sensitive to changes in some of the data streams, those thresholds can be set such that the probability of exceeding the threshold(s) is greater in those data streams than in the others. For the purposes of the simulations that follow in this chapter, there was no reason to favor one data stream over another, so all the thresholds were set such that the probability of false alarm was equal for all data streams.

Univariate CuSum Algorithm

The CuSum is a sequential hypothesis test for a change from a known in-control density f_0 to a known alternative density f_1. The algorithm monitors the statistic S_i, which satisfies the recursion

$$S_i = \max(0, S_{i-1} + L_i), \tag{1}$$

where the increment L_i is the log likelihood ratio

$$L_i = \log \frac{f_1(X_i)}{f_0(X_i)}.$$

The algorithm stops and concludes that $X_i \sim f_1$ at time $\hat{\tau} = \inf\{i : S_i > h\}$ for some prespecified threshold h that achieves a desired ARL under the given in-control distribution.

If f_0 and f_1 are normal distributions with means μ and $\mu + \delta$, respectively, and unit variances, then (1) reduces to

$$S_i = \max(0, S_{i-1} + (X_i - \mu) - k), \tag{2}$$

where $k = \delta/2$. This is the form commonly used, even when the underlying data is only approximately normally distributed. For the DC hospital data we examined, the log transformed data was generally very close to normally distributed, so we applied (2) to $\log(X_i)$. Note that k may be set to values other than $\delta/2$ and frequently users specify a value for k rather than the mean of f_1. What is relevant to the performance of the CuSum is that when the process shifts to a state where $\mathrm{E}(X_i) > \mu + k$, then the expected value of the increment$X_i - \mu - k$ is positive and the CuSum S_i tends to increase and subsequently exceed h relatively quickly.

Note that, since the univariate CuSum is "reflected" at zero, it is only capable of looking for departures in one direction. If it is necessary to guard against both positive and negative changes in the mean, then one must simultaneously run two CuSums, one of the form in (2) to look for changes in the positive direction, and one of the form

$$S_i = \max(0, S_{i-1} - (X_i - \mu) - k),$$

to look for changes in the negative direction. For the syndromic surveillance problem, we are only interested in looking for increases in rates, so we only use (2).

As with the univariate Shewhart, multiple univariate CuSum algorithms must be applied, one to each data stream, for the multivariate syndromic surveillance problem. As with the univariate Shewhart algorithms, for the purposes of our simulations, there was no reason to favor one data stream over another, so all the thresholds were set such that the probability of false alarm was equal in all data streams and so that the resulting expected time to false alarm for the combined set of univariate algorithms was equal to the expected time to false alarm of the multivariate algorithm.

Multivariate Shewhart Algorithm (Modified Hotelling's T^2)

Hotelling [Hot47] introduced the T^2 (sometimes referred to as the χ^2) algorithm. For multivariate observations $\mathbf{X}_i \in \mathbb{R}^d$, $i = 1, 2, \ldots$, compute

$$T_i^2 = \mathbf{X}_i' \Sigma^{-1} \mathbf{X}_i,$$

where Σ^{-1} is the inverse of the covariance matrix. The algorithm stops at time $\hat{\tau} = \inf\{i : T_i > h\}$ for some prespecified threshold h.

We refer to this as a multivariate Shewhart algorithm since it only looks at data from one period at a time. Like the original univariate Shewhart $\bar{X}$ algorithm, because it only uses the most recent observation to decide when to stop, it can react quickly to large departures from the in-control distribution, but will also be relatively insensitive to small shifts. Of course, it also requires that the covariance matrix is known or well-estimated.

For the syndromic surveillance problem, it is desirable to focus the T^2 algorithm on the detection of increases in incident rates. We accomplish that by modifying the stopping rule for the T^2 so that it meets two conditions: (1) $T_i > h$ and (2) $\mathbf{X}_i \in \mathcal{S}$, where $\mathcal{S}$ is a particular subspace of $\mathbb{R}^d$ that corresponds to disease outbreaks, for example an increase in one or more data streams.

For the purposes of the syndromic surveillance simulations, we defined $\mathcal{S}$ as follows. Choose values $s_1, s_2, \ldots, s_d$ such that

$$\int_{x_1=s_1}^{\infty} \int_{x_2=s_2}^{\infty} \cdots \int_{x_d=s_d}^{\infty} f_o(\mathbf{x}) d\mathbf{x} \approx 0.99,$$

and then define $\mathcal{S} = \{x_1 > s_1, x_2 > s_2, \ldots, x_d > s_d\}$.

For example, consider an in-control distribution following a bivariate normal distribution with some positive correlation, so that the probability contour for the density of f_0 is an ellipse with its main axis along a 45-degree line in the plane. Then you can think about $\mathcal{S}$ as the upper right quadrant that almost encompasses the 99% probability ellipse.

The idea of using this region for $\mathcal{S}$ is that if f_1 represents a shift in the mean vector in any direction corresponding to an increase in one or more of the data streams, then the modified T^2 algorithm will have an increased probability of alarming, which should result in a decreased expected time to alarm. On the other hand, if f_1 represents a condition where the mean vector corresponds to a decrease in one or more of the data streams, then the probability of alarming will decrease and the algorithm will have less of a chance of producing an alarm.

Multivariate CuSum Algorithm (Modified Crosier's MCuSum)

The abbreviation MCuSum, for multivariate CuSum, is used here to refer to the algorithm proposed by Crosier [Cro88] that at each time i considers the statistic

$$\mathbf{S}_i = (\mathbf{S}_{i-1} + \mathbf{X}_i - \boldsymbol{\mu})(1 - k/C_i), \text{ if } C_i > k, \tag{3}$$

where k is a statistical distance based on a predetermined vector $\mathbf{k}$, $k = \{\mathbf{k}'\Sigma^{-1}\mathbf{k}\}^{1/2}$ and $C_i = \{(\mathbf{S}_{i-1} + \mathbf{X}_n - \boldsymbol{\mu})'\Sigma^{-1}(\mathbf{S}_{i-1} + \mathbf{X}_i - \boldsymbol{\mu})\}^{1/2}$. If $C_i \leq k$, then reset $\mathbf{S}_i = \mathbf{0}$. The algorithm starts with $\mathbf{S}_0 = \mathbf{0}$ and sequentially calculates

$$Y_i = (\mathbf{S}_i'\Sigma^{-1}\mathbf{S}_i)^{1/2}.$$

It concludes that $X_i \sim f_1$ at time $\hat{\tau} = \inf\{i : Y_i > h\}$ for some threshold $h > 0$.

Crosier proposed a number of other multivariate CuSum-like algorithms but generally preferred (3) after extensive simulation comparisons. Pignatiello and Runger [PR90] proposed other multivariate CuSum-like algorithms as well, but found that they performed similarly to (3).

It is worth noting that Crosier derived his algorithm in an ad hoc manner, not from theory, but found it to work well in simulation comparisons. Healy [Hea87] derived a sequential likelihood ratio test to detect a shift in a mean vector of a multivariate normal distribution that is a true multivariate CuSum. However, while we found Healy's algorithm to be more effective (had shorter ARLs) when the shift was to the precise f_1 mean vector, it was less effective than Crosier's for detecting other types of shifts, including mean shifts that were close to but not precisely the specific f_1 mean vector.

In this application we prefer Crosier's algorithm to Healy's since it seems to be more effective at detecting a variety of departures from the in-control mean vector and the types of shifts for the syndromic surveillance problem are not well-defined. That is, if we knew the type of departure to look for, we could design a detection algorithm that would have more power to detect that specific signal. However, given that the types of signals will vary, we have opted for Crosier's method because it is robust at detecting many types of departures well.

We also prefer Crosier's formulation for the syndromic surveillance problem as it is easy to modify to look only for positive increases. In particular, in our simulations, when $C_i > k$ we bound $\mathbf{S}_i$ to be positive in each data stream by replacing (3) with $\mathbf{S}_i = (S_{i,1}, \ldots, S_{i,d})$ where

$$S_{i,j} = \max[0, (S_{i-1,j} + X_{i,j} - \mu_j)(1 - k/C_i)],$$

for $j = 1, 2, \ldots, d$.

2.4 Performance Comparisons via Abstract Simulations

Before evaluating the performance of the methods using actual data, we compared their performance using simulated data from normal and multivariate normal distributions. The purpose of these simulations was to:

1. Compare and contrast the performance of the methods under known, ideal conditions;
2. Gain some insight into how they performed as the dimensionality of the data changed; and
3. Reach some preliminary conclusions about how best to implement the algorithms for the real data.

In these simulations, we compared the performance by average run length, first setting the ARL under the in-control distribution (i.e., $\mathrm{E}_\infty(T)$, the expected time to false alarm) equally, and then comparing the ARL performance under numerous out-of-control distributions resulting from various shifts in the mean vector at time 0 (i.e., $\mathrm{E}_0(T)$).

For example, Fig. 4 illustrates the improved performance of the modified T^2 algorithm and the modified MCuSum regardless of dimensionality and size of (a positive) mean shift. Here (and in the other figures in this section) the in-control distribution is a six-dimensional multivariate normal centered at the zero vector with unit variance in all the dimensions and covariance $\varrho = 0.3$ between all the dimensions; that is, the in-control distribution is

$$f_0 = N\left(\begin{pmatrix} 0 \\ 0 \\ 0 \\ 0 \\ 0 \\ 0 \end{pmatrix}, \begin{pmatrix} 1 & 0.3 & 0.3 & 0.3 & 0.3 & 0.3 \\ 0.3 & 1 & 0.3 & 0.3 & 0.3 & 0.3 \\ 0.3 & 0.3 & 1 & 0.3 & 0.3 & 0.3 \\ 0.3 & 0.3 & 0.3 & 1 & 0.3 & 0.3 \\ 0.3 & 0.3 & 0.3 & 0.3 & 1 & 0.3 \\ 0.3 & 0.3 & 0.3 & 0.3 & 0.3 & 1 \end{pmatrix}\right).$$

The out-of-control distributions are the same as the in-control distributions but with components of the mean vector shifted as indicated on the horizontal axis for the number of dimensions shown in the key. So, for example, the darkest line is for a mean vector that was shifted in all six dimensions from 0.0 — no shift — on the left to 3.4 on the right.

The vertical axis in Fig. 4 is the difference (Δ) between the ARL for the unmodified algorithm and the modified algorithm for a given mean vector shift (measured only at the values indicated on the horizontal axis). Positive values indicate the modified algorithm had a smaller ARL and so performed better, so that for a particular out-of-control condition the modified algorithm had a shorter time to alarm. A difference of 0 at mean shift $= 0.0$ indicates that the false-alarm rates (equivalently, the in-control ARLs) were set equally for each algorithm before comparing the expected time to alarm for various out-of-control mean vector shifts (within the bounds of experimental error, where a sufficient number of simulation runs were conducted to achieve a standard error of approximately 2.5 on the estimated in-control ARLs).

Figure 4 shows, as expected, that the modified algorithms perform better than the original algorithms at detecting positive shifts regardless of whether the shift occurs in one dimension, in all the dimensions, or in some number of dimensions in-between, and for all magnitudes of shift. As the number of dimensions experiencing a shift of a given size increases, the modified algorithms do considerably better. However, for the largest shifts, the performance of the original algorithms approaches that of the modified algorithms.

Not shown here, the results for other low-to-moderate values of ϱ, from $\varrho = 0$ to $\varrho = 0.9$, are very similar. Only for large ϱ and small shifts in a low number of dimensions does the original MCuSum algorithm best the modified algorithm. However, in our actual data the covariances between chief complaints, both within and between hospitals, whether aggregated or not, tended to be quite low, generally less than 0.1 and never greater than 0.3.

A further benefit of the modified algorithms, at least in terms of syndromic surveillance, is that they will not alarm if incidence rate(s) decrease. While a decrease in rates might be interesting to detect for some purposes, for the purpose of syndromic surveillance such detection would constitute a false alarm. In addition, because these multivariate algorithms only look for positive shifts, they can be directly compared to multiple one-sided univariate algorithms operating simultaneously.

Given that the modified T^2 performs better than the original T^2 for this problem, Fig. 5 focuses the performance of the modified T^2 as compared to six one-sided Shewhart algorithms operating simultaneously. The comparison is shown in two different ways: in terms of the distance of a shift measured in the direction of one or more of the axes ("on axes" in the left graph), or in terms of the distance of a shift "off axes" (right graph). At issue is that the univariate algorithms are direction specific, meaning they are designed to look for shifts along the axes. The multivariate algorithms are direction invariant, meaning they are just as effective at detecting a shift of distance x whether the shift occurs in the direction of one or more axes ("on axes") or in some other direction ("off axes").

The left-side graph of Fig. 5, constructed just like Fig. 4, shows that six simultaneous univariate Shewharts are more effective (have shorter ARLs) than the modified T^2 when the shift occurs on axes. At best, for large shifts, the ARL of the modified T^2 is equivalent to the multiple univariate Shewharts, and for smaller shifts (roughly > 0.0 to 1.0) the multiple univariate Shewharts are clearly better.

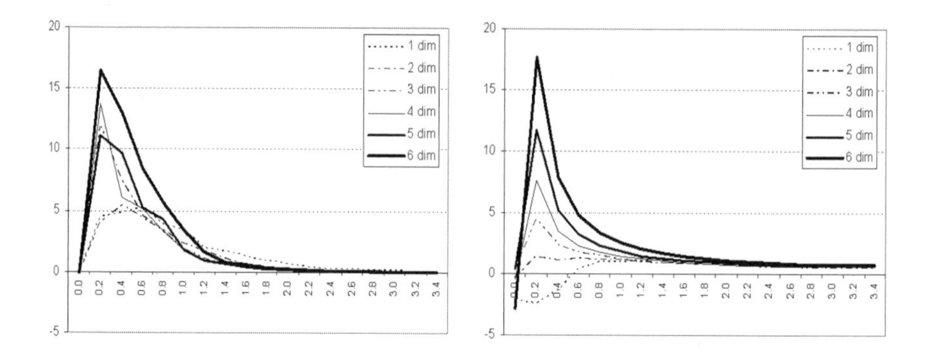

Fig. 4. Performance comparison of the T^2 and MCuSum algorithms versus their modified counterpart algorithms. The modified algorithms (T^2 on the left and MCuSum on the right) perform better than the original algorithms at detecting positive shifts regardless of whether the shift occurs in one dimension, in all the dimensions, or in some number of dimensions in-between, and for all magnitudes of shift.

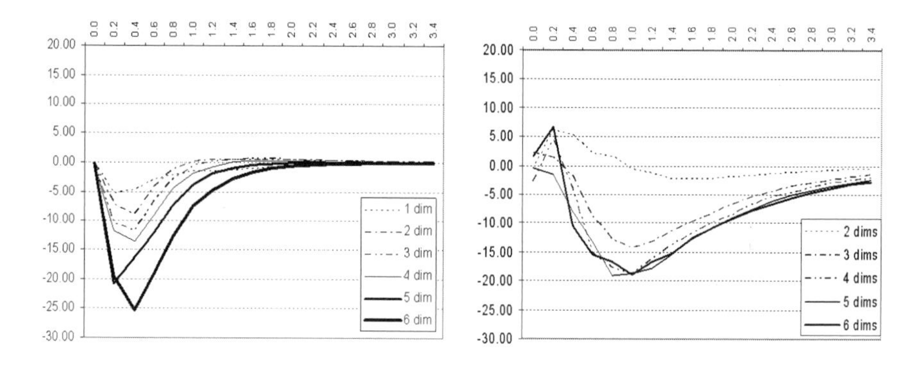

Fig. 5. Performance comparison of the modified T^2 algorithm versus multiple simultaneous univariate Shewhart algorithms for $\varrho = 0.3$. The multiple simultaneous Shewhart algorithms generally have smaller ARLs except for small "off axis" shifts.

The graph on the right side of Fig. 5 is constructed differently. The horizontal axis of this graph shows the *distance* of the shift, where the shift is in the number of dimensions indicated in the key, and was constructed so that the projection of the shift onto the axes for those dimensions was equal. That is, for a shift of distance l in n dimensions, the mean vector component for each of the affected dimensions shifted from 0 under f_0 to $l/\sqrt{n}$ under f_1 (and where in the other 6–n dimensions, the mean vector components remain unchanged at 0).

This type of shift is the most extreme off-axis type of shift (meaning for a given distance l, the maximum projection on the nonzero axes was the smallest) and here we see a result similar to Fig. 5, except that the modified T^2 does better than the simultaneous univariate Shewhart algorithms for very small shifts.

Why is the distinction between the two types of shifts (on-axes versus off-axes) relevant? Well, if each of the types of bioterrorism events to be detected will manifest itself in the data being monitored as a separate increase in one of the data streams, such as ER admit counts for a particular chief complaint, then thinking about and optimizing the detection algorithm to look specifically for shifts along the axes makes sense. On the other hand, for a bioterrorism event that will manifest itself as changes in a number of dimensions of the data being monitored, such as with less specific health data that in combination may increase, it makes sense to provide for an event that manifests itself more like a latent variable and hence appearing most strongly in some off-axes direction.

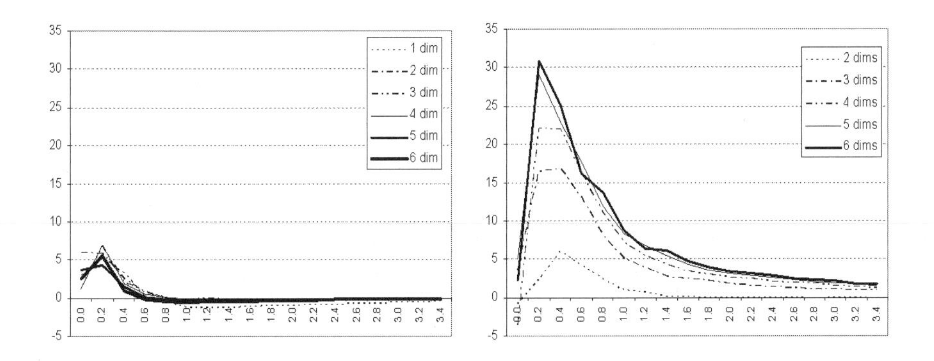

Fig. 6. Performance comparison of the modified MCuSum algorithm versus multiple simultaneous univariate CuSum algorithms. The modified MCuSum tends to have smaller ARLs whether the shift is along the axes (left) or whether the shift is off the axes (right).

Given that the goal is a robust methodology to guard against either possibility, the results for the simultaneous univariate Shewharts versus the modified T^2 are mixed. However, the results for the modified MCuSum versus simultaneous univariate CuSums presented in Fig. 6 differ in that the modified MCuSum is generally better than the simultaneous univariate CuSums regardless of whether the shift is on- or off-axis. In particular, in the left graph of Fig. 6 the modified MCuSum performance is substantially better for small shifts (roughly > 0.0 to 0.5 or so), equivalent for large shifts (roughly > 3.0), and only marginally degraded for other shifts, with an ARL difference of less than 1. As expected, in the right graph of Fig. 6 the modified MCuSum performance is better than or, for very large shifts, equivalent to the simultaneous univariate CuSums.

Though not shown here, these results also hold for a range of low to moderate correlations, from $\varrho = 0$ to $\varrho = 0.6$. Hence, these results would tend to indicate that the modified MCuSum would be preferable to simultaneous univariate CuSums for detecting a variety of types of mean shifts. What remains, then, is a comparison of the modified MCuSum to either the multiple univariate Shewharts or the modified T^2 in those scenarios where each does better.

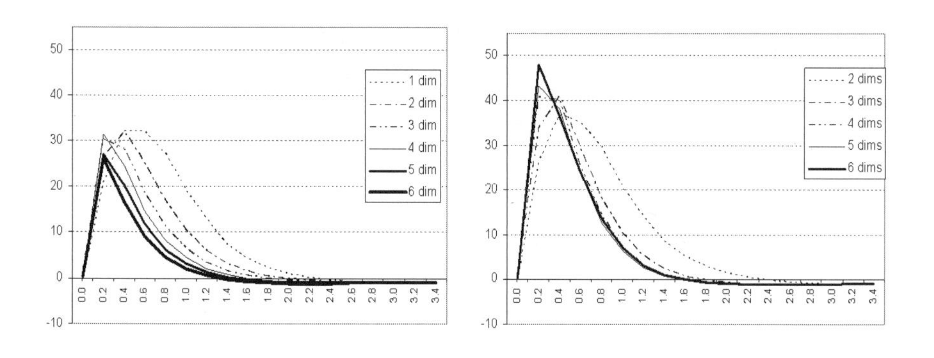

Fig. 7. Performance comparison of the modified MCuSum algorithm to multiple simultaneous univariate Shewhart algorithms (left) and to the modified T^2 algorithm (right). Whether the shift is on-axes or off-axes, the modified MCuSum algorithm performs better than the preferred Shewhart-type algorithm.

Figure 7 provides this comparison: to the simultaneous univariate Shewharts when the shift is on-axes (left graph) and to the modified T^2 when the shift is off-axes (right graph). In both cases, the modified MCuSum algorithms' performance is better. Our conclusion, then, is a preference for the modified MCuSum, at least in these simulations for a jump change in the mean vector of multivariate normal distributions with moderate covariance. In the next section, then, we further examine the performance of these methods using real data and more realistic shifts to evaluate the performance of the algorithms.

3 A Simulation Study Using DC ER Syndromic Surveillance Data

One would expect that a properly designed multivariate algorithm would be more effective — particularly, more sensitive and timely when the false-positive rate is controlled — than standard univariate methods. However, as the previous section demonstrated, some multivariate methods are better than others, and there are situations in which simultaneous univariate algorithms are preferable. Furthermore, since the evaluations in the SPC literature tend to focus on a jump change in the mean, as did the evaluations in the preceding section, it does not necessarily follow that those results will directly apply to the syndromic surveillance problem in which the mean will likely change in some monotonically increasing fashion. Hence, to evaluate the univariate and multivariate algorithms described in Sect. 2, we also conducted a simulation study on data from the DC Department of Health ER syndromic surveillance system and then evaluated how the algorithms performed under a series of outbreak scenarios.

3.1 Data and Methods

As baseline data for our simulation study we used data on the daily number of ER admissions for four syndromic group "chief complaints" (unspecified infection, rash, respiratory complaints, and gastrointestinal complaints) from seven Washington, DC, hospitals with relatively complete data. Of the eight syndromic groups available, these four were chosen because they are the most common and, in univariate analyses, are most effective at detecting disease outbreaks. The data on the resulting 28 data streams (4 syndromic groups x 7 hospitals) span the period of September 2001 through May 2004 (with missing data imputed as required [SJF04] to simplify the comparisons of the detection algorithms).

This data provides the naturally occurring incident rates and variation in the hospital ERs for the four syndromic groups. We then "seeded" these data in various ways, meaning we added extra cases to the data, to simulate a bioterrorism event. In the base case, Scenario A, we seeded the data adding 1 additional observation on day τ, 2 additional on day $\tau+1$, and so on up to 10 on day $\tau+9$ for each of the 28 data streams resulting in a total of 1,540 extra cases over 10 days. Scenario A is intended to represent a bioterrorism event that manifests itself in multiple ways across the entire population. Hence, all of the chief complaints increase in all the hospital ERs.

In contrast, we defined Scenario C to represent a situation in which the outbreak shows up in one syndromic group only, so we only seeded the "unspecified infection" syndromic group only for all seven hospitals adding 1 additional observation on day τ, 2 additional on day $\tau + 1$, and so on up to 10 on day $\tau + 9$ (for a total of 385 extra cases).

Since the total number of cases added in Scenario C is only one-quarter of that of the base scenario, we also constructed Scenario CA in which the seed was increased to 4 on day τ, 8 on day $\tau + 1$, and so on, resulting in 220 extra unspecified infection cases in each of seven hospitals, which is a total of 1,540 extra cases over 10 days. Hence, like Scenario C, Scenario CA represents an event that manifests itself in only one syndromic group but with the magnitude of Scenario A.

Scenarios D and DA repeat this with a focus on hospitals rather than syndromic groups. In Scenario D we seeded the data adding 1 additional observation on day τ, 2 additional on day $\tau + 1$, and so on up to 10 on day $\tau + 9$ for every syndromic group but in only one medium-sized hospital. In Scenario DA, the seed was increased to 7 on day τ, 14 on day $\tau + 1$, and so on, resulting in 385 extra cases in each syndromic group in only one hospital. So, Scenarios D and DA represent an outbreak in a smaller geographic region, with Scenario D being of a smaller magnitude and Scenario DA having the magnitude of Scenario A.

These five scenarios were chosen to represent the extremes of a range of ways in which a real bioevent might occur. (As the gap in the naming convention suggests, we investigated other scenarios as well, but do not present them here.) Some might regard Scenario C, in which the outbreak is concentrated in only one syndromic group, as the most likely of the scenarios. However, we expect that any real outbreak will look like some combination of these scenarios, so detection algorithms that work well across the test scenarios are likely to be effective in actual practice.

Given these scenarios, we then compared the performance of the algorithms described in Sect. 2.3 and a trend-adjusted CuSum (see Stoto et al. [SSM04] for additional detail) applied in two ways. First, we applied simultaneous univariate algorithms or one multivariate algorithm to the individual 28 data streams, setting the detection threshold empirically so that the probability of an alarm outside the flu season (i.e., the false-alarm rate) was 1%. Second, as an alternative to reduce the dimensionality of the problem, we first summed the total number of cases across all hospitals in each of the four syndromic groups and then applied either simultaneous univariate algorithms or a multivariate algorithm to the resulting four data streams (again setting the false-alarm rates equal at 1%).

To carry out the simulation we began by setting $\tau = 1$ and adding the appropriate seed on days 1 through 10 of the dataset. We repeated this setting $\tau = 2$ and adding the appropriate seed on days 2 through 11, and so on, until we had created 970 alternative datasets. We then applied the detection algorithms to each alternative dataset and calculated the proportion of times that each algorithm alarmed on day τ through $\tau + 9$, the first day of the simulated bioevent to the 9th day after the simulated bioevent, to estimate the sensitivity of the detection algorithm. Because we expect performance to differ by season, the results are calculated separately for the flu season (defined as December 1–April 30) and the rest of the year.

3.2 Results

Figure 8 compares the performance of the modified MCuSum ("MV" in the key) and simultaneous univariate CuSum methods ("Z" in the key) outside of the flu season for all five scenarios — just one summary result of the many simulations we ran. Unlike the graphs presented in the previous section, showing estimated ARLs for various changes in the mean, Fig. 8 plots the probability of detection (which can be interpreted as an estimated probability of alarm) for each algorithm under each scenario by day of the outbreak.

Note first that the probability of detection on day 0, that is, the day before the outbreak begins, is 1% for each detection algorithm, the false-alarm rate we set. Focusing first on Scenario A (in which the seed appears in all 28 data streams), the results show that in 18% of the sample datasets the simultaneous univariate CuSum algorithms (dashed line with open circles) alarm on day 2 of the outbreak, increasing to 67% on day 3 and 100% on day 4 and higher. In this scenario, the modified MCuSum (solid line with open circles) does slightly better. The probabilities of alarming on days 2, 3, and 4 are 36%, 93%, and 100%, respectively.

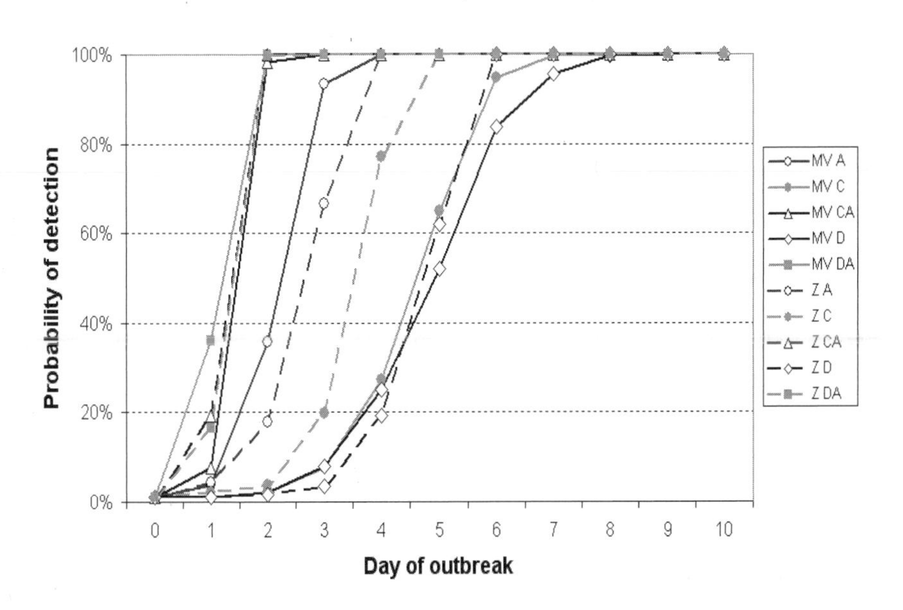

Fig. 8. Comparison of the modified MCuSum ("MV" in the key, solid lines in the graph) and the simultaneous univariate CuSums ("Z" in the key and dashed lines in the graphs) in terms of probability of outbreak detection by day of outbreak for Scenarios A, C, CA, D, and DA previously defined.

In Scenario C (closed circles), in which the simulated outbreak is concentrated in only one syndromic group and consequently involves only one-quarter the number of cases of Scenario A, both detection algorithms not surprisingly do less well. On day 4 of the simulated outbreak the univariate CuSums have only a 77% probability of alarming, and the modified MCuSum only a 27% probability of alarming. The results for Scenario CA (triangles), however, show that most of the reason for the poorer performance is that there are fewer excess cases. In this scenario the univariate and multivariate CuSum algorithms have 100% and 98% probabilities of alarming, respectively, by day 2. Note that in Scenario C by day 4 a total of 10 excess cases of unspecified infection have been seen in each hospital, and in Scenario CA there are 40 excess cases. The average daily number of such cases in the baseline data is less than 1 for two of the hospitals, between 3 and 6 for four hospitals, and over 30 for one hospital in the analysis.

Scenarios D and DA, in which the outbreak is concentrated in only one hospital, show similar results. In Scenario D, which involves only one-seventh the number of cases as Scenario A (open circles), the univariate CuSums alarm probability reaches 62% only on day 5, and the modified MCuSum only reaches 52% on that day. With the same number of cases as in Scenario A, the performance of both algorithms improves under Scenario DA (squares). Both reach a 100% alarm rate by day 2.

Comparing the performance of the two CuSum algorithms across these scenarios, it is difficult to conclude that one is better than the other. The modified MCuSum does noticeably better than the univariate CuSums in Scenario A (solid versus dashed lines with open circles, respectively), but worse in Scenario C (closed circles). In the other scenarios their performance is similar.

To summarize this type of comparison for all of the algorithms we tested, we calculated a performance index, defined as Σ_iProb(detection on day i) for $i = 1$ to 10. This is essentially the area under the curve in Fig. 8. In Scenario A, for instance, the performance index for the modified MCuSum is 8.34 and for the univariate CuSums is 7.90. In Scenario C, the situation is reversed: 5.99 versus 7.04. To get a sense of the range of the performance index, the best performance represented in Fig. 8 is the modified MCuSum in Scenario DA, with a performance index of 9.37, and the worst is the modified MCuSum in Scenario D, with a performance index of 5.68.

Table 1 displays the performance index for 12 detection algorithms for the five scenarios. "Z" results are univariate analyses operating on all 28 data streams, and "C" results sum the syndromic group data over seven hospitals (resulting in four data streams). In both cases we investigate simultaneous univariate Shewhart (1a and 3a), CuSum (1b and 3b), and trend-adjusted CuSum (1c and 3c) algorithms. "MV" results are multivariate algorithms, and "CMV" results are for multivariate algorithms but with the data summed across hospitals as in the C results: Hotelling's T^2 (5a, 6a), the modified T^2 (5b, 6b), and the modified MCuSum (7, 8).

Table 1. Comparison of performance of the univariate and multivariate algorithms for the five scenarios using a performance index of Σ_iProb(detection on day i) for $i = \tau$ to τ+9. The "Z" results (1a, 1b, and 1c) are simultaneous univariate algorithms operating simultaneously on all 28 series. The "C" results (3a, 3b, and 3c) are also simultaneous univariate algorithms operating on syndromic group data summed over the seven hospitals. "MV" results (5a, 5b, and 7) and "CMV" results (6a, 6b, and 8) are multivariate algorithms operating on the 28 data streams and the summed four data streams, respectively. These include Hotelling's T^2 (5a and 6a), the modified T^2 (5b, 6b), and the modified MCuSum (7 and 8)

	Scenario				
Performance Indices	A	C	CA	D	DA
"Z" Algorithms					
1a – Shewhart	6.34	5.76	9.18	3.30	9.34
1b – CuSum	7.90	7.04	9.20	5.89	9.17
1c – Trend-adjusted CuSum	1.64	0.68	9.34	0.24	9.34
"C" Algorithms					
3a – Shewhart	8.03	4.78	9.08	0.67	8.03
3b – CuSum	2.00	0.16	0.94	0.49	2.00
3c – Trend-adjusted CuSum	8.22	5.97	9.08	0.62	8.22
"MV" Algorithms					
5a – Hotelling's T^2	6.71	3.27	8.71	0.87	9.03
5b – Modified T^2	7.40	4.30	8.22	1.87	9.03
7 – Modified MCuSum	8.34	5.99	9.07	5.68	9.37
"CMV" Algorithms					
6a – Hotelling's T^2	7.63	0.92	7.68	0.29	7.63
6b – Modified T^2	8.00	1.35	7.76	0.52	8.00
8 – Modified MCuSum	8.43	0.58	6.42	1.75	8.43

As displayed in Table 1, these results suggest that no one or two detection algorithms clearly dominate the others across all five of the scenarios tested. However, the two best are the simultaneous univariate CuSums and modified MCuSum algorithms (Z-1b and MV-7), which are the focus of Fig. 8. Each has a performance index in the 8 to 10 range for scenarios A, CA, and DA, but in the 5 to 7 range for scenarios C and D.

Pooling data across hospitals is a common way to analyze multiple data streams, the rationale being that the signal is more likely to emerge above the random variability. Our results, however, suggest that at least for the scenarios we used, algorithms operating on the pooling data (the C and CMV results) were less effective than those same algorithms operating on the unpooled data (Z and MV results).

Among the unpooled data for the simultaneous univariate algorithms (the Z results), the standard CuSum algorithm (1b) performs at least as well and usually better than the Shewhart algorithm (1a) and the trend-adjusted CuSum (1c). That the CuSum performs better than the Shewhart algorithm should be expected since the CuSum is better at detecting small changes and, in our scenarios, the outbreaks all begin with relatively small increases early on. However, in contrast, with the pooled data (C results) the standard CuSum (3b) performs substantially less well than the alternatives (3a and 3c).

Stoto et al. [SJF04] extend these results by investigating other detection algorithms and performance outside the flu season and perform various sensitivity analyses.

It should be noted that these results are potentially sensitive to many arbitrary choices that had to be made in the details of the detection algorithms tested and the design of the simulation. The performance of CuSum methods, for instance, depends on the choice of the parameter k, and may be better or worse for fast- or slow-growing outbreaks. The CuSum also depends on the estimated mean count $\hat{\mu}_0$ used as the baseline to calculate departures for each series. The trend-adjusted CuSum method depends on the weighting parameter λ in the exponentially weighted moving average.

In addition, we chose to set the false-alarm rate to 1% outside the flu season, which we arbitrarily defined as December 1–April 30; a different set of dates may have given different results. Our simulated outbreaks used seeds of the same size in every hospital, ignoring substantial variability in the background ER admission rates; again, a different and possibly more realistic choice might lead to different results. Finally we should note that the results also depend on the particular dataset used as the baseline for the simulation. The results are likely to apply to similar data in the future, but may be different for syndromic surveillance systems in cities other than Washington, DC.

These results show roughly similar performance for the simultaneous univariate CuSum and modified MCuSum algorithms, with one better than the other or both having similar performance characteristics depending on the scenario. In contrast, the abstract simulations in Sect. 2.4 show that the modified MCuSum has a clear advantage when the shift to be detected is "off-axes" and seems to show some performance improvements over the simultaneous univariate CuSum algorithm even when the shifts are on-axes. Whether these differences are the result of the simulation choices (jump change in the mean versus gradual increase, for example) or some other factor or factors remains to be determined.

However, some conclusions are clear:

- CuSum and CuSum-like algorithms are preferable to Shewhart and Shewhart-like algorithms for syndromic surveillance applications.
- For multivariate algorithms, appropriately modifying the algorithms to look only for increases in rates, such as we did in Sect. 2.3, provides additional detection power in syndromic surveillance applications.

- When designing, implementing, and comparing syndromic surveillance algorithms it is critical to ensure the appropriate thresholds are chosen to achieve a common aggregate false-alarm rate.
- While the CuSum algorithms generally performed better than the others we evaluated, unless the bioevent is so large so as to be obvious, a syndromic surveillance system will take some time to detect the incident — likely on the order of 2 to 5 days, depending on the size of the incident, for a system using data similar to what we have evaluated here.

4 Discussion

Out of concern about the possibility of bioterrorist attacks, many health departments throughout the United States and elsewhere are energetically developing and implementing a variety of syndromic surveillance systems. Our analyses suggest that while these systems may be valuable, their effectiveness for this purpose has not yet been demonstrated, and health departments ought to be cautious in investing in this area and take the time and effort to evaluate the performance of proposed systems in their own setting.

The central problem is that syndromic surveillance has been sold on the basis that it is able to detect outbreaks hours after people begin to develop symptoms, but our analyses suggest that unless the number of people affected is exceptionally large, it is likely to be a matter of days before enough cases accumulate to trigger detection algorithms. Of course, if the number of people coming to emergency departments is exceptionally large, sophisticated detection systems are simply not needed — the incident will be obvious. Further, the window (in terms of number of excess cases and time) between what is reasonably detectable with a syndromic surveillance system, and what is obvious, may be small.

Although an increasing number of statistically sophisticated detection algorithms have been developed, there is a limit to their efficacy. More generally, detection algorithms can be tuned to particular types of outbreaks (e.g., those that are geographically focused), but are only effective if the terrorists choose a matching method of exposing people. Moreover, as Stoto, Schonlau, and Mariano [SSM04], Reingold [Rei03], and others have pointed out, the value of an alarm system is limited by what happens when the alarm goes off. Simply knowing that there are an excess number of people with flulike symptoms is not enough, in itself, to initiate or guide a public health response.

Syndromic surveillance systems, however, can serve other public health purposes. The information technology that has been developed in many cities and states is truly impressive, and many health departments have worked hard to build relationships with hospitals and other entities in their communities to get access to data. The resulting systems and relationships would have additional value for detecting food-borne disease and other outbreaks. For many public health issues, for instance, knowing what is happening in a matter of days rather than weeks or months would indeed be a major advance for state and local health departments. During the cryptosporidium outbreak in Milwaukee in 1993, for instance, a syndromic surveillance system would have made health officials aware of the outbreak weeks/months before they actually were [MNG98].

Indeed, syndromic surveillance might prove to be most useful in determining the arrival of influenza in a community each year and in helping to determine whether pandemic flu has emerged. Nationally, influenza surveillance is based on a network of sentinel physicians who report weekly on the proportion of their patients with influenzalike symptoms, plus monitoring deaths attributed to influenza or pneumonia in 122 cities. Laboratory analysis to determine whether a case is truly the flu, or to identify the strain, is only rarely done [CDC04b]. Whether the flu has arrived in a particular state or local area, however, is largely a matter of case reports, which physicians often do not file. Pandemic influenza, in which an antigenic shift causes an outbreak that could be more contagious and/or more virulent, and to which few people are immune by virtue of previous exposure, is a growing concern [WW03]. Syndromic surveillance of flulike symptoms might trigger more laboratory analysis than is typically done and hasten the public health response.

References

[Alt85] Alt, F. B. 1985. "Multivariate quality control." In *Encyclopedia of statistical science*, edited by S. Kotz and N. L. Johnson, Volume 6. New York: John Wiley & Sons.

[BBM04] Buckeridge, D. L., H. Burkom, A. Moore, J. Pavlin, P. Cutchis, and W. Hogan. 2004. "Evaluation of syndromic surveillance systems — design of an epidemic simulation model." *Morbidity and Mortality Weekly Report* 53 (Supplement): 137–143.

[Bue04] Buehler, J. W. 2004. "Review of the 2003 National Syndromic Surveillance Conference – Lessons learned and questions to be answered." *Morbidity and Mortality Weekly Report* 53 (Supplement): 18–22.

[CDC04a] Centers for Disease Control and Prevention. 2004. "Framework for evaluating public health surveillance systems for early detection of outbreaks; recommendations from the CDC Working Group." *Morbidity and Mortality Weekly Report* 53 (RR-5): 1–13.

[CDC04b] Centers for Disease Control and Prevention. 2004. Overview of influenza surveillance in the United States fact sheet. http://www.cdc.gov/flu/weekly/pdf/flu-surveillance-overview.pdf accessed January 28, 2005.

[CF99] Chang, J. T., and R. D. Fricker, Jr. 1999. "Detecting when a monotonically increasing mean has crossed a threshold." *Journal of Quality Technology* 31:217–233.

[Cro88] Crosier, R. B. 1988. "Multivariate generalizations of cumulative sum quality control schemes." *Technometrics* 30:291–303.

[Hea87] Healy, J. D. 1987. "A note on multivariate CuSum procedures." *Technometrics* 29:409–412.

[Hen04] Henning, K. J. 2004. "What is syndromic surveillance?" *Morbidity and Mortality Weekly Report* 53 (Supplement): 7–11. Syndromic Surveillance: Reports from a National Conference, 2003.

[Hot47] Hotelling, H. 1947. "Multivariate quality control — Illustrated by the air testing of sample bombsights." In *Techniques of statistical analysis*, edited by C. Eisenhart, M. W. Hastay, and W. A. Wallis, 409–412. New York: McGraw-Hill.

[Lor71] Lorden, G. 1971. "Procedures for reacting to a change in distribution." *Annals of Mathematical Statistics* 42:1897–1908.

[Mon85] Montgomery, D. C. 1985. *Introduction to statistical quality control*, 2nd ed. New York: John Wiley & Sons.

[MNG98] Morris, R. D., E. N. Naumova, and J. K. Griffith. 1998. "Did Milwaukee experience waterborne cryptosporidiosis before the large documented outbreak in 1993?" *Epidemiology* 9:264–270.

[NM95] Nomikos, P., and J. F. MacGregor. 1995. "Multivariate SPC charts for monitoring batch processes." *Technometrics* 37:41–59.

[Pag54] Page, E. S. 1954. "Continuous inspection schemes." *Biometrika* 41:100–115.

[PR90] Pignatiello, J. J., Jr., and G. C. Runger. 1990. "Comparisons of multivariate CuSum Charts." *Journal of Quality Technology* 3:173–186.

[Rei03] Reingold, A. 2003. "If syndromic surveillance is the answer, what is the question?" *Biosecurity and Bioterrorism: Biodefense Strategy, Practice, and Science* 1 (2): 1–5.

[Rob59] Roberts, S. W. 1959. "Control chart tests based on geometric moving averages." *Technometrics* 1:239–250.

[Rob66] Roberts, S. W. 1966. "A comparison of some control chart procedures." *Technometrics* 8:411–430.

[She31] Shewhart, W. A. 1931. *Economic control of quality of manufactured product.* Princeton, NJ: D. van Nostrand Company.

[Shi63] Shiryayev, A. N. 1963. "On optimum methods in quickest detection problems." *Theory of Probability and its Applications* 8:22–46.

[Shi73] Shiryayev, A. N. 1973. *Statistical sequential analysis.* Providence, RI: American Mathematical Society.

[Sie04] Siegrist, D. 2004. "Evaluation of algorithms for outbreak detection using clinical data from five U. S. cities." Technical Report, DARPA Bio-ALERT Program.

[SP04] Siegrist, D., and J. Pavlin. 2004. "Bio-ALERT biosurveillance detection algorithm evaluation." *Morbidity and Mortality Weekly Report* 53 (Supplement): 152–157.

[Sta04] Stacey, D. 2004, November. Simulating pharmaceutical sales and disease outbreaks based on actual store sales and outbreak data. *2004 Syndromic Surveillance Conference, Boston, MA.*

[SJF04] Stoto, M. A., A. Jain, R. D. Fricker, Jr., J. O. Davies-Cole, S. C. Washington, G. Kidane, C. Glymph, G. Lum, and A. Adade. 2004, November. Multivariate methods for aberration detection: A simulation study using DC ER data. *2004 Syndromic Surveillance Conference, Boston, MA.*

[SSM04] Stoto, M. A., M. Schonlau, and L. T. Mariano. 2004. "Syndromic surveillance: Is it worth the effort?" *Chance* 17:19–24.

[WTE01] Wagner, M., F. C. Tsui, J. Espino, V. Dato, D. Sittig, R. Caruana, L. McGinnis, D. Deerfield, M. Druzdzel, and D. Fridsma. 2001. "The emerging science of very early detection of disease outbreaks." *Journal of Public Health Management and Practice* 7:50–58.

[WW03] Webby, R. J., and R. G. Webster. 2003. "Are we ready for pandemic influenza?" *Science* 302:1519–1522.

[WMC03] Wong, W. K., A. Moore, G. Cooper, and M. Wagner. 2003. "Bayesian network anomaly pattern detection for disease outbreaks." Edited by T. Fawcett and N. Mishra, *Proceedings of the Twentieth International Conference on Machine Learning.* Menlo Park, CA: AAAI Press, 808–815.

[WN85] Woodall, W. H., and M. M. Ncube. 1985. "Multivariate CuSum quality control procedures." *Technometrics* 27:285–292.

A Spatiotemporal Analysis of Syndromic Data for Biosurveillance

Laura Forsberg[1], Caroline Jeffery[2], Al Ozonoff[3] and Marcello Pagano[4]

[1] Department of Biostatistics, Harvard School of Public Health, lforsber@hsph.harvard.edu
[2] Department of Biostatistics, Harvard School of Public Health, cjeffery@hsph.harvard.edu
[3] Department of Biostatistics, Boston University School of Public Health, aozonoff@hsph.harvard.edu
[4] Department of Biostatistics, Harvard School of Public Health, pagano@hsph.harvard.edu

1 Introduction

The word *surveillance* is a conjunction of the French root *sur* and the Latin *vigilare* to yield the meaning: to watch over. It is easier to terrorize the ignorant than the well-informed, so vigilance, and in particular active surveillance, is an important component of counterterrorism. This chapter shows how we can apply particular statistical methods to the task of biosurveillance to improve our ability to withstand bioterrorism.

History has taught us that biological weapons have been used throughout time and across the world. For example, as early as the sixth century B.C., the Assyrians poisoned the wells of their enemies with rye ergot, and Solon of Athens during the siege of Krissa poisoned the water supply with the purgative hellebore. But sometimes the effects of the biological weapons were much more widespread: during the siege of Caffa in 1347 the Tartar forces of Kipchak khan Janibeg, after laying siege to the Genoese city for a year, were decimated by the plague (*Pasteurella pestis*) that had infected northern China. As a parting gift, Janibeg catapulted infested corpses into the city. The Italians were of course infected, but hoping to escape, four Genoese ships sailed for home and brought the plague to Europe. It is unknown how many individuals were killed by the plague, but some estimate that as many as 40% of the total population were killed, with some areas affected even more severely.

These modes of attack persisted through the ages. Even as recently as World War II, the Japanese Unit 732 under the direction of Shiro Ishii spread the bubonic plague in northern China [Har02], so it does not seem that we have seen the last of this kind of warfare. Indeed, as recently as 1984 (and as close to home as Oregon) members of the Rajneeshee, an Indian religious cult, contaminated salad bars in The Dalles with *Salmonella typhimurium* in an attempt to keep local citizens from voting. This resulted in over 750 people being poisoned and 40 being hospitalized, but a disturbing aspect of this attack is that the public health system in place at the time did not detect that a bioterrorist attack had taken place. It was over one year after the poisoning, when some cult members bragged about their actions, that it came to light that the poisoning had been intentional.

Undoubtedly it is best to prevent bioterrorist attacks, but if that is not possible, then it is beneficial to detect them as soon as possible, in the hope that this would allow the early introduction of remedial action. That the current state of public health alertness is not optimal was proven further by the cryptosporidium infection of the water supply in Milwaukee in 1993. One of the two water treatment plants was contaminated, but this was not recognized until at least two weeks after the onset of the outbreak [MHP94]. As a result, it is estimated that more than 400,000 people were afflicted, including over 100 deaths. It is of some comfort that this was not due to a bioterrorist attack, but troubling that it took so long to admit to such a major change: according to the *Morbidity and Mortality Weekly Report* published by the Centers for Disease Control and Prevention (CDC), "the outbreak ... in Milwaukee was the largest documented waterborne disease outbreak in the United States since record keeping began in 1920." As the CDC further notes, the fact that "the extent to which waterborne disease outbreaks are unrecognized and underreported is unknown" is disturbing.

The detection and reporting of outbreaks can be placed in the classical statistical hypothesis testing framework: the null hypothesis is that normalcy reigns, and the alternative is that there is an outbreak. The two types of possible errors are not reporting an outbreak on the one hand and raising a false alarm on the other. Assuming daily (or perhaps more frequent) observations, since the process is ongoing we must also consider the timeliness of detection. The data available for analysis depends on the situation; we concentrate here on syndromic surveillance.

Syndromic surveillance, as the term is normally used, refers to observing patients with particular syndromes that are chosen because of their association with the early symptoms of certain diseases. It may also include several nontraditional data sources, such as nurse hotline call volume or over-the-counter (OTC) pharmaceutical sales; however, we restrict our attention here to the automated collection of emergent disease cases, grouped into syndromes. Examples of such syndromes include influenzalike illness (ILI), upper respiratory infections (URIs), or gastrointestinal (GI) syndrome. Syndromic surveillance is attractive in its potential for early detection of disease outbreaks, since syndromes precede definitive diagnosis of disease, but the method suffers in its lack of specificity for any particular disease.

Traditional methods typically consider the number of patients, but as the outbreak of anthrax at Sverdlovsk [MGH94] proves, knowing also where the patients were located when they were afflicted is a very powerful piece of evidence. Another example is the Milwaukee outbreak mentioned above. Milwaukee at the time had two water plants, one servicing predominantly the northern part of the city and the other the southern part of the city. During a particular time period, the excessive number of those patients reporting to the city's hospitals with gastrointestinal complaints were predominantly from the southern part of the city, the part serviced by the polluted water plant. Thus we argue that knowing where the patients originate may prove to be useful information (we have shown [OFB04] that there is an increase in the power of detection if we not only consider the temporal aspect of the series of patients reporting to hospitals, but also consider their spatial distribution). So we demonstrate in the remainder of this chapter how to combine the two streams of information: the number of patients and their addresses. As noted in a recent review [BMS04], "the routine application of space-time analytic methods may detect aberrations in bioterrorism surveillance data with greater sensitivity, specificity, and timeliness," an idea that is supported by previous work [OFB04, OBP05].

2 Temporal Surveillance

Temporal surveillance of syndromic data depends first on establishing a baseline pattern of behavior, then detecting departures from this baseline (usually an excess in the number of patients). For example, on a daily basis let $N_i, i = 1, \ldots, T$, denote the time series of the number of patients registered with a particular set of syndromes over a particular time period of length T. Figure 1 is a plot of four years of respiratory syndrome data from Cape Cod, Massachusetts. These data represent all patients reporting to three local hospitals and who complain of upper respiratory syndromes. Syndrome assignments were made on the basis of ICD-9 codes, according to the commonly used ESSENCE-II classifications [LBP04]. To make the graph more comprehensible, we aggregate the number of patients over an unweighted seven-day moving average. This has the further advantage of eliminating a day-of-week effect, and we use this as our basic time series. Although we lose some precision in not modeling at the daily level (for example we do not account for public holidays, a predictable deviation from the behavior of other days), we have chosen this path for the sake of clarity of presentation. It is important to note that aggregation or moving averages ordinarily result in a loss of timeliness. For this reason, a more sophisticated modeling effort would be required for a sensitive and timely system. However, for the modeling aims that we choose to demonstrate here, the moving average will suffice and allows for a more clear illustration of techniques.

There is a clear seasonal pattern that is typical of respiratory disease, with a pronounced spike in the "flu season" months of December, January, and February. The timing and magnitude of this spike change from year to year, posing a challenge for establishing a seasonal baseline. This may result

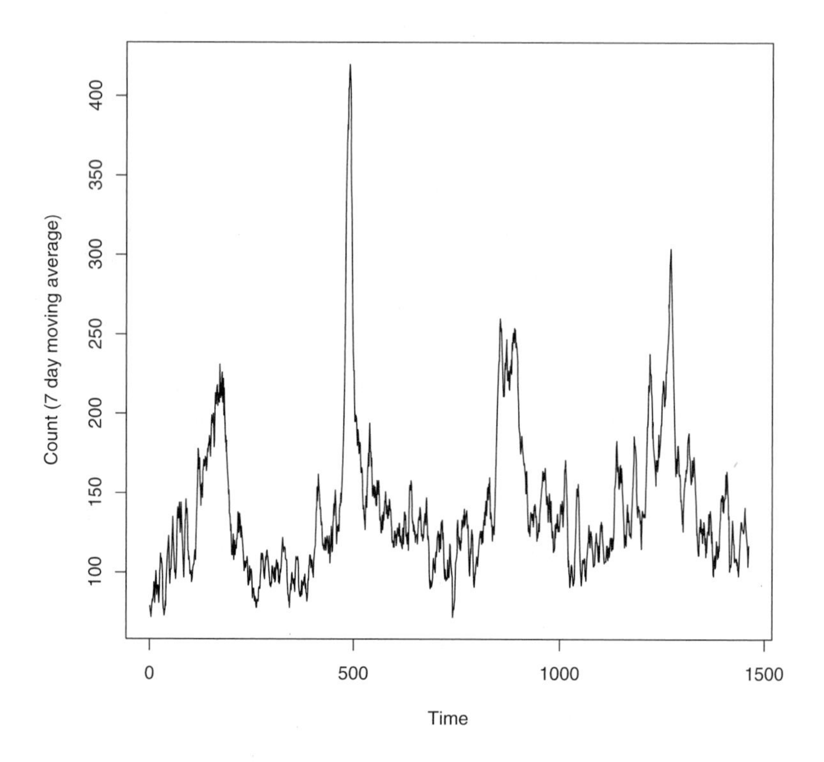

Fig. 1. Four years of syndromic data from Cape Cod, MA.

in poor model fit for years that are atypical. Moreover, the choice of historical data used to estimate baseline may influence estimates of model parameters. Despite research into stochastic models that are flexible enough to accommodate these challenges [Mad05, RCS03], the standard approach used by CDC and others [BBF05, RM03, HTS03] is similar to what we describe here.

A typical surveillance approach attempts to model the process, then performs a daily prediction of what should be expected to compare with what is observed. The traditional approach (used by CDC for influenza surveillance) is Serfling's method [Ser63]. This is an ordinary regression of the form

$$N_t = \mu + \alpha t + \beta \cos\left(\frac{2\pi t}{365} + \theta\right) + \varepsilon_t.$$

Although this method is adequate for capturing the seasonal component of baseline, it is not suitable for making predictions because it does not take into account the autocorrelation of the individual observations. Moreover, by ignoring the pronounced spikes in the winter months, this method is prone to sounding an alarm at the beginning of every flu season. While this may serve its purpose for detecting the onset of an epidemic of influenza, it is deficient if the goal is to detect an additional signal (such as a bioterror attack) superimposed on the spikes that commonly occur in the winter months.

To improve upon Serfling's method, and to simulate a genuine prospective surveillance system, we use the first year of data to fit a simple seasonal model. This model consists of a sinusoidal term of period one year, reflecting the annual cycle of respiratory disease. An additional three indicator terms for the months of December, January, and February are also included to account for the increased volume during influenza season. There are several cautionary points when following such a regression modeling approach.

Our model is

$$N_t = \mu + \delta_t + \sin\left(\frac{2\pi t}{12}\right) + \eta_t,$$

where

$$\delta_t = \begin{cases} \mu_{jan}, & \text{if January} \\ \mu_{feb}, & \text{if February} \\ \mu_{dec}, & \text{if December} \\ 0, & \text{otherwise.} \end{cases}$$

These terms were fit to the first year's data using a standard linear regression model. The residuals from this model are autocorrelated, and so we fit an autoregressive (AR) model [BD02] to the residuals. Here we consider models of the form

$$\eta_t = \alpha_1 \eta_{t-1} + \cdots + \alpha_k \eta_{t-p} + \varepsilon_t,$$

where η_t is the seasonal residual (after subtracting the seasonal mean as fitted above from the observed data). Note that we omit the intercept from the model here since we work under the assumption that the departures from seasonal baseline are mean zero. Importantly, the remaining residuals ε_t form a new time series whose elements are not correlated. Parameter estimation as well as order selection is easily handled in most statistical packages.

Fitting this model to the first year's data, we simulated prospective temporal surveillance by extending the seasonal component forward in time and applying the AR model to the observed residuals to obtain one-step-ahead prediction for each day. The residuals from these predictors form the basis for a test statistic that detects departures from baseline. Figure 2 shows the remaining three years of residuals. In practice, model parameters would be estimated more frequently than once every four years; however, for the purposes of illustration we have restricted attention to one round of model-fitting.

Having created residuals that are independent over time, we can now use known techniques to monitor the series. For example, the CuSum control process is intended for independent data; this approach has been followed successfully in a syndromic surveillance setting [MKD04], and weighting modifications along these lines has also been suggested [RPM03]. But one should not stop here, as there is still other information potentially available in the

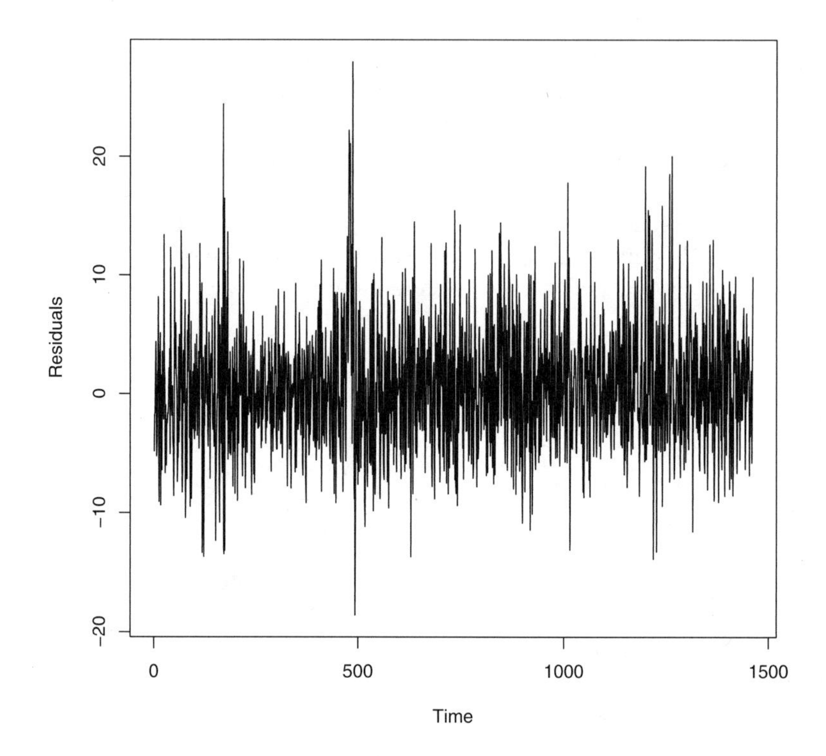

Fig. 2. Residuals after removing seasonal and AR components.

data. For example, if patients' addresses are available, we may wish to exploit that information. The following section considers one such statistic designed to study the spatial distribution of the data.

3 The *M*-Statistic

In conjunction with monitoring the temporal series, one should expect to gain greater power to detect outbreaks or abnormal patterns in disease incidence by also investigating the spatial distribution of the locations of the cases. Several statistics have been proposed for the detection of spatial anomalies (for a review, see Brookmeyer and Stroup [BS04] and Kulldorff [Kul98]). Many of these statistics were designed for retrospective surveillance to detect spatial clusters of diseases associated with long incubation periods, usually cancers [AB96]. As described in the introduction, recent events have led to a growing interest in methods of spatial surveillance that can be performed in real time to detect outbreaks and unusual patterns in diseases with a short incubation period. In what follows we describe the M-statistic, which is useful and effective for implementation in the detection of spatial anomalies and that works well in conjunction with the temporal monitoring described in the previous section.

Let $\mathbf{X} = \{X_1, \ldots, X_n\}$ be the locations of n cases on the plane and let $\mathbf{d} = \{d_1, \ldots, d_{\binom{n}{2}}\}$ be the $\binom{n}{2}$ interpoint distances, or the distances between pairs of points in $\mathbf{X}$. Experience has shown that the distribution of these distances between hospital patients is remarkably stable over time [OBP05]. This would suggest that we can empirically construct an expected, or null, distribution of the distances and expect that it would be stable through time for a given space. Therefore, to detect spatial anomalies, one might consider some measure of the difference between the observed distribution of interpoint distances for any $\mathbf{X}$ and the expected distribution of interpoint distances. There are many metrics that could be used for such a comparison. Several of these were explored by Bonetti and Pagano [BP05], leading to the adoption of one of these methods, the M-statistic.

The M-statistic calculates deviations of the empirical cumulative distribution function (ecdf) of the distances from the expected cdf via a Mahalanobis type distance. This approach has several appealing features that will now be described. One such feature is that we can rely upon the theory of U processes in the development of this statistic. Consider the ecdf given by

$$\begin{aligned} F_n(d) &= \frac{1}{n^2} \sum_{i=1}^{n} \sum_{j=1}^{n} 1(d(X_i, X_j) \leq d) \\ &= \left(\frac{n-1}{n}\right) \binom{n}{2}^{-1} \sum_{i<j} 1(d(X_i, X_j) \leq d) + \frac{1}{n} \\ &= \left(\frac{n-1}{n}\right) U_n(d) + \frac{1}{n}, \end{aligned}$$

where $U_n(d) = \binom{n}{2}^{-1} \sum_{i<j} 1(d(X_i, X_j) \leq d)$ is a U statistic of order two, with symmetric kernel $\psi(i, j; d) = 1(d(X_i, X_j) \leq d)$ [Hoe48, Lee90]. One approach to measuring the difference between the cdfs is to discretize the distribution of the distances such that the ecdf and expected cdf can be represented by $k \times 1$ vectors of cumulative probabilities such that the kth entry is 1. A statistic can be formulated as

$$M^* = (F_n(\mathbf{b}) - F(\mathbf{b}))^\top \Sigma_F^- (F_n(\mathbf{b}) - F(\mathbf{b})),$$

where Σ_F^- is the generalized inverse of the rank $k-1$ variance covariance matrix of $F_n(\cdot)$ and $\mathbf{b}$ is the $(k-1) \times 1$ vector of the bin cutoffs used to discretize the cdfs. In Bonetti and Pagano [BP05], the authors prove that as

n goes to infinity, $F_n(\cdot)$ converges weakly to a multivariate normal random variable with zero mean vector and variance covariance matrix

$$\Sigma_F = \{\sigma_{a,b}\},\ \sigma_{a,b} = \text{Cov}[1(d(X_1, X_2) \le d_a, d(X_1, X_3) \le d_b)],\ a, b = 1, \ldots, k.$$

Then,

$$\sigma_{a,b} = E\{1(d(X_1, X_2) \le d_a, d(X_1, X_3) \le d_b)\} \\ -E\{1(d(X_1, X_2) \le d_a)\}\, E\{1(d(X_1, X_3) \le d_b)\}.$$

One can also consider using the successive differences of the cdf to perform the same test. Therefore, in place of $F_n(\cdot)$, we consider
$\mathbf{f}_n = (F_n(b_1), F_n(b_2) - F_n(b_1), \ldots, F_n(b_{k-1}) - F_n(b_{k-2}), 1 - F_n(b_{k-1}))$.
Then, we can consider the test statistic

$$\tilde{M} = (\mathbf{f}_n - \mathbf{f})^\top \Sigma^- (\mathbf{f}_n - \mathbf{f}),$$

where Σ^- is the generalized inverse of the variance covariance matrix of $\mathbf{f}_n$. As $\mathbf{f}_n$ is a linear transform of $F_n(\cdot)$, $\mathbf{f}_n$ is also normal with zero mean vector and variance covariance matrix that is the appropriate transform of Σ_F. Additionally, $\tilde{M}$ converges weakly to a chi-square random variable with $k-1$ degrees of freedom (the rank of Σ) [BP05]. Implementation of this statistic requires estimation of Σ, as we are usually unable to specify its exact form. We describe two methods of estimation. The first method relies on the theory of U statistics. Accordingly, the covariance matrix of $F_n(\mathbf{b})$ can be estimated consistently by

$$\widehat{\sigma}_{a,b} = 4\left\{ \frac{1}{\binom{n}{3}} \sum_{1 \le i < j < k \le n} h(\mathbf{X}_i, \mathbf{X}_j, \mathbf{X}_k; q_a, q_b) \right. \\ \left. - \left[\frac{1}{\binom{n}{2}} \sum_{1 \le i < j \le n} 1(d(\mathbf{X}_i, \mathbf{X}_j) \le q_a) \right] \left[\frac{1}{\binom{n}{2}} \sum_{1 \le i < j \le n} 1(d(\mathbf{X}_i, \mathbf{X}_j) \le q_b) \right] \right\},$$

where

$$h(\mathbf{X}_i, \mathbf{X}_j, \mathbf{X}_k; q_a, q_b) = 6^{-1} \sum_{\varrho} \left[1(d(\mathbf{X}_{\varrho_1}, \mathbf{X}_{\varrho_2}) \le q_a, d(\mathbf{X}_{\varrho_1}, \mathbf{X}_{\varrho_3}) \le q_b)\right],$$

is the symmetrized kernel computed over the collection $\varrho = \{(\varrho_1, \varrho_2, \varrho_3)\}$ of the six permutations of the indices (i, j, k) (see Bonetti and Pagano [BP05]). In the calculation of this estimator, for efficiency the triple sum should be implemented as a single loop by making use of (fast) matrix multiplications for the inner sums. An example of this appears in Forsberg et al. [FBJ05]. To estimate the covariance matrix of $f_n(\mathbf{b})$, an appropriate linear transform can be applied to $\hat{\Sigma}_F$.

The second estimator, which we use in the examples below, is calculated by resampling the historical data. This will be described in greater detail below.

In practice we estimate $f_n(\mathbf{b})$ by the observed counts in each bin and calculate

$$M = (\mathbf{o} - \mathbf{e})^T S^- (\mathbf{o} - \mathbf{e}),$$

where $\mathbf{o}$ and $\mathbf{e}$ are the $k \times 1$ vectors of observed and expected counts and S^- is the generalized inverse of the estimate of the $k \times k$ variance covariance matrix of $\mathbf{o}$. We use resampling methods to implement this statistic, as the asymptotic results described above converge slowly. These resampling methods will now be described.

First, it is necessary to determine a method of discretizing the interpoint distance distribution, so as to represent it by a vector of cell counts. We prefer to bin the data into equiprobable bins, to avoid sparse cell counts and to increase the power [OBF05]. Power is maximized by choosing k to be approximately 50% to 75% of n [FBP05]. The cutoff points for the bins are selected by sampling from historical data to obtain an estimate of the null distribution of the distances. The cutoff points, or bin breaks, are determined as the $\{\frac{1}{k}, \ldots, \frac{k-1}{k}\}$ quantiles of this distribution.

Next, we obtain an estimate of the variance covariance matrix of the cell counts, S. Again this is obtained by resampling m times from historical data and computing

$$S = \frac{1}{m} \sum_{i}^{m} (\mathbf{o}_i - \mathbf{e})(\mathbf{o}_i - \mathbf{e})^T,$$

where $\mathbf{o}_i$ is the vector of observed cell counts from the ith sample and $\mathbf{e}$ are the expected cell counts ($e_j = \binom{n}{2}/k, j = 1, \ldots, k$) for the k cells.

Once the S matrix has been estimated, it is possible to get an estimate of the null distribution of the M-statistic via resampling. Again we resample from historical data and obtain the null distribution of M. From this we can extract a $(1 - \alpha)100\%$ critical value for M.

These calculations can be computationally intensive and care must be taken in the number of samples to take in calculating the bin breaks, S and the null distribution of M, as well as the number of bins, k. However, when done correctly [BFO03, OFB04], we obtain all the needed information for the computation of a statistic that has been shown to be powerful for detecting spatial anomalies [OBF05]. This statistic has many features that make it particularly useful for syndromic surveillance. For instance, this statistic is capable of detecting a wide array of spatial anomalies that would create an aberration in the interpoint distance distribution. Further, the M-statistic is easily adapted to the data source on hand, for example exact address locations, census tracts, zip codes, or any other form of location data. An additional strength of the M-statistic is the ability to combine it with N, the number of cases, and obtain a bivariate statistic that simultaneously considers temporal and spatial behavior, as we next show.

4 Bivariate Spatiotemporal Surveillance

An appealing theoretical aspect of the M-statistic is its asymptotic independence from the number of cases observed. Thus we would expect that a temporal test statistic would remain approximately independent from the values of the spatial test statistic M. To verify this result when applied to real data, we examined the joint distribution of M and the residuals obtained from our temporal model (Sect. 2). Figure 3 is a scatterplot of the joint test statistics for N and M, showing approximate independence (correlation $\varrho = 0.047$, 95% CI (-0.004,0.098)).

A bivariate test statistic for the joint test statistics can be formulated in a number of ways. To use the information available, i.e., both the number of patients and their locations, we need a bivariate test statistic. We are motivated by the asymptotic independence of M and N, but we need to know the particular alternative hypotheses against which we wish to have power. Here we choose a compromise hypothesis that will guard against a shift in both the number of patients and an increase in M. This seems to work well for a number of alternative hypotheses.

The joint values of N and M provide us with additional information regarding the nature of a disease outbreak. Referring to Fig. 1, we see that the flu season in year 2 was especially large in magnitude. The case volume far exceeded predictions and thus we can consider this an outbreak of respiratory disease in Cape Cod. The corresponding values for M, however, do not indicate any deviation from spatial baseline. Thus we learn that the outbreak is pandemic to the region and not specific to a particular location within Cape Cod.

Conversely, there is a period of roughly three weeks (days 709–728) where the case volume is essentially as predicted. However, high values of M on those days indicate a localized outbreak that is not sufficient in extent to qualify as a temporal aberration. This type of outbreak might be a candidate for investigation by local public health authorities. It is also worth noting that a local release of a biological agent (e.g., anthrax, smallpox) might exhibit a similar spatial pattern. Using distance-based methods of mapping (Sect. 5), Fig. 7 illustrates the local nature of this outbreak. The dynamic nature of spatiotemporal patterns of disease is more dramatically illustrated when viewed sequentially (e.g., as images projected together to form a movie). Although it is not possible to display animated images here, there is more to learn when considering these patterns dynamically (see Ozonoff et al. [OBF04] for further discussion).

5 Locating Clusters

Having used the methods described above, suppose that in a particular instance we decide that the null hypothesis of normalcy should be rejected. The

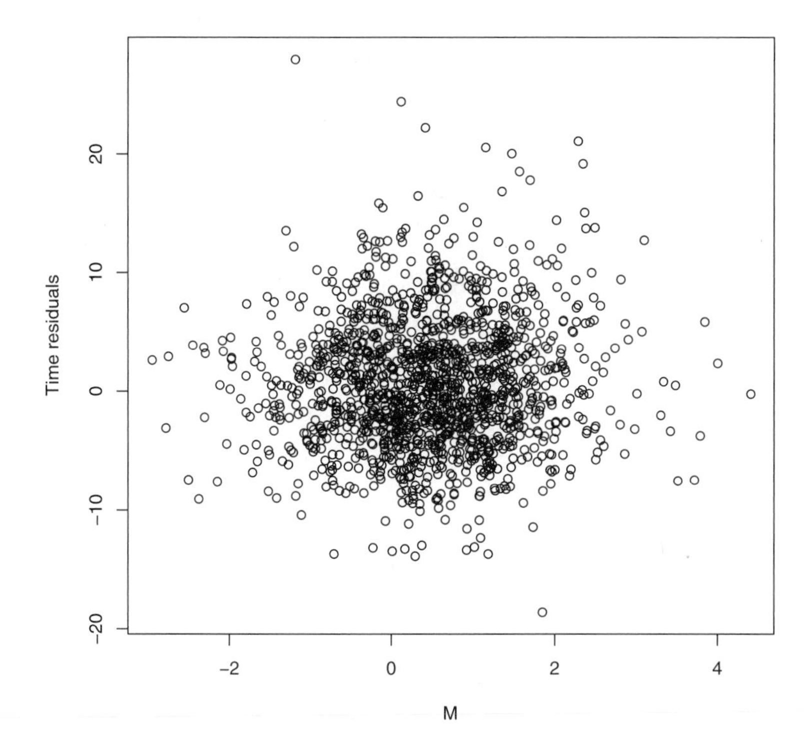

Fig. 3. Plot of spatial statistic (M) versus temporal residuals.

next step then is to study the available spatial data to determine whether the deviation from the null hypothesis is due to an exogenous cluster(s), or not, and if it is, to locate and describe the cluster(s). Here we outline such a detection method that is based on using the distribution of the distances between points, in keeping with the methods in the rest of this chapter.

We concentrate on the plane, but these methods extend directly to higher-dimensional spaces [GP05]. Consider a closed region R that contains the observed points. Construct C a circle surrounding R. For $N > 1$, divide the circumference of C into N equal arcs described by the points on the circumference: $(c_1, c_2, ..., c_N)$. We label these as observing points because from each c_i we observe the distribution of the distances to points in R and compare these with what we expect to see under the null distribution. We then aggregate these results over the N points to identify possible clusters in R.

For illustration, suppose that on a given day we observe n cases in R, located at points $(x_1, x_2, ..., x_n)$, and assume, for the sake of discussion, that all n observed locations are different (continuous data). We are interested in knowing whether $(x_1, x_2, ..., x_n)$ is a random sample from the null distribution of points in R. So choose an observing point c_i, $i \in \{1, ..., N\}$, and an observation x_j, $j \in \{1, ..., n\}$, and let d_{ij} be the Euclidean distance between c_i and x_j. Suppose the distribution of these distances under the null hypothesis is $F_i(d)$ with associated density function $f_i(d)$. Define the associated estimator $\widehat{F}_i(d) = \frac{1}{n}\sum_{k=1}^{n} I(d_{ik} \leq d)$, where $I(.)$ is the Heaviside function $I(x) = 0$ if $x < 0$ and $= 1$ if $x \geq 0$.

The $F_i(d)$ are either given or we can estimate them to any degree of accuracy via resampling or Monte Carlo methods applied to the null distribution of distances between c_i and points in R, for each i.

For a given distance d_{ij} and a fixed constant $0 < k < 1$, and letting A be the subset of R intersecting with the annulus bounded by radii $(d_{ij} - \frac{h}{2})$ and $(d_{ij} + \frac{h}{2})$, we next find h as the smallest value such that

$$\int_A f_i \mathrm{d}\mu = \frac{1}{k},$$

where $\mathrm{d}\mu$ is the ordinary Lebesgue measure on the Euclidean plane. The quantity $1/k$ represents the proportion of points in A under the null. To estimate it from the observed $(x_1, \ldots, x_n)$, we look at the proportion of observations at a distance from c_i within h of d_{ij} (see Fig. 4):

$$\widehat{F}_i(d_{ij}+h) - \widehat{F}_i(d_{ij}-h) = \frac{1}{n}\sum_{k=1}^{n} I(x_k \in A) = \frac{1}{n}\sum_{k=1}^{n} I(d_{ij}-h < d_{ik} \leq d_{ij}+h).$$

We can then calculate the deviation from the null by defining

$$\gamma_{ij} = \sqrt{n}\left(\frac{1}{n}\sum_{l=1}^{n} I(d_{ij} - h < d_{il} \leq d_{ij} + h) - \frac{1}{k}\right).$$

Notice that by decomposing γ_{ij} into $\sqrt{n}(\widehat{F}_i(d_{ij} + h) - F_i(d_{ij} + h)) - \sqrt{n}(\widehat{F}_i(d_{ij} - h) - F_i(d_{ij} - h))$ we see that for a fixed d_{ij}, $\gamma_{ij} \sim N(0, \sigma^2)$ as $n \to \infty$, assuming that the observed points are distributed according to f.

Hence γ_{ij} incorporates whether there are too many or too few observations in the region of R covered by the strip. Of course, this will not tell us alone whether x_j is part of a cluster, yet we can aggregate similar information when looking from other observing points. For that, define $\Gamma : \mathbb{R} \to \mathbb{R}$ by

$$\Gamma(x_j) = \frac{1}{N}\sum_{i=1}^{N} \gamma_{ij}.$$

In Fig. 5, we look at x_j from five different observing points. Each strip covers a fraction of the density of distances from the considered observing point under the null ($1/k$ in each case). The strips overlap more and more as they get closer to x_j, hence the estimation of the deviation from the null will be more

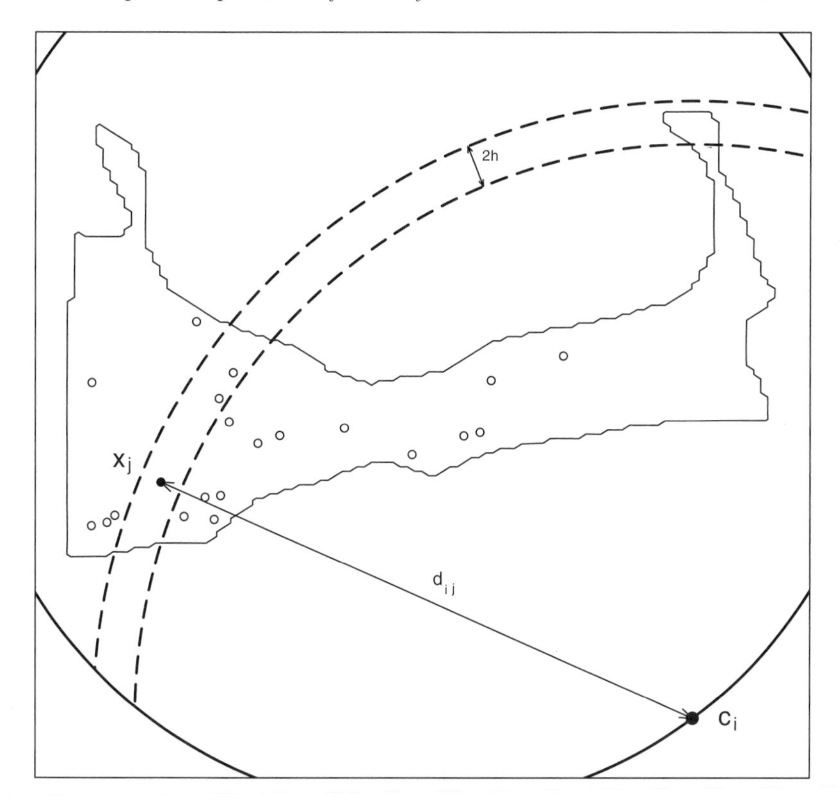

Fig. 4. Estimating deviation from null.

precise as we increase the number of points on C. The same process can be repeated for all other observations and in fact Γ can be estimated at all points in R.

This method was applied to respiratory syndrome data from Cape Cod, Massachusetts. In particular we consider the large outbreak during the flu season of year 2. As mentioned in Sect. 4, the case volume exceeds by far the expected, but the M-statistic detects no spatial aberration. Figure 6 gives a representation of the estimation of Γ during that period. We use 40 observing points circumscribing the Cape Cod region. In the figure, triangles correspond to case locations for the time period studied. The void region in the western part of Cape Cod corresponds to Otis Air Force Base, where military personnel receive care within the Air Force system and thus no patients are represented from the base in this dataset. The values of Γ throughout the Cape are displayed with different levels of gray, with darker shades corresponding to higher values. Confirming the results from the bivariate analysis above, we can see that there is no significant deviation from the null spatially,

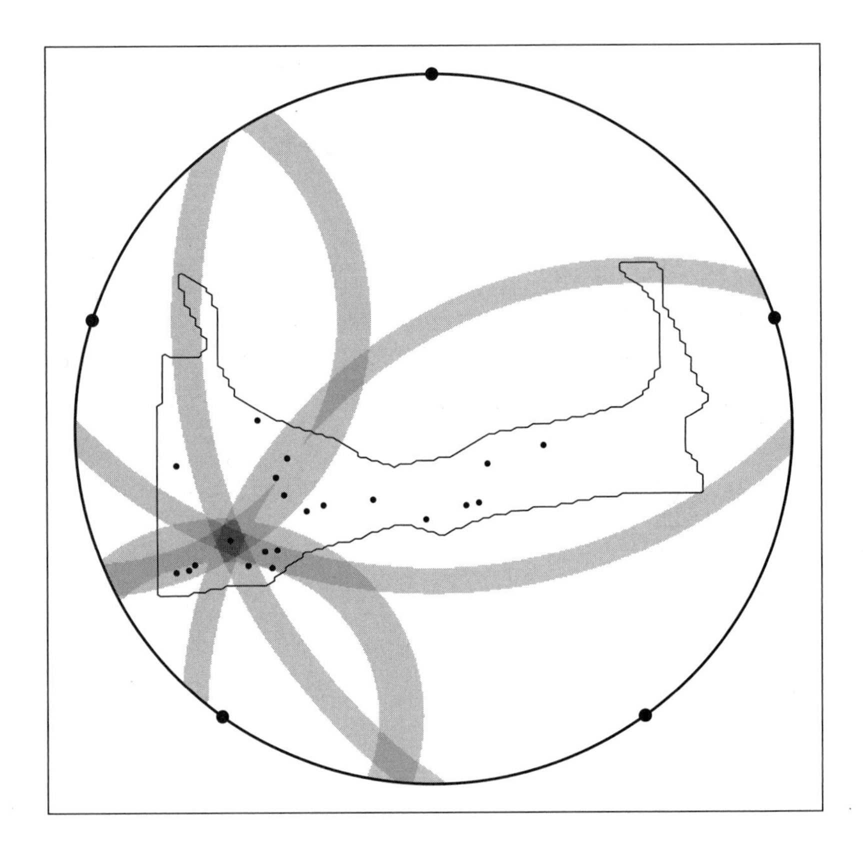

Fig. 5. Aggregating over all observing points.

hence the increase in the report of flu cases was distributed evenly throughout the Cape according to the null.

Conversely, during the three-week period between days 709 and 728, the M-statistic indicates a localized outbreak, which was not detected as a temporal aberration. In Fig. 7 we see that Γ takes higher values in the eastern portion of Cape Cod. The region delimited by the black solid contour indicates a significant spatial aberration.

Coming back to the method presented, and letting some of the parameters approach infinity, a few observations can be made. In Fig. 5 we can see that the five strips do not cover the whole region R. Part of the observed points are not taken into account when calculating $\Gamma(x_j)$. We can increase N, the number of fixed points, and also increase k so that all the strips intersect only around x_j while remaining large enough to include all the observed data. In fact we can rewrite $\Gamma(x_j)$ as follows: first we assume that k is large enough so that the strips are close to being arcs (that can be achieved if we suppose that we can sample as much or as little as possible from the null, i.e., the null distribution

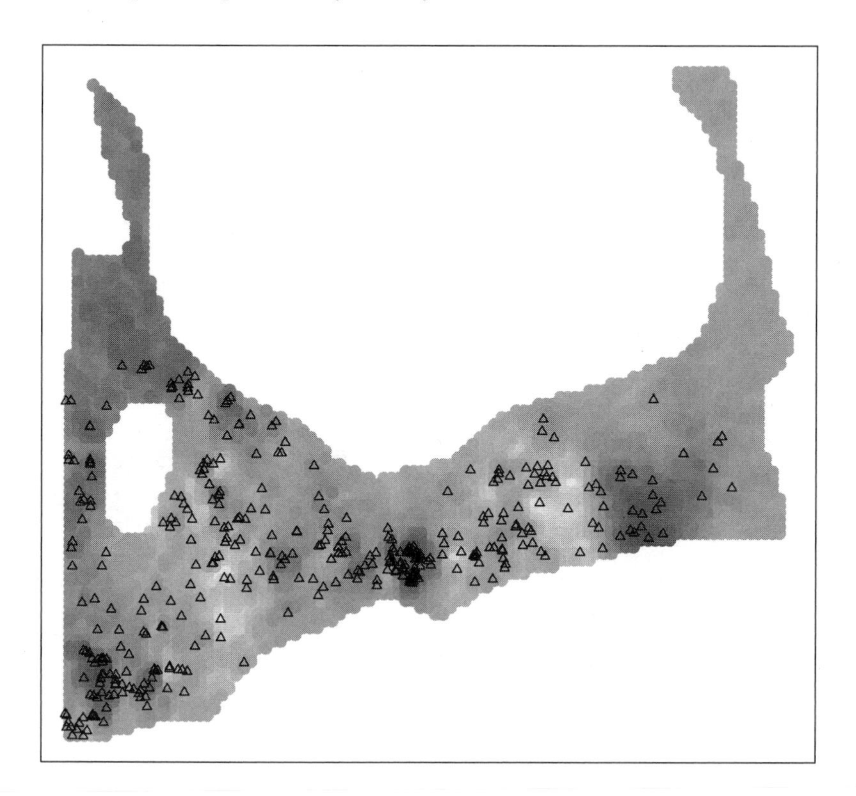

Fig. 6. Regionwide outbreak of respiratory illness on Cape Cod during year 2 of the study period. This outbreak displays no significant spatial deviations.

is continuous in R); also we assume N large enough so that each observation $x_{j'}$ other than x_j is contained in exactly two strips (both corresponding to the two observing points intersecting C and the right bisector of $[x_j, x_{j'}]$). Letting ν_{ij} be the number of observations in strip i corresponding to d_{ij}, we can rewrite $\Gamma(x_j)$ as follows:

$$\Gamma(x_j) = \frac{1}{N}\sum_{i=1}^{N}\sqrt{n}\left(\frac{\nu_{ij}}{n} - \frac{1}{k}\right) = \frac{\sqrt{n}}{Nn}\left(\sum_{i=1}^{N}\nu_{ij}\right) - \frac{\sqrt{n}}{k}.$$

Now given that all the strips only intersect at x_j, the ν_{ij} will sum to $N + 2(n-1)$. Hence

$$\Gamma(x_j) = \frac{\sqrt{n}}{Nn}(N + 2n - 2) - \frac{\sqrt{n}}{k}.$$

As mentioned earlier we can also estimate Γ for every point in R, and so for $x \in R \backslash \{x_1, ..., x_n\}$,

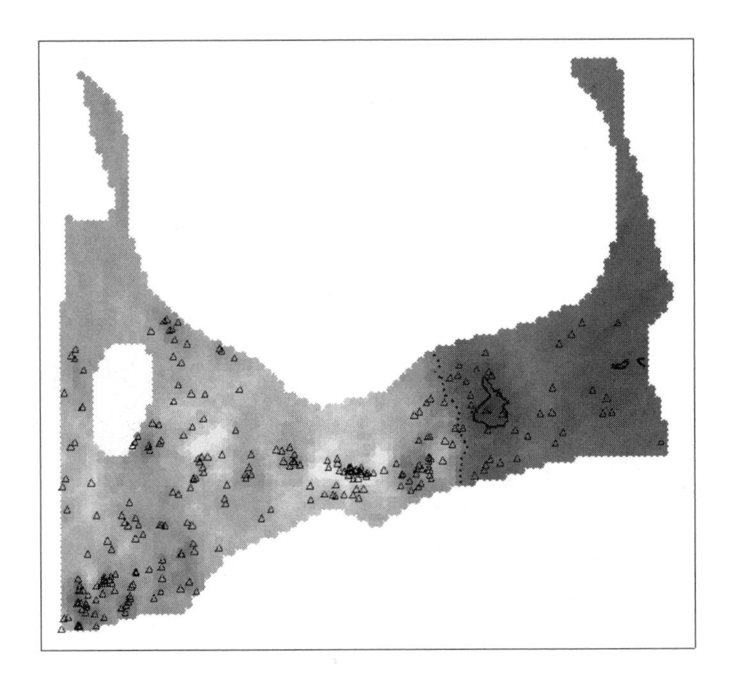

Fig. 7. A localized outbreak of respiratory illness (days 709–728).

$$\Gamma(x) = \frac{\sqrt{n}}{Nn}(N + 2n) - \frac{\sqrt{n}}{k}.$$

Letting $k \to \infty$ and $N \to \infty$, for all $x \in R$ we have $\Gamma(x) \to \frac{1}{\sqrt{n}}$. Note that as $k \to \infty$ and $N \to \infty$, the strips become arcs and there are infinitely many of them. They all intersect at x and cover the rest of the region twice.

6 Conclusion

We have shown in this chapter not only how to look at the number of patients entering into the surveillance system, but also to consider from whence they came. The statistic used to measure the goodness-of-fit of the spatial distribution we consider is the *M*-statistic. This statistic has the advantage that its distribution, for large numbers of patients, is approximately independent of the number of patients, so it affords us a two-dimensional statistic (one spatial and one temporal) to simultaneously evaluate departures from normalcy.

The M-statistic is based on the distances between patients. We use the fact that in practice this distance distribution is stationary and deviations from this stationarity would indicate a disturbance that might prove worth investigating.

As research into distance-based methods and other methods of spatiotemporal methods continues, visualization and disease mapping will become increasingly important since it provides public health officials and decision-makers additional information, possibly on a real-time or near real-time basis. Our approach to disease mapping, a logical outgrowth of our work on distance-based methods, can be seen as one approach to this important problem.

Acknowledgments

Research supported in part by grants from the National Institutes of Health RO1-AI28076, T32-AI07358 and National Library of Medicine RO1-LM007677.

References

[AB96] Alexander, F. E., and P. Boyle, eds. 1996. *Methods for investigating localized clustering of disease.* Lyon, France: International Agency for Research on Cancer.

[BFO03] Bonetti, M., L. Forsberg, A. Ozonoff, and M. Pagano. 2003. "The distribution of interpoint distances." In *Bioterrorism: Mathematical modeling applications in homeland security*, edited by H. T. Banks and C. Castillo-Chavez, 89–109. Philadelphia, PA: SIAM.

[BP05] Bonetti, M., and M. Pagano. 2005. "The interpoint distance distribution as a descriptor of point patterns, with an application to cluster detection." *Statistics and Medicine* 24:753–773.

[BMS04] Bravata, D. M., K. M. McDonald, W. M. Smith, C. Rydzak, H. Szeto, D. L. Buckeridge, C. Haberland, and D. K. Owens. 2004. "Systematic review: Surveillance systems for early detection of bioterrorism-related diseases." *Annals of Internal Medicine* 140:910–922.

[BBF05] Brillman, J. C., T. Burr, D. Forslund, E. Joyce, R. Picard, and E. Umland. 2005. "Modeling emergency department visit patterns for infectious disease complaints: Results and applications to disease surveillance." *BioMed Central Medical Informatics and Decision Making* 5 (1): 4.

[BD02] Brockwell, P. J., and R. A. Davis. 2002. *Introduction to time series and forecasting*, 2nd ed. New York: Springer-Verlag.

[BS04] Brookmeyer, R., and D. Stroup. 2004. *Monitoring the health of populations: Statistical principles and methods for public health.* Oxford, UK: Oxford University Press.

[FBJ05] Forsberg, L., M. Bonetti, C. Jeffery, A. Ozonoff, and M. Pagano. 2005. "Distance-based methods for spatial and spatiotemporal surveillance." In *Spatial and syndromic surveillance for public health*, edited by A. B. Lawson and K. Kleinman, 133–152. Hoboken, NJ: Wiley.

[FBP05] Forsberg, L., M. Bonetti, and M. Pagano. 2005. "The choice of the number of bins for the M-statistic." In preparation.

[GP05] Graham, D., and M. Pagano. 2005. "Dimension reduction of HIV genotype measures with application to modeling the relationship between genotype and phenotype." In preparation.

[Har02] Harris, S. H. 2002. *Factories of death.* New York: Routledge.

[Hoe48] Hoeffding, W. 1948. "A class of statistics with asymptotically normal distributions." *Annals of Mathematical Statistics* 9:293–325.

[HTS03] Hutwagner, L., W. Thompson, G. M. Seeman, and T. Treadwell. 2003. "The bioterrorism preparedness and response Early Aberration Reporting System (EARS)." *Journal of Urban Health* 80 (Suppl. 1): i89–i96.

[Kul98] Kulldorff, M. 1998. "Statistical methods for spatial epidemiology: Tests for randomness." *GIS and Health*, pp. 49–62.

[Lee90] Lee, A. J. 1990. *U-statistics.* New York: Marcel Dekker.

[LBP04] Lombardo, J. S., H. Burkom, and J. Pavlin. 2004. "ESSENSE II and the framework for evaluating syndromic surveillance systems." *Morbidity and Mortality Weekly Report* 53 (Supplement): 159–165.

[MHP94] MacKenzie, W. R., N. J. Hoxie, M. E. Proctor, M. S. Gradus, K. A. Blair, D. E. Peterson, J. J. Kazmierczak, D. G. Addiss, K. R. Fox, and J. B. Rose. 1994. "A massive outbreak in Milwaukee of cryptosporidium infection transmitted through the public water supply." *New England Journal of Medicine* 331:161–167.

[Mad05] Madigan, D. 2005. "Bayesian data mining for health surveillance." In *Spatial surveillance for public health*, edited by A. B. Lawson and K. Kleinman, 203–221. Hoboken, NJ: Wiley.

[MGH94] Meselson, M., J. Guillemin, M. Hugh-Jones, A. Langmuir, I. Popova, A. Shelokov, and O. Yampolskaya. 1994. "The Sverdlovsk anthrax outbreak of 1979." *Science* 266:1202–1208.

[MKD04] Miller, B., H. Kassenborg, W. Dunsmuir, J. Griffith, M. Hadidi, J. D. Nordin, and R. Danila. 2004. "Syndromic surveillance for influenzalike illness in an ambulatory care network." *Emerging Infectious Diseases* 10:1807–1811.

[OBP05] Olson, K. L., M. Bonetti, M. Pagano, and K. D. Mandl. 2005. "Real time spatial cluster detection using interpoint distances among precise patient locations." *BioMed Center Medical Informatics and Decision Making* 5:19.

[OBF04] Ozonoff, A., M. Bonetti, L. Forsberg, C. Jeffery, and M. Pagano. 2004. "The distribution of interpoint distances, cluster detection, and syndromic surveillance." *Proceedings of the Joint Statistical Meetings.* American Statistical Association, 420–422.

[OBF05] Ozonoff, A., M. Bonetti, L. Forsberg, and M. Pagano. 2005. "Power comparisons for an improved disease clustering test." *Computational Statistics and Data Analysis* 48:679–684.

[OFB04] Ozonoff, A., L. Forsberg, M. Bonetti, and M. Pagano. 2004. "A bivariate method for spatiotemporal syndromic surveillance." *Morbidity and Mortality Weekly Report* 53 (Supplement): 61–66.

[RCS03] Rath, T., M. Carreras, and P. Sebastiani. 2003. "Automated detection of influenza epidemics with hidden Markov models." Edited by M. R. Berthold, H. J. Lenz, E. Bradley, R. Kruse, and C. Borgelt, *Advances in intelligent data analysis V. 5th International Symposium on Intelligent Data Analysis, IDA 2003, Berlin, Germany, August 28–30, 2003.* New York: Springer, 521–531.

[RM03] Reis, B. Y., and K. D. Mandl. 2003. "Time series modeling for syndromic surveillance." *BioMed Central Medical Informatics and Decision Making* 3:2.

[RPM03] Reis, B. Y., M. Pagano, and K. D. Mandl. 2003. "Using temporal context to improve biosurveillance." *Proceedings of the National Academy of Sciences* 100:1961–1965.

[Ser63] Serfling, R. E. 1963. "Methods for current statistical analysis of excess pneumonia-influenza deaths." *Public Health Reports* 78:494–506.

Part IV

Modeling

Modeling and Simulation for Defense and National Security

Wendy L. Martinez

Office of Naval Research, martinwe@onr.navy.mil

1 Introduction

Modeling is an essential (and sometimes the only) problem-solving tool in defense and national security applications. Of course, modeling is important and useful to industry as well, partly because contractors are doing much of the actual defense work (research and development, combat support, security, etc.), but also because industrial applications are similar to those in defense in their structure and requirements. Thus, this group of papers will be of interest to a wide audience.

The term *modeling* can mean many things and has broad applicability, as we will see from the four papers in this section. Modeling can mean transforming various behaviors, rules, phenomena, etc. into computer codes for the purposes of simulating some complex (or maybe simple) system. Modeling also pertains to the creation and specification of the rules and mathematical expressions that govern the phenomena of interest. Finally, when we carry out regression, probability density estimation, classification, or some other estimation method, then we are also doing modeling, because we are estimating (or creating) a model of some process or phenomenon.

2 Summary of Papers

The first paper describes the underlying models for a simulation of smallpox transmission in a large city. It is called EpiSims, and it was developed at the Los Alamos National Laboratory by expanding an existing simulation package that models the movement of individuals in large metropolitan areas. The authors describe the original simulation program called TRANSIMS (Transportation Analysis Simulation System), the models for spreading an infectious disease, and how this can be used to provide information to policymakers.

This paper illustrates an essential application of modeling and simulation. It is not possible to conduct physical experiments in the spread of infectious diseases, such as smallpox, HIV-AIDS, hemorrhagic fevers, etc., or the outcome of a chemical, biological, radiological, nuclear (CBRN) attack. In other words, we cannot expose a population to a disease or chemical attack and see what happens. Thus, to understand the problem and possible outcomes, we have to rely on computer simulations and the models that are used to build them. The goal of the EpiSims experiment had to do with understanding the various courses of action in the case of a smallpox attack, such as mass vaccination, targeted vaccination in conjunction with quarantining, and self-isolation.

The EpiSims paper also addresses many important issues in modeling and simulation. First, this is an interesting application of the reuse of existing models and how we can adapt them to other problems. While the initial version of EpiSims models the spread of smallpox, the creators made the simulation and disease models somewhat modular, so other diseases and biological attacks could be implemented easily. Second, they have an excellent description of how such models are built. The authors went to subject-matter experts and to the literature on smallpox and present a thoughtful discussion of the limitations of the models. Finally, this application shows how modeling, simulation, and statistics can help decision-makers.

The second paper describes a modeling methodology that would also help decision-makers understand what to do in the case of a suspected biological or chemical attack. The focus is on modeling the concentration field of a contaminant in a building. The subsequent model would be used to help make decisions regarding sampling the site for the presence of a contaminant and decontamination. The authors describe several methods for modeling spore dispersal, such as computational fluid dynamics, multizone modeling, and geostatistical modeling techniques. They also show the need for a multidisciplinary approach to this problem and the important role that statisticians play.

The next paper addresses a type of modeling many statisticians are familiar with: regression. The authors look at the issue of privacy and security when decision-makers need to combine or make use of data that are stored in multiple, distributed databases. The ability to make connections and to discover threats using all available information is critical to homeland defense. This is one of the lessons learned from the September 11 attacks. However, due to privacy, security, and domain issues, owners of the databases do not want to openly share their data.

Using regression as an application, the authors illustrate how one can perform secure regression without actually integrating or knowing all of the data. They show secure multiparty computation in both a horizontal (agencies have the same attributes on disjoint sets of records) and a vertical (agencies have different attributes on the same records) fashion.

The final paper is similar to the second paper in that it describes the issues involved with modeling physical phenomena. In particular, they discuss passive detectors of radiation emitted from cargo or passenger vehicles. This is an important problem for national security and counterterrorism, since potential weapons include nuclear and radiological devices hidden in our vast transportation system. One of the interesting issues brought up in this paper concerns threat models. This is an important consideration with modeling and problem-solving, in general. If we are not modeling the correct threat, then we are solving the wrong problem. For example, if a denial-of-service cyber-attack is not a viable threat, we should really be investigating and modeling something else.

3 Research Directions in Modeling

In addition to the main topic, many important research issues are presented in these papers:

1. Estimating normal background signals and system behavior.
2. The impact of model uncertainty on the results.
3. Modeling complicated phenomena (e.g., infectious disease spread, transport of contaminants in an occupied building).
4. Computational issues when integrating databases.
5. The adaptation of existing models and simulations for other uses.

However, as with most interesting and relevant topic areas, there are many open research directions and problems.

One very important area is the modeling of human behavior based on culture and other demographic information. Such models could be used to predict enemy and terrorist behavior and to enhance our decision-making tools. Such models exist in the social and psychological disciplines, but efforts should be made to translate these into computational models. Research in this area will also help in establishing and modeling appropriate threat scenarios.

Researchers can help ensure our national security by developing new methods and technologies in statistical data mining for the automatic integration of disparate types of data. Faced with massive amounts of data generated by persistent sensors and surveillance, decision-makers need tools that not only integrate data from diverse sources and of different types, but also provide an estimate of significance. We need to expand our notions of what types of data provide information and develop models that exploit nontraditional data sources (databases, open-source, Web-based, symbolic, images, etc). Statistical data mining techniques for discovering trends in enemy activities, links among objects, and hidden models of behavior will be needed to support intent analyses and to develop courses of action. In this context, the exploitation of textual data seems particularly promising and necessary. To realize the full potential of processing web-based and open-source information we must be able to convert free-form text to other structures that allow us to employ our computational tools and techniques. As an example of recent work that applies graph theoretic techniques and visualization to knowledge discovery and to the creation of information threads based on collections of documents, see Solka et al. [SBW05].

As with any modeling, users and researchers must be concerned about validation (how accurately a model represents the real world) and verification (a model implementation is correct) of models. This can be very difficult to do when the models involve the spread of disease, a CBRN attack, or human behavior. Also, the use of sensitivity analysis and residual analysis are often neglected areas of research in modeling.

Models and simulations are just some of the tools in our toolbox, and they should be combined with others for maximum effect. Thus, more research needs to be done on methodologies for combining experimental data, data from models and simulations, and expert opinion. In applications such as these, we need to be concerned about understanding the effect of model errors and how these propagate through the system, especially when sources are combined. Additionally, it would be interesting to develop techniques and methods for experimental design, when our experiment includes multiple sources, such as observed and simulated data.

In conclusion, I offer the following websites for those who want more information on modeling and simulation in defense:

- The Defense Modeling and Simulation Office (DMSO) has a website at `https://www.dmso.mil/public/`. This is an excellent entry point for all modeling and simulation in the Department of Defense.
- For more information on verification and validation, see the website at `http://vva.dmso.mil/`.
- The EpiSims paper shows how important it is to reuse models and simulations. To facilitate this, the Department of Defense provides a searchable repository of modeling and simulation resources. This can be found at `http://www.msrr.dmso.mil/`. Agencies that provide information to this repository include the Department of Defense, Missile Defense Agency, Defense Intelligence Agency, and others.
- For information on Department of Defense research thrusts in data mining and modeling, readers are encouraged to visit the following websites. For Navy programs, see `http://www.onr.navy.mil`, Air Force programs can be found at `http://www.afosr.af.mil`, and Army research in these areas is described at `http://www.aro.army.mil`. The Defense Advanced Research Projects Agency, Information Exploitation Office (IXO), supports efforts also. Visit `http://www.darpa.mil` for more information.

Reference

[SBW05] Solka, J. L., A. C. Bryant, and E. J. Wegman. 2005. "Data mining and data visualization." In *Handbook of statistics*, edited by C. R. Rao, E. J. Wegman, and J. L. Solka, Volume 24. Amsterdam: Elsevier.

Modeling and Parameterization for a Smallpox Simulation Study

Sarah Michalak[1] and Gregory Wilson[2]

[1] Statistical Sciences Group, Los Alamos National Laboratory, michalak@lanl.gov
[2] Statistical Sciences Group, Los Alamos National Laboratory, gdwilson@lanl.gov

1 Introduction

In the wake of September 11, 2001, and the anthrax mailings that followed it, the U.S. government and other organizations had a keen recognition of the need for emergency preparedness and strategies for disaster mitigation. In particular, concern focused on the potential for additional terrorist attacks, including the threat of an attack using a biological agent like smallpox, botulism, or plague. Decision-makers had a renewed interest in learning how they could best respond to such a terrorist attack, and scientists and public health experts wanted to contribute to the effort to answer such questions.

Before September 11, 2001, various role-playing simulations had been used to study responses to incidents involving bioterrorism (see, for example, Inglesby et al. [IGO01] and O'Toole et al. [OMI02]). Although playing these game scenario-type exercises was instructive, they required extensive planning and resources even if only played once. Furthermore, this approach did not facilitate the investigation of multiple scenarios. In the post-September 11, 2001, context, detailed information that could inform decision-making was needed, and the need for it was urgent. In such a setting, computer simulation models could efficiently provide decision-makers with information about the strengths and weaknesses of different response strategies that could be used in the event of a bioterrorist attack.

The foundation of such an epidemiological simulation model had been developed by Los Alamos National Laboratory (LANL) researchers. This foundation was the Transportation Analysis Simulation System (TRANSIMS) [BBB99] that was created to model the mobility of individuals in large metropolitan areas. During its development, researchers realized that TRANSIMS could form the basis of an epidemiological simulation tool that could model the spread of disease in urban areas. In the aftermath of September 11, 2001, the U.S. Office of Homeland Security (OHS) asked LANL to develop this new technology, named EpiSims, to study response strategies that might be used if there were a bioterrorist attack involving smallpox. As part of this effort, individuals in the LANL Statistical Sciences Group reviewed and parameterized some of the models that underlay EpiSims and researched and parameterized possible response strategies. This proved challenging as the existing literature didn't contain all of the information needed to use EpiSims to study response strategies in the event of a bioterrorist attack using smallpox. Moreover, OHS required a quick turnaround, so the results presented to them could not be based on the in-depth analysis one would ideally perform.

A key objective of the work for OHS was to avoid using models that were only applicable to smallpox and instead to ensure that they depended on parameter values that could be adjusted so that the resulting disease would behave like different biological agents. In particular, the U.S. Centers for Disease Control and Prevention (CDC) had identified six biological agents that could likely be used as bioterrorist weapons: anthrax, botulism, certain hemorrhagic fevers, plague, smallpox, and tularemia. The disease and transmission models described below are designed to be flexible so that they may be used to model multiple diseases such as these. But for purposes of providing information to OHS, the focus was on smallpox as LANL developed end-to-end capability for EpiSims.

Following September 11, 2001, a number of researchers published work presenting mathematical or simulation models of smallpox. Specifically, Bozzette et al. [BBB03], Eichner [Eic03], Epstein et al. [ECC02], Halloran et al. [HLN02], Kaplan et al. [KCW02], Kaplan [Kap04], Kretzschmar et al. [KVW04], Legrand et al. [LVB03], Meltzer et al. [MDL01], and Nishiura and Tang [NT04] used such models to assess different response strategies in the event of a bioterrorist attack using smallpox, with Bozzette et al. [BBB03] and Epstein et al. [ECC02] also considering preevent vaccination strategies. Ferguson et al. [FKE03] provided an overview of some of the issues inherent in modeling smallpox and a review of the models presented in Bozzette et al. [BBB03], Halloran et al. [HLN02], Kaplan et al. [KCW02], and Meltzer et al. [MDL01]. While some studies assumed homogeneous mixing of the population [KCW02, Kap04, LVB03, MDL01, NT04], others modeled different transmission mechanisms based on the type of contact between an infectious individual and a susceptible individual [BBB03, Eic03, KVW04], or structured populations of individuals that incorporated different types of contact [ECC02, HLN02]. However, to our knowledge no other existing model is grounded on a structured population of individuals as large as or as detailed as that which underlies EpiSims via the TRANSIMS technology. In addition, EpiSims explicitly incorporates two manifestations of smallpox (ordinary smallpox, which has a 30% mortality rate, and hemorrhagic smallpox, which is almost always fatal) and two subgroups (pregnant women and certain individuals with HIV infections) presumed to suffer from more severe manifestations of smallpox at a higher rate than the general population.

These smallpox models, including EpiSims, reflect varying levels of detail. In any given situation, the granularity of the chosen model could depend on the timeframe and/or resources available for completing the study of interest, the level of detail required for the analysis, or the characteristics of the disease under study. The necessity of providing results within a short timeframe or via limited resources can require use of a simple model that may be fit quickly and easily. In terms of the results of interest, if one wants to study transmission among different subpopulations that are defined by region or demographic characteristics, those subpopulations and their contact patterns must be reflected in the model. As an example of the importance of disease characteristics, the transmission of some diseases may be more contact-pattern dependent than for others. For example, transmission of a disease in which infectious individuals are immobilized and close contact is required for transmission is much more dependent on contact patterns in the locations of the infectious individuals (most likely their homes or health care facilities) than would be transmission of a disease in which infectious individuals are mobile and close contact is not required for transmission. Finally, if information about the accuracy of results given varying times and costs to produce them is available, this may inform model choice as well.

This paper describes TRANSIMS, the work the authors and their collaborators performed in the development of EpiSims, the information EpiSims simulations provided to decision-makers in OHS, and some example output that is representative of that from an EpiSims simulation. A companion paper [MW05] discusses this work for a general audience. The work presented in this paper is our own and that of our collaborators on the EpiSims project.

2 TRANSIMS

Previous to our EpiSims work, TRANSIMS had been used to create a model of the activity, including mobility, patterns of the roughly 1.6 million people who travel in the metropolitan Portland, OR, area on a typical weekday. In using TRANSIMS to model the activity patterns in Portland, a population of synthetic individuals was constructed that matched the actual Portland population at the census block group level in terms of certain demographic variables including household size, age of head of household, and annual household income. (A census block group is a cluster of blocks that usually includes about 250–550 dwelling units.) Next, each individual was assigned a set of daily activities that described his or her activities in Portland on a typical weekday. For example, a person may take a child to day care, go to work, go out to lunch, leave work, go to the grocery store, pick another child up from sports practice, and then return home. Such daily activities were based on daily activity surveys collected from individuals in the actual Portland population. Once daily activities such as these had been assigned to each synthetic individual, locations for the activities were determined. TRANSIMS includes roughly four locations per city block, which are used to approximate the locations at which individuals perform different activities. Next, each synthetic individual was assigned a mode of transportation to be used in traveling from one activity to the next. With this, the transportation flow of the synthetic Portland population resulted. In building the model, iteration at each step occurred until the result was sufficiently close to that of the actual Portland population before moving on to the next step in the modeling process. Once fitted, TRANSIMS provided a second-by-second snapshot of the locations of all of the 1.6M synthetic individuals in virtual Portland. The idea behind EpiSims is to overlay an infectious disease model on the resulting activity patterns so that researchers can observe how the disease spreads through the population as infectious individuals infect susceptible individuals during daily activities involving typical contact patterns.

The development of EpiSims required the addition of several capabilities to TRANSIMS. These included abilities to model (1) the transmission of the disease of interest from an infectious individual to a susceptible individual via a transmission model, (2) the progress of the disease within an infected individual via a disease model, and (3) response strategies that might be used in the event of a terrorist attack or other outbreak. Since certain demographic information about the synthetic individuals in the TRANSIMS population was available, the disease and transmission models could incorporate parameters that depended on relevant individual-level variables such as age, immune status, and other health-related variables. The response strategies to be investigated were modeled as functions of parameters that reflected public compliance with the requests of public health officials, readiness to respond to an outbreak of smallpox, and the extent to which the response strategy could realistically be fully implemented.

3 Disease and Transmission Models for EpiSims

EpiSims was envisioned as a general framework for modeling infectious diseases, and initial disease and transmission models for EpiSims had been developed previous to our EpiSims work and the interest in using EpiSims to model smallpox [EGK04]. These models were functions of parameters that could be set to reflect the transmission and disease course characteristics of different infectious diseases. However, it was unclear whether they could be used to model smallpox.

In the initial EpiSims disease model synthetic individual i in the TRANSIMS population has a "disease load" $L_i(t)$ that varies with time t during the course of his infection with the disease under study. The disease load describes the progress of the disease within the individual, with greater values indicating a sicker individual. In addition, an individual's infectiousness at a given time depends on his disease load at that time. Specified values of the disease load correspond to an individual who has become infected with the disease of interest, is experiencing symptoms of the disease, is sick at home, is capable of infecting others, or has just died as a result of the disease. Changes in an individual's disease load are modeled as piecewise exponential.

The initial EpiSims transmission model describes how susceptible individuals are infected with the disease of interest. In this model, the environment in a given location has a disease load associated with it that results from individuals at the location who shed disease load into it or from contamination of the location. In particular, individual i sheds a fraction of his disease load R_i^s into his location per unit time. Thus, infectious individual i with a constant disease load $L_i(t)$ sheds $L_i(t) \times R_i^s$ disease particles into the environment per unit time. If susceptible individual j is in an environment with a positive disease load, then he absorbs a fraction of the available load in the environment, R_j^a, per unit time.

Determining whether these disease and transmission models could be used to model smallpox involved a review of the available smallpox literature. Based on how these models would need to function in the EpiSims machinery, a list of bioagent factors that covered a range of characteristics relevant to the disease model, the transmission model, and the response strategies was developed. Using the bioagent factors to guide the literature review helped ensure that it would uncover all of the aspects of smallpox that EpiSims needed to incorporate. Thus, the suitability of the existing disease and transmission models for modeling smallpox could be assessed.

The bioagent factors covered characteristics such as the agent type (virus or bacteria), transmission mechanisms, the virulence of the agent, existence of different disease manifestations, symptoms, diagnostic confounds, treatments, whether a vaccine existed for the agent, survival outside the host, and the course of the disease in the host. For a broad range of agents that might be used in a bioterrorist attack, much of this information was available from online CDC sources (for current information, see `http://www.bt.cdc.gov/agent/agentlist.asp`). In addition, a recent series of consensus documents published in the *Journal of the American Medical Association* outlined what was known about various agents from a medical and public health perspective and their potential use as bioterrorist weapons, with Henderson et al. [HIB99] a consensus statement concerning smallpox. The literature review also drew heavily on Fenner et al. [FHA88] and a few other sources.

The literature review led to the development of a diagrammatic disease model that was used in conjunction with the piecewise exponential disease model. The diagrammatic disease model was intended to be general to the extent that it depended on parameters that could be set so that the resulting disease course would reflect different diseases. The diagrammatic disease model, shown in Fig. 1, describes the course of the disease within an individual from his exposure to the disease through recovery or death. In the case of smallpox, an individual is initially exposed to the disease when he has absorbed enough smallpox virions[3] to develop a smallpox infection. Next, based on the individual's characteristics, a smallpox infection may develop. Individuals who do not develop a smallpox infection would include those who have recently received a successful smallpox vaccination. If a smallpox infection is present, it may be a clinical infection or a subclinical infection, with subclinical infections those in which the individual has evidence of a smallpox infection based on a diagnostic test, but does not develop any other symptoms. An individual who has been previously vaccinated may develop a subclinical case of smallpox. If the individual develops a clinical case of smallpox, any of several manifestations may be present. Possible manifestations of smallpox include ordinary smallpox; modified smallpox and variola sine eruptione, which are less severe than ordinary smallpox; and two very severe forms, hemorrhagic smallpox and flat-type smallpox, which are almost always fatal. Our simulations for OHS included ordinary smallpox and hemorrhagic smallpox. Depending on the manifestation of smallpox, a different disease course ensues that includes an asymptomatic incubation period and, in the case of ordinary and several other manifestations of smallpox, a prodromal period in which the individual suffers from a high fever and other symptoms followed by the development and progression of smallpox pustules. Lastly, the individual may either recover or die from his smallpox infection. As indicated in Fig. 1, an individual's disease course may depend on relevant health and background characteristics such as his vaccination history and immunocompromisation status. An individual's disease course informs his infectiousness.

The piecewise exponential disease model and the diagrammatic disease model worked in concert in EpiSims. The diagrammatic disease model was used to determine whether an individual developed a smallpox infection and, if so, the manifestation of smallpox from which he suffered and the particulars of his disease course such as the lengths of the incubation and prodromal periods, whether he recovered or died, and if he died, the point in his disease course at which he died. This information then determined how the disease load $L_i(t)$ of the individual progressed with time, with this progression modeled using the piecewise exponential disease load model. Since an individual's infectiousness depends on his disease load, its parameterization needed to reflect how infectiousness varies with disease course for different disease manifestations.

The first step in parameterizing the EpiSims disease and transmission models was to expand the review of the available information about smallpox, with an emphasis on uncovering data and information relevant to these models. In terms of the disease models, Fenner et al. [FHA88] included many datasets that provided information that linked disease manifestation and/or disease outcome (recovery or death) to demographic variables such as age, gender, pregnancy status, and vaccination history. Detailed descriptions of

[3] A virion is a complete virus particle capable of causing infection.

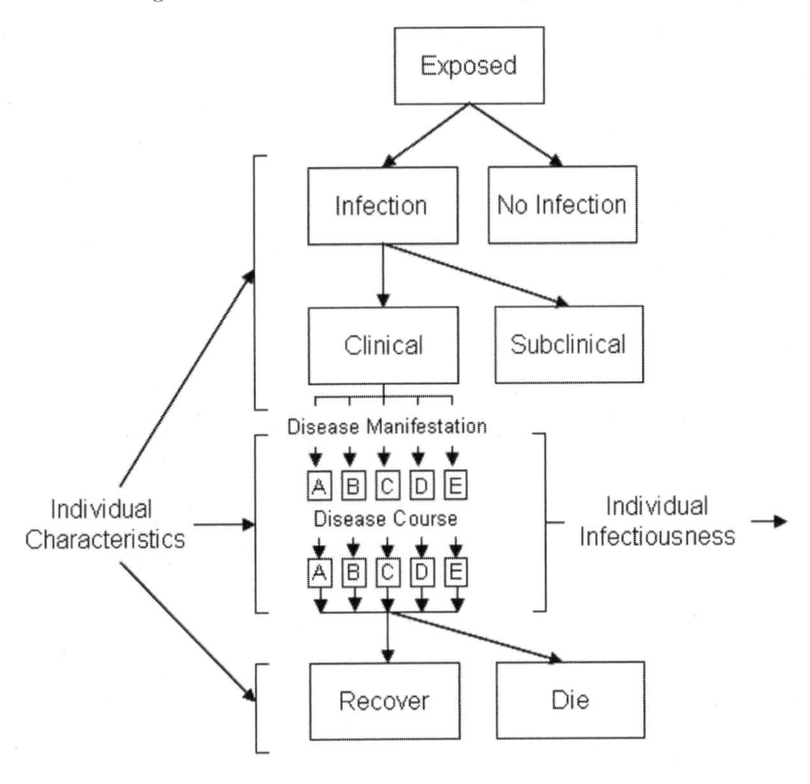

Fig. 1. Diagrammatic disease model. This model describes the course of smallpox from an individual's exposure to it through his eventual outcome, which may be either recovery or death.[4]

the disease courses for different manifestations of smallpox were also available in this source.

Less information was available for modeling individual disease loads and the transmission of smallpox from an infectious individual to a susceptible individual. In Fenner et al. [FHA88, p. 191], it is reported that "transmission was usually direct, by the implantation of infective droplets on to the nasal, oral, or pharyngeal mucous membrane, or the alveoli of the lung, or less commonly indirect, as an airborne infection or from fomites."[5] It was believed that an individual was most infectious during his first week with pustules, with transmission almost never occurring before the pustules appeared and

[4] This figure is reprinted with permission from *Chance*. Copyright 2005 by the American Statistical Association. All rights reserved.

[5] A fomite is an object that may serve as a means of transmitting a disease from one individual to another.

infectivity decreasing substantially during the later stages of the disease that followed the first week of pustules [FHA88, p. 189]. From an epidemic management perspective, an individual was isolated until after the last scab resulting from a pustule had separated from his body; however, the consensus appeared to be that actual infectiousness had declined substantially before that time [FHA88, p. 185].

Fenner et al. [FHA88] also included information about the conditions under which smallpox was more or less likely to be transmitted. Closer contact made transmission more likely. For example, a susceptible individual was more likely to be infected by an infectious individual if he shared a room with the smallpox victim than if he merely shared a dwelling unit with him [FHA88, p. 191]. Transmission was even less likely if the susceptible individual merely lived in the same housing compound as the infectious individual [FHA88, p. 191]. Some of the information in Fenner et al. [FHA88] about the conditions under which smallpox might be transmitted from an infectious individual to a susceptible individual seemed contradictory. For example, they report "especially in India, long-distance movements by train or bus of patients suffering from smallpox, with an overt rash, used to occur frequently, yet infection of casual fellow-travelers was rare indeed — so rare that instances of it were deemed worthy of special report" [FHA88, p. 191]. However, in another case, a visitor to a hospital housing an individual with smallpox contracted smallpox despite spending only 15 minutes in the hospital in areas away from the isolation corridor where the smallpox patient was being housed [FHA88, p. 192]. As discussed below, such discrepancies required resolution, and CDC researchers offered expert opinion that was crucial to this resolution.

Once this information had been compiled, it was compared to that needed to parameterize EpiSims for smallpox. In terms of the disease models, the available literature yielded estimates of many parameter values. However, for other subjects such as the likely disease courses and mortality probabilities for immunosuppressed individuals, additional information was needed. Specifically, individuals with different types of immunosuppression (i.e., those with HIV infections, those undergoing chemotherapy, and recent organ transplant recipients) were not prevalent in the population when smallpox was endemic, so information about their likely response to smallpox infection was largely unavailable. In terms of the transmission model, the available information helped ground our understanding of smallpox transmission mechanisms and could guide development of the transmission model. However, the detailed understanding of the conditions under which smallpox transmission was possible that was needed for our modeling was not uncovered in the literature. Since EpiSims was intended to model smallpox transmission by utilizing the social contact patterns provided by TRANSIMS, the probability of smallpox transmission given a distance between an infectious individual and a susceptible individual, the duration and other characteristics of their contact, and the disease stage of the infectious individual needed to be specified. Determination of whether fomites were a substantial source of transmission was also needed so that they could be incorporated in our transmission model if necessary.

Next, a list of information required to parameterize EpiSims for smallpox was compiled, and each item was assigned a priority based on the availability of data and information pertaining to it. Consultation with several smallpox experts at the CDC followed.

The CDC smallpox experts were invaluable in terms of their ability to provide information useful for parameterization of the transmission model. For example, they advised that after an infectious person left an area, any fomites were practically uninfectious. Thus, EpiSims could reasonably exclude fomites as a method of transmission of smallpox. They also affirmed that, absent other factors such as centralized ventilation systems that may cause smallpox virions to travel farther than they otherwise would, smallpox transmission typically requires close ($\leq$ 6.5 feet or 2 meters) contact. They also provided information about the duration of contact typically necessary for smallpox to be transmitted from one person to another given close contact. To account for transmission that occurs via close contact and for transmission that might occur at greater distances, large locations in EpiSims were divided into sublocations. In EpiSims, transmission of smallpox occurred more quickly within a sublocation than between sublocations.

For the disease models, the CDC smallpox experts reaffirmed and as necessary revised our understanding of the smallpox disease course for different smallpox manifestations and furnished information about the likely course of smallpox in immunosuppressed individuals. They further advised that our simulations should include pregnant women because smallpox tends to manifest more severely in pregnant women than in most other individuals. Our study also incorporated the subgroup of individuals with HIV infections who were likely to suffer from the more severe manifestations of smallpox. Since neither of these populations was included in the original TRANSIMS population, their frequency in the metro Portland, OR, population needed to be estimated so that they could be reflected in the population of synthetic TRANSIMS individuals.

Following this exchange with the CDC smallpox experts, our parameter values were finalized with additional consultation as needed. Parameterization of the disease models followed a two-step process. First, parameter values for the diagrammatic disease model were developed. Once the diagrammatic disease model was parameterized, it was translated into the language of the disease load model so that individuals were symptomatic, staying home sick, infectious, and recovered or dead at appropriate times in their disease courses and so that an individual's infectiousness varied appropriately with his disease manifestation and disease course. The disease load thresholds for an infected individual, a symptomatic individual, an individual who was staying home sick, an individual who was infectious, and an individual who had just died were set to common values for all individuals, with an individual's disease load $L_i(t)$ varying appropriately given his disease manifestation. The disease models were stochastic so that different individuals with the same manifestation of smallpox might experience different disease courses, e.g., different lengths of the incubation and prodromal periods.

For the transmission model, R_i^s varied across individuals to reflect the different levels of infectiousness that correspond to different disease manifestations and R_i^a varied across individuals to reflect how the length of contact required for a susceptible individual to become infected with smallpox might vary.

With these parameter values, EpiSims could be implemented for smallpox and its course through the TRANSIMS model of the Portland, OR, population and the efficacy of different response strategies could be observed.

4 Response Strategies

Early in the project, different response strategies that could be investigated in the simulation study for OHS were outlined. These included mass vaccination, targeted vaccination combined with quarantine, and self-isolation. Mass vaccination, as the name implies, is a response strategy in which the entire population is vaccinated as quickly as possible. Targeted vaccination combined with quarantine is a response strategy based on techniques used during the smallpox eradication campaign. Under this strategy, an individual suffering from smallpox is interviewed to determine the individuals with whom he has had contact while sick. Following the interview, the individual with smallpox is quarantined to prevent the infection of further individuals and his household members and contacts are located, vaccinated, and placed under quarantine. Vaccination within several days of smallpox exposure may prevent infection with smallpox or reduce the severity of a resulting smallpox infection, including the probability that an infected individual dies. Thus, strategies that include a vaccination component can be effective at controlling a smallpox outbreak and reducing the severity of symptoms of those who contract smallpox. In addition, targeted vaccination focuses vaccination resources on those individuals most in need of them. Under the self-isolation strategy, individuals isolate themselves in their homes, thus reducing the number of individuals they may infect with smallpox. This strategy was included to assess the efficacy of individuals isolating themselves without any other overt public health response. Information about such a strategy could be useful in the event that public health resources were not immediately available to implement one of the other responses, but could broadcast public health announcements encouraging voluntary self-isolation. Finally, a baseline case in which there was no response to the presence of smallpox in the community was included.

The response strategies also required parameterization. Our simulations needed to reflect the amount of time between the initial detection of smallpox in the community and the implementation of the response strategy, the efficacy with which the response strategy could be implemented, and the extent of compliance with it. For example, under a mass vaccination campaign, it is unlikely that every individual in the population will be vaccinated; some will refuse vaccination or never be found. Similarly, not all individuals will comply with a request to go to a central quarantine facility.

Infectious disease specialists at the CDC provided expert opinion on issues such as the amount of time it might take to mount a public health response to an outbreak of smallpox, the willingness of individuals to comply with a contact tracer requesting information about people with whom he has had contact, and the percentage of contacts an individual might remember based on the duration and proximity of contact. This information was reflected in the response strategies used in the simulation study for OHS.

5 Results for Decision-makers

Information related to the relative efficacy of and the resources required to implement the different response strategies was of interest to OHS. For the OHS study, each simulation ran for 70 days, with day 1 being the day on which smallpox virions were released and the initial cohort of individuals was infected. Each simulation was summarized statistically and graphically. Total counts and maximum daily counts of individuals who were infected, symptomatic, infectious, at home sick, vaccinated, in a centralized quarantine facility, or who recovered or died were presented as were graphical daily counts of individuals who became infected, were infectious, recovered, died, were vaccinated, or were in a centralized quarantine facility.

Outcomes that describe the course of smallpox through the population, e.g., counts of individuals who became infected, symptomatic, or who died or recovered during the course of the outbreak, provide information about the severity of the smallpox outbreak, while information pertaining to vaccination, isolation, and disease outcomes (recovery or death) could guide resource planning. For example, the total number of vaccinations given provides information about the total number of vaccine doses needed to implement the response strategy under a given scenario, and the maximum daily number of individuals vaccinated provides information about the number of individuals trained in smallpox vaccination techniques necessary to implement the response strategy under the scenario. The total number of people who die from smallpox provides information about both the magnitude of the smallpox outbreak and the need for resources, e.g., mortuary facilities.

Figure 2 displays example results similar to those that might result from an EpiSims simulation. As in the results forwarded to OHS, they cover the initial 70 days that smallpox is present in the TRANSIMS population, with day 1 being the day on which the initial cohort of individuals is infected with smallpox. The upper plot in Fig. 2 describes the course of smallpox in the population, with the line denoted by triangles displaying the number of people who are infectious each day and the line denoted by plus signs presenting the number of people who become infected each day. The day is indicated on the horizontal axis of the graph. The lower graph in Fig. 2 displays the number of people vaccinated each day as part of the targeted vaccination and quarantine response strategy considered in these example results.

As indicated in the upper plot, the scenario begins with a cohort of 2300 individuals who are infected with smallpox on day 1. The relationship between the daily counts of infectious individuals and the daily counts of individuals who became infected with smallpox in the upper plot in Fig. 2 indicates how smallpox spreads and then is contained by the targeted vaccination and quarantine strategy. The first wave of smallpox is indicated by the first peak in each of the two lines on the upper plot. During the beginning of the first wave of smallpox, each infectious individual infects more than one susceptible individual on average. However, when the targeted vaccination and quarantine program is instituted, a dramatic decrease in the number of people infected per day follows. During the second wave of the smallpox outbreak, indicated by the second peak in the number of infectious individuals, very few susceptible individuals are infected, with further spread of the disease stopped by roughly day 55. Thus, the targeted vaccination and quarantine strategy is effective in stopping the spread of smallpox in this example.

The lower graph in Fig. 2 provides information about the targeted vaccination program used to contain smallpox in these example results. Because the targeted vaccination strategy involves vaccinating contacts of individuals with smallpox, the trend in the daily counts of vaccinated individuals mimics the daily counts of infectious individuals. However, many more people are vaccinated than become infectious, indicating that on average an individual suffering from smallpox has contact with more than one susceptible individual while he is sick. Exactly 12,000 individuals are vaccinated on the peak day of the vaccination program, with 133,000 people vaccinated during the course of the smallpox outbreak. Information such as this could be used to guide resource planning related to the necessary number of doses of vaccine and individuals trained to give smallpox vaccinations required to implement this hypothetical targeted vaccination and quarantine strategy.

6 Conclusion

To provide information to OHS within its desired timeframe, the work discussed herein, the steps necessary to implement it in EpiSims, and a 167-page report detailing the initial EpiSims smallpox simulations were completed over six months starting in late 2001 and ending in mid-2002. The results of the study emphasized that self-isolation, or the voluntary isolation of individuals at home when they are sick, can be an important means of controlling the disease either as part of a broader response strategy or on its own; that the amount of time it takes to mount a response to the presence of smallpox in the community is important, particularly if there is no self-isolation of individuals

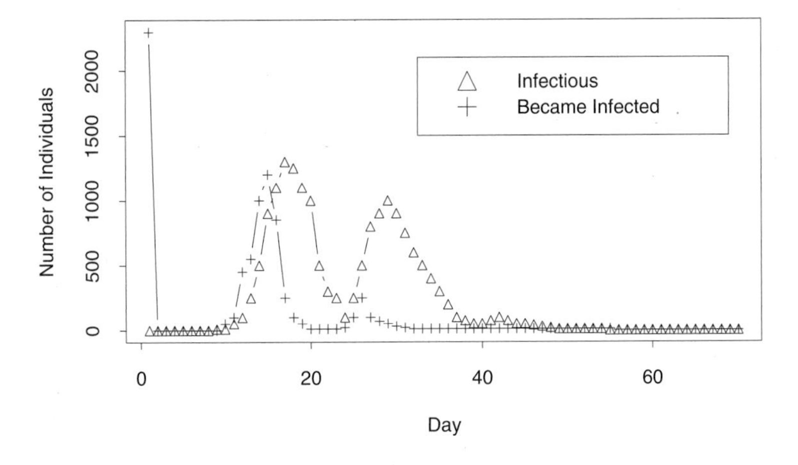

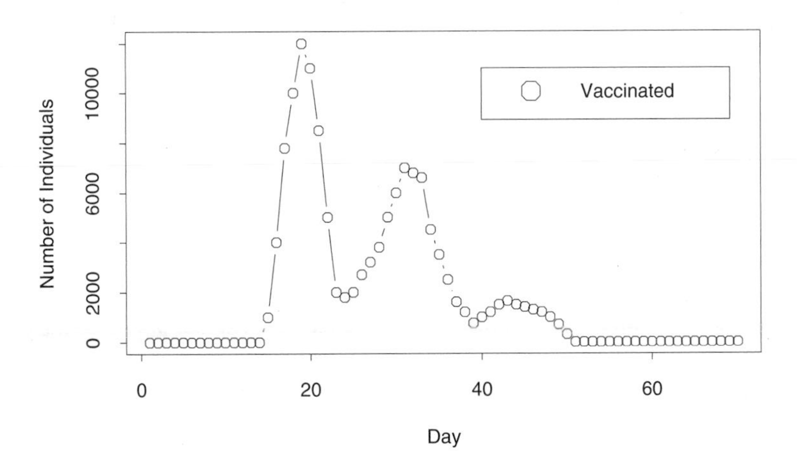

Fig. 2. Example output representative of that from an EpiSims smallpox simulation. In this hypothetical example, the response is targeted vaccination and quarantine. The upper plot describes the course of smallpox in the population over its first 70 days using daily counts of individuals who are infectious and of individuals who become infected with smallpox. The lower plot provides the number of individuals vaccinated each day during the same time period.

in the population; and that both mass vaccination and targeted vaccination and quarantine can be efficacious responses, with the more effective strategy depending on the characteristics of the incident.

Given the timeframe for this study, many of the statistical issues inherent in this work could not be grappled with, despite a profound appreciation of their importance. Key statistical issues not addressed in this paper include computer model validation or rigorous ascertainment of whether the TRANSIMS and EpiSims models produce realistic results; sensitivity analyses that investigate the sensitivity of the simulation results to the specification and parameterization of the models underlying them; the design of computer experiments involving simulations such as those performed for the OHS smallpox study; parameter estimation via the combination of data and expert opinion; the careful quantification of uncertainty so that simulation results incorporate replication error and the uncertainty inherent in the parameter values used for the study; and the use of Monte Carlo variance reduction techniques to produce more precise estimates of the quantities under investigation.

Since the completion of the initial EpiSims study for OHS, the EpiSims technology has been further developed. Eubank et al. [EGK04] discuss how the contact patterns modeled by TRANSIMS could be used in developing response strategies. EpiSims has also been chosen as one of the modeling systems in the National Institutes of Health's Modeling of Infectious Disease Agent Study (MIDAS) that supports development of computation disease models that are accessible to a wide range of researchers.

References

[BBB99] Barrett, C., R. Beckman, K. Berkbigler, K. Bisset, B. Bush, S. Eubank, J. Hurford, G. Konjevod, D. Kubicek, M. Marathe, J. Morgeson, M. Rickert, P. Romero, L. Smith, M. Speckman, P. Speckman, P. Stretz, G. Thayer, and M. Williams. 1999. "TRANSIMS Volume 0 — Overview." Los Alamos Unclassified Report LA-UR-99-1658, Los Alamos National Laboratory.

[BBB03] Bozzette, S., R. Boer, V. Bhatnagar, J. Brower, E. Keeler, S. Morton, and M. Stoto. 2003. "A model for smallpox-vaccination policy." *New England Journal of Medicine* 348:416–425.

[Eic03] Eichner, M. 2003. "Case isolation and contact tracing can prevent the spread of smallpox." *American Journal of Epidemiology* 158:118–128.

[ECC02] Epstein, J., D. Cummings, S. Chakravarty, A. Derek, R. Singa, and D. Burke. 2002. "Toward a containment strategy for smallpox bioterror: An individual-based computational approach." Working paper 31, Brookings Institute Center on Social and Economic Dynamics. http://www.brook.edu/dybdocroot/es/dynamics/papers/bioterrorism.pdf.

[EGK04] Eubank, S., H. Guclu, A. Kumar, M. Marathe, A. Srinivasan, Z. Toroczkai, and N. Wang. 2004. "Modelling disease outbreaks in realistic urban social networks." *Nature* 429:180–184.

[FHA88] Fenner, F., D. Henderson, I. Arita, Z. Jezek, and I. Ladnyi. 1988. *Smallpox and its eradication.* Geneva: World Health Organization.

[FKE03] Ferguson, N., M. Keeling, W. Edmunds, R. Gani, B. Grenfell, R. Anderson, and S. Leach. 2003. "Planning for smallpox outbreaks." *Nature* 425:681–685.

[HLN02] Halloran, M., I. Longini, A. Nizam, and Y. Yang. 2002. "Containing bioterrorist smallpox." *Science* 298:1428–1432.

[HIB99] Henderson, D. A., T. V. Inglesby, J. G. Bartlett, M. S. Ascher, E. Eitzen, P. B. Jahrling, J. Hauer, M. Layton, J. McDade, M. T. Osterholm, T. O'Toole, G. Parker, T. Perl, P. K. Russell, and K. Tonat. 1999. "Smallpox as a biological weapon — Medical and public health management." *Journal of the American Medical Association* 281 (22): 2127–2137.

[IGO01] Inglesby, T., R. Grossman, and T. O'Toole. 2001. "A plague on your city: observations from TOPOFF." *Clinical Infectious Diseases* 32:436–445.

[Kap04] Kaplan, E. 2004. "Preventing second-generation infections in a smallpox bioterror attack." *Epidemiology* 15:264–270.

[KCW02] Kaplan, E., D. Craft, and L. Wein. 2002. "Emergency response to a smallpox attack: the case for mass vaccination." *Proceedings of the National Academy of Sciences* 99:10935–10940.

[KVW04] Kretzschmar, M., S. van den Hof, J. Wallinga, and J. van Wijngaarden. 2004. "Ring vaccination and smallpox control." *Emerging Infectious Diseases* 10:832–841.

[LVB03] Legrand, J., C. Viboud, P. Boelle, A. Valleron, and A. Flahault. 2003. "Modelling responses to a smallpox epidemic taking into account uncertainty." *Epidemiology and Infection* 132:19–25.

[MDL01] Meltzer, M. I., I. Damon, J. W. LeDuc, and J. D. Millar. 2001. "Modeling potential responses to smallpox as a bioterrorist weapon." *Emerging Infectious Diseases* 7:959–969.

[MW05] Michalak, S., and G. Wilson. 2005. "Reconsidering smallpox." *Chance* 18:38–43.

[NT04] Nishiura, H., and I. Tang. 2004. "Modeling for a smallpox-vaccination policy against possible bioterrorism in Japan: The impact of long-lasting vaccinal immunity." *Journal of Epidemiology* 14:41–50.

[OMI02] O'Toole, T., M. Mair, and T. Inglesby. 2002. "Shining light on "Dark Winter"." *Clinical Infectious Diseases* 34:972–983.

Approaches to Modeling the Concentration Field for Adaptive Sampling of Contaminants during Site Decontamination*

William K. Sieber[1], James S. Bennett[2], Abera Wouhib[3], Joe Fred Gonzalez, Jr.[4], Myron J. Katzoff[5], and Stanley A. Shulman[6]

[1] National Institute for Occupational Safety and Health, Centers for Disease Control and Prevention, WSieber@cdc.gov
[2] National Institute for Occupational Safety and Health, Centers for Disease Control and Prevention, JBennett@cdc.gov
[3] National Center for Health Statistics, Centers for Disease Control and Prevention, AWouhib@cdc.gov
[4] National Center for Health Statistics, Centers for Disease Control and Prevention, JGonzalez@cdc.gov
[5] National Center for Health Statistics, Centers for Disease Control and Prevention, MKatzoff@cdc.gov
[6] National Institute for Occupational Safety and Health, Centers for Disease Control and Prevention, SShulman@cdc.gov

1 Introduction

Following the anthrax outbreaks of October–December 2001, there has been an increasing awareness of the potential for biological and chemical contaminant releases in a closed environment such as an office building. Characterizing the spread and distribution of such contaminants so as to minimize worker exposure to them, and to maximize the effectiveness of field procedures used in removing contaminants, has been a priority. During the anthrax outbreaks, for example, extensive environmental sampling was done by the Centers for Disease Control and Prevention (CDC)/National Institute for Occupational Safety and Health (NIOSH) with over 100 individuals involved in taking approximately 10,000 environmental samples. Between 4% and 50% of environmental samples taken from a given site were positive, depending on location sampled [SMW03]. Thus there was a high degree of variability in sample results depending on sampling location. Further, many sites continue to undergo remediation to remove all traces of *Bacillus anthracis*. If optimal sampling locations could be determined where bacteria were most likely present, and if

* The findings and conclusions in this report are those of the authors and do not necessarily represent the views of the National Institute for Occupational Safety and Health or the National Center for Health Statistics.

environmental sampling were done at these locations, it is possible that the cost of complete building remediation could be significantly reduced. Decontamination efforts, including those continuing ones in postal facilities, are not only associated with the direct and indirect economic impacts of current sampling procedures, but also with the psychological costs such as:

- Loss of faith in management claims about safety of the work environment and subsequent strained relations between labor and management.
- Postal system processing delays and imbalances in the use of institutional resources.
- Loss of work time, wasteful use of medical services, and lawsuits.

Various approaches may be used to characterize the spread and distribution of contaminants in a building. Two such approaches are computational fluid dynamics (CFD) and multizone (MZ) modeling. In CFD simulations, contaminant release information (such as from a mail sorting machine or digital bar code sorting machine) is used to form a set of boundary conditions. The equations for conservation of mass, momentum, and energy for dispersion of the contaminant (gas, liquid droplets, or solid particles) throughout the environment are solved iteratively on a high-resolution mesh until the imbalances in the equations (the residuals) decrease to an acceptably small number (iterative convergence). MZ modeling is a coarser rendering of the conservation laws and is more practical for buildings because it is less computationally burdensome. It is a network model dividing the space into zones that may also have architectural relevance such as rooms. The zones are connected by an airflow path that is often the heating, ventilation, and air conditioning (HVAC) system.

Real transport of contaminants in occupied spaces is a complicated phenomenon, made up of knowable and unknowable features, particularly when trying to reconstruct a past event. Dispersion may occur via mechanisms that can be modeled deterministically: airflow in a ventilation system, air movement due to pressure differences between areas, temperature gradients, and via activities that are difficult or impossible to characterize, such as office mail delivery and sorting, or foot traffic.

Another approach, initially developed for identification of metal ore deposits, has been through adaptation of geostatistical models for identification of concentrations of environmental contaminants. CFD modeling of the concentration field of a contaminant is a deterministic approach while the geostatistical approach is a probabilistic one.

It has been suggested that, once a concentration field for the contaminant is known, an algorithm could be developed so as to optimize the taking of further environmental samples by intensively sampling locations where there would likely be a high probability of extreme concentrations of contaminant present. Such an algorithm could implement taking additional environmental samples in locations adjacent to those where extreme levels of contaminant were present as determined in previously sampled locations. Such a process using information from previous sampling locations to determine the next sampling location is called adaptive sampling [KWG03a, KWG03b].

In this paper, we consider approaches to modeling of the concentration field in a building following release of a contaminant and subsequent environmental sampling procedures for decontamination of that site. We first consider dispersion characteristics and concentration levels of contaminants that might be released in the workplace. We then consider models of the concentration field using probabilistic approaches such as geostatistical or spatial techniques, and deterministic ones such as CFD. A procedure for site characterization and decontamination incorporating adaptive sampling is introduced. Finally, investigation of contaminant dispersion in a room is proposed using mathematical modeling of tracer gas concentrations in a ventilation chamber, and limitations and advantages of the modeling process are discussed.

2 Dispersion of Contaminants in the Workplace

2.1 Example — Dispersion of Anthrax Spores

Investigators have determined that *Bacillus anthracis* spores contained in a letter arriving at a postal facility can be aerosolized during the operation and maintenance of high-speed, mail-sorting machines, potentially exposing workers and possibly entering HVAC systems [CDC02c]. Spores could be transported to other locations in the facility through the ventilation system, by airflow differentials between work areas, or during everyday work activities as shown in Fig. 1. In a building such as a postal processing and distribution facility, factors such as the use of compressed air to clean work surfaces, movement of personnel and carts throughout the building, and a large-volume building with multiple doors being opened and closed may also alter ventilation patterns and change contaminant transport patterns. Indeed, among environmental samples taken at postal processing and distribution facilities by investigators from the NIOSH, positive samples for *Bacillus anthracis* were obtained at diverse locations including furniture and office walls. Some of these locations were some distance away from the mail-sorting machine (Table 1) [SMW03]. Because of the potentially ubiquitous distribution of contaminants, recommendations for protecting building environments from airborne chemical, biological, or radiological (CBR) attack have been developed [NIO02]. These include maintenance and control of the building ventilation and filtration systems, isolation of areas where CBR agents might enter the building (lobbies, mailrooms, loading docks, and storage areas), and control of pressure/temperature relationships governing airflow throughout the building.

Environmental Sampling for Anthrax

Recognized techniques for collecting environmental samples to detect the presence of *Bacillus anthracis* include wipe, swab, and vacuum surface sampling techniques [CDC02b]. Environmental samples are cultured and then tested for the presence of the bacillus [CDC02a]. Often only binary outcomes (presence/absence of anthrax spores) are reported following culture, although continuous measures such as concentration (spores per area sampled) may sometimes be reported [SHT02].

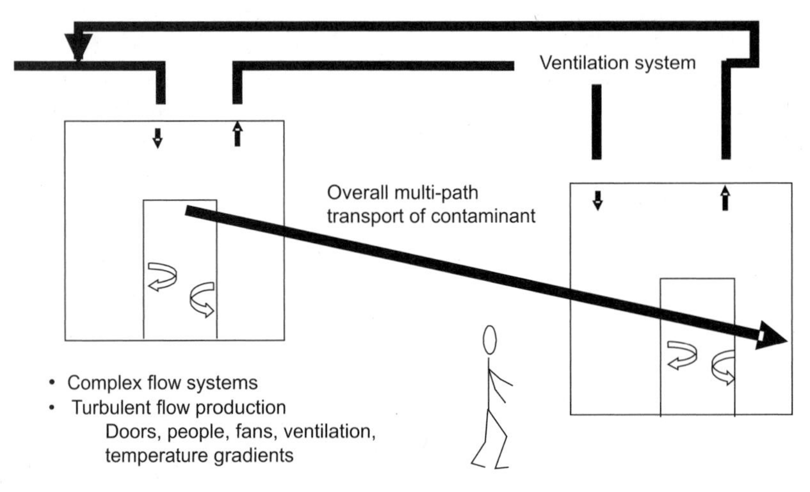

Fig. 1. Possible contaminant transport mechanisms.

Table 1. Anthrax environmental sampling results by location at postal processing and distribution facilities*

Sampling Location	# of Samples	# and Percent Positive
Mail-sorting devices	435	151 (34.7%)
Other postal machines/equipment	288	59 (20.5%)
Office furniture	49	20 (40.8%)
Office equipment	32	8 (25.0%)
Ventilation system	26	10 (38.5%)
Windows	24	11 (45.8%)
Mailbag/pouch/box	16	13 (81.3%)
Wall/wall boxes	14	2 (14.3%)
Floor	5	2 (40.0%)

*Surface samples taken by NIOSH investigators.

2.2 Contaminant Distributions

The behavior of contaminants in occupational settings has been studied extensively [LBL77, Rap91]. Concentrations of contaminants such as aerosols or dusts in occupational and environmental samples are commonly considered to follow a lognormal distribution with probability density function:

$$p(c) = \frac{e^{-(\log((c-\theta)/m))^2/(2\sigma^2))}}{(c-\theta)\sigma\sqrt{2\pi}} \qquad c \geq \theta; m, \sigma > 0, \tag{1}$$

where c is the contaminant concentration, σ is the shape parameter, θ is the location parameter, and m is the scale parameter. Typically we consider $\theta = 0$ and $m = 1$ for the standard lognormal distribution:

$$f(c) = \frac{e^{-(\log(c))^2/(2\sigma^2))}}{(c\sigma\sqrt{2\pi})} \qquad c \geq \theta; \sigma > 0. \tag{2}$$

The lognormal distribution is completely determined by the median or geometric mean (GM) and geometric standard deviation (GSD). Conditions appropriate for the occurrence of lognormal distributions in environmental and occupational data are [LBL77]:

- The concentrations cover a wide range of values, often several orders of magnitude.
- The concentrations lie close to a physical limit (zero concentration).
- The standard deviation of the measured concentration is proportional to the measured concentration.
- A positive probability exists of very large values (or data "spikes") occurring.

It should be noted that the distribution of contaminants in a building environment following an airborne CBR attack might not necessarily follow a standard lognormal distribution considered above, however, due to outside factors affecting dispersion. Such factors could include airflow in the HVAC system, airflow in the occupied space driven by the ventilation system, air movement due to pressure differences between room areas and temperature gradients, and general office activities such as mail delivery, sorting, or general foot traffic (see Fig. 1).

A hypothetical lognormal distribution of anthrax spores in environmental samples collected using a high-efficiency particulate air (HEPA) vacuum in work areas of a postal facility is shown in Fig. 2. The curve in Fig. 2 was developed using measured values of $c = 8.93$ (GM of spore concentrations) and $\sigma = 1.87$ (GSD of spore concentrations).

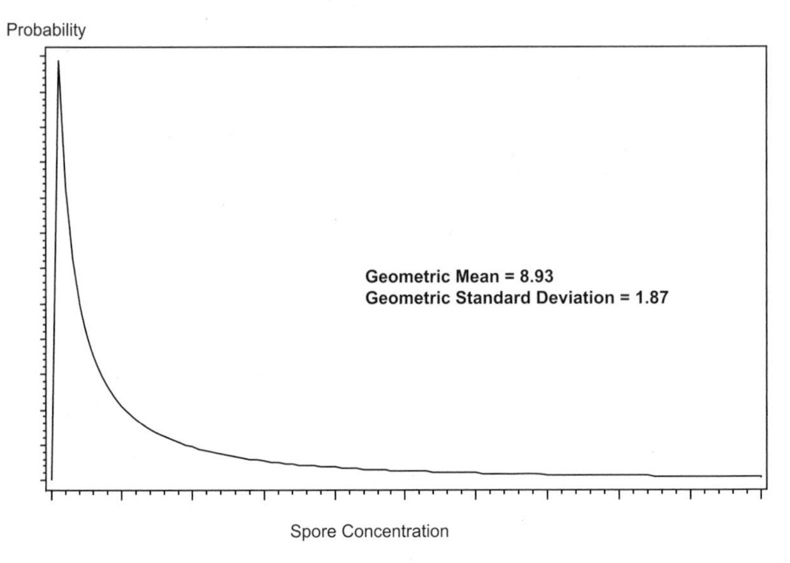

Fig. 2. Hypothetical lognormal distribution of anthrax spore concentrations in work areas of a postal facility.

3 Remediation: How Clean Is Clean?

The desired concentration levels of contaminants may determine methods used during remediation, and thus the cost of remediation of any facility. Thus the extent of the cleanup effort may well be determined by contaminant levels deemed to be acceptable by governments, workers, and employers. For some agents any presence is unacceptable. This is the case for anthrax spores, for which there are currently no occupational or environmental exposure standards, resulting in the massive remediation efforts, closings of certain postal facilities, and large numbers of samples taken to test for the presence or removal of anthrax following the anthrax investigations of October–December, 2001. It should be remembered that even with large numbers of samples, uncertainty still may exist since the limits of detection of present sampling and analytical methods for anthrax spores are unknown and because of limitations on the number of locations tested. In addition, there are currently no validated sampling and analytical methods specifically for *Bacillus anthracis* in environmental samples.

Some evidence on lethal contaminant levels comes from event outbreaks, epidemiological case–control studies, and animal studies. For the anthrax example, information about the quantity of spores needed for health effects comes from an accidental release of anthrax spores in Sverdlovsk, Russia, in 1979 [Gui99]. From epidemiological analysis, the number of spores calculated to cause infection in half the exposed population (LD_{50}) was found to be between 8,000 and 40,000 spores, and the typical incubation period was 2–6 days with an illness duration of 3–5 days [BB02, BB03]. Extrapolation from animal data suggests that the LD_{50} is 2,500 to 55,000 inhaled spores [IOH02]. Inhalation anthrax is often fatal, whereas cutaneous anthrax usually is not. Such information has been used to model the extent of an anthrax epidemic, and to determine the role public health measures such as vaccination may play in minimizing an epidemic [BB03, JE01, MGH94].

Alternatives may exist for deciding that a building is "clean" other than requiring complete removal of a contaminant. Statistically, a building might be declared "clean" following remediation efforts: (a) if the upper 99% confidence limit for the average concentration (spore count) is less than a threshold determined by experiment, or (b) if the probability of obtaining a determination greater than the threshold is small enough, say less than α. In setting appropriate limits, it should also be remembered that the probabilities in (a) and (b) will be affected by environmental sampling and analytical limitations in spore detection. Therefore high inferential limits and low threshold values are desirable. In the anthrax example where any presence is unacceptable, the threshold value would be 0 spores.

It should be mentioned that in certain cases, such as during a first response effort where the presence of contaminant is known and the sources must be quickly identified, it may be that an entirely statistically based sampling strategy may be undesirable or impractical. This case is in contrast to ones in which characterization and estimation of contaminant levels are needed, for which the modeling approaches discussed here may be used. In first response cases, the rationale for clearance sampling might be based on an empirical approach (observation, good practices). Since sample size and surface sampling areas cannot be calculated based on known risk, sampling in such situations might better be determined based on practical experience, observation, and good industrial hygiene practices, guided by statistical considerations.

4 Approaches to Modeling the Concentration Field

4.1 Probabilistic Methods

Geostatistical models originally developed in the mining industry for identification of high concentrations of materials have been adapted for identification of high concentrations of environmental contaminants [WE92]. Examples of models that may be of interest for these data include kriging models that make use of variogram estimates. Suppose that the measurement of a contaminant at location $\mathbf{s}_i$ is $Z(\mathbf{s}_i)$. For a shift value of $\mathbf{h}$, a model of the form

$$\mathrm{Var}(Z(\mathbf{s}_i + \mathbf{h}) - Z(\mathbf{s}_i)) = 2\gamma(\mathbf{h}) \tag{3}$$

has been found useful. The quantity $2\gamma(\mathbf{h})$ on the right-hand side is called the variogram. For the prediction problem, measurements are taken at $\{\mathbf{s}_i, i = 1, 2, \ldots, n\}$. The measurements are assumed to follow the model $Z(\mathbf{s}_i) =$

$\mu(\mathbf{s}_i)+\delta(\mathbf{s}_i)$, for a nonrandom systematic component $\mu(\mathbf{s}_i)$ and error $\delta(\mathbf{s}_i)$. The predictor is $P(Z) = \Sigma_{i=1,2,\ldots,n}\lambda_i Z(\mathbf{s}_i)$, where $\Sigma_{i=1,2,\ldots,n}\lambda_i = 1$ [Cre93]. The optimal predictor in ordinary kriging minimizes the mean square prediction error over the class of linear predictors by using the variogram of the data and assuming that $\mu(\mathbf{s}_i)$ is constant. There are many variants of this procedure, including use of the logarithms of the measurements Z, application of robust procedures, kriging on ranks, and replacement of the mean $\mu(\mathbf{s}_i)$ by a linear combination of known functions.

In the applications of interest here, geostatistical methods could be used as follows. After an initial phase of sampling in a contaminated building, these methods could be used to predict locations of high levels of contaminant, and these locations could then be used in a second sampling phase. An application of this sort has been described by Englund and Heravi [EH94]. In the present work, differences from that study are the use of these methods in an indoor environment, with biological contaminants, which are distributed in three dimensions. The inclusion of knowledge of ventilation systems, building geometry, and effect of human occupation makes this new application a considerable challenge.

4.2 Computational Fluid Dynamics

CFD is a powerful deterministic model of contaminant transport that requires solving conservation laws in scenarios such as rooms. One approach, the control volume method, involves division of the physical space of the room into discrete control volumes called cells. The partial differential equations (Navier–Stokes equations) that govern fluid motion are integrated over each control volume to form simplified algebraic equations. To illustrate using ϕ for a general scalar variable such as mass, a momentum component, or energy, the continuum form of the general steady-state conservation equation is [BFK03, BCS03, Flu98]

$$\oint \varrho\phi\vec{\mathbf{v}} \bullet \mathrm{d}\vec{\mathbf{A}} = \oint \Gamma_\phi \bigtriangledown \phi \bullet \mathrm{d}\vec{\mathbf{A}} + \int S_\phi \mathrm{d}V, \tag{4}$$

where

ϱ = fluid density
$\vec{\mathbf{v}}$ = velocity vector
ϕ = scalar variable
$\vec{\mathbf{A}}$ = surface area vector
S_ϕ = source of ϕ per unit volume
Γ_ϕ = diffusivity for ϕ
V = cell volume

When this partial differential equation is discretized in the control volume method, it becomes

$$\sum_{f}^{N_{faces}} v_f \phi_f A_f = \sum_{f}^{N_{faces}} \Gamma_\phi (\nabla \phi)_n A_f + S_\phi V, \tag{5}$$

where

f = cell face
N_{faces} = number of faces enclosing cell
ϕ_f = value of ϕ convected through face f
v_f = mass flux through face f
A_f = area of face f
$(\nabla \phi)_n$ = magnitude of the gradient of ϕ normal to face f
V = cell volume

The algebraic equations are then solved iteratively, starting at the boundaries of the physical space. The calculations are repeated until the conservation laws are satisfied to an acceptable degree in each cell. The degree of imbalance, called the "residual," in the conservation law is computed as the difference between the value of a variable in a cell and the value that would be expected based on what is flowing into and out of that cell via adjacent cells. A global measure of the imbalance in the conservation law is formed by summing the residuals for all cells, then dividing by the sum of the variable in all the cells. This quantity is termed a "normalized residual."

Sensitivity of CFD models to boundary conditions (e.g., building geometry and HVAC parameters) represents both strengths and limitations of the method. While accurate reconstructions of contaminant transport and deposition are practical in simple environments, real buildings and the activities therein are often prohibitively complex and uncertain. A simplified approach for airflow in buildings, known as MZ modeling, treats building zones as nodes and the HVAC system as links in a network model [DW02, pp. 131–154]. Sandia National Laboratories has developed this method intensely, beginning in 1995, in response to the sarin gas release in Tokyo's subway system. Their technology has been refined now to the point where building blueprints (CAD (computer-aided design) drawings) are used to more rapidly develop flow models that incorporate detailed physics of contaminant behavior. Such information helps optimize detector placement, emergency response strategies, and decontamination tactics [SNL03]. It is also possible to combine a local CFD model with a building-scale MZ model. Sextro et al. [SLS02], for example, modeled the spread of anthrax in buildings using a combination of MZ and CFD modeling.

The results of these numerical models are a predicted concentration field that can be treated as a population, from which sparser samples can be drawn, for the evaluation of statistical approaches to estimating a complete field. The fidelity of the numerically predicted fields must be assessed before they can be used as a meaningful population of concentration values. The authenticity of CFD solutions can be decomposed into two areas, verification and validation. The American Institute of Aeronautics and Astronautics defines verification as "the process of determining that a model implementation accurately represents the developer's conceptual description of the model and the solution to the model." Validation is defined as "the process of determining the degree to which a model is an accurate representation of the real world from the perspective of the intended uses of the model" [OT02]. The following verification and validation tests would be performed.

Spatial Discretization. The solutions are brought to convergence using the second-order upwind scheme. Also, the results from the original grid are compared to a new grid where the number of cells has increased by a factor of 8 in regions of large gradients. This grid convergence test will be evaluated using Roache's grid convergence index (GCI):

$$\mathrm{GCI} = F_S \left| \frac{f_2 - f_1}{1 - r^p} \right|, \tag{6}$$

where $F_S = 3$ is a safety factor designed to approximate errors in convergence in the coarse-grid and fine-grid studies, f_2 is the coarse-grid solution, f_1 the fine-grid solution, r the ratio of cell sizes, and p the order of the discretization [Roa98, pp. 107–136].

Solution Convergence. The solution process is iterated until the normalized residuals for each conservation equation are less than 10^{-3}. The second-order solution will require adjustment of the under-relaxation parameters in order to converge to this level.

Experimental Measurements. A CFD simulation of a space that is also characterized experimentally provides a basis for comparison of variables such as velocity and concentration. Concentration is a useful endpoint variable to determine whether contaminant transport has been simulated accurately. If not, the velocity field agreement may be used to understand how to improve the contaminant transport simulation.

The verifications and validations should be looked at as a set. Experimental measurements by themselves are not a sufficient yardstick of CFD accuracy, given that experiments also have error and that agreement between CFD and experiment can occur by chance.

5 Adaptive Sampling for Site Characterization and Decontamination

In general, adaptive sampling refers to sampling designs in which the procedure for selecting sites or units to be included in a second sample may depend on values of interest observed or predicted from previous samples. With this in mind, the primary application of an adaptive sampling procedure is to capture as many of the units of interest as possible in the sample based on a well-established linking mechanism. In the anthrax example, the main purpose of the adaptive sampling procedure is to identify units of a given building with concentration of spores above the threshold for remediation.

As with MZ modeling, the discretization of a building into adaptive sampling units may be done room by room or by another rule with contaminant transport relevance, such as connection via the ventilation system. It is desirable to choose units small enough to be relatively homogeneous with respect to the level of contamination within, while not making the total number of units impractically large.

To maximize the number of sampled units with concentration above the threshold, the procedure could be composed of several rounds of adaptive sampling depending on the outcomes during the survey. At the first round, an initial sample of n units is taken, which could be any probability sample, for instance, a simple random sample, a stratified random sample, or a systematic sample. The sampling units could be defined as a volumetric grid of spatial units of uniform dimensions for the building, which are then indexed. It may be convenient to have this grid coincide with an equal or lower frequency subset of the numerical grid, though the numerical grids are often not spatially uniform. Interpolation between grid points then becomes necessary. If the concentration population was formed by a MZ model, the sampling units can coincide with the zones. We then examine each unit in the initial sample to determine if it contains "above the threshold level of spores." When it does, we add units that are connected by some linking mechanism, such as a ventilation system, ordinary foot traffic, interoffice mail distribution route, or adjacency to other units that may contain the spores above the threshold. In other words, if adjacency is the linking mechanism and there are grid units in the initial sample that contain spore concentrations in excess of a prescribed threshold, add grid units adjacent to each contaminated sample grid unit according to a fixed "neighboring units" pattern. Continue to add units adjacent to these grid units in excess of the predefined threshold value until no other units need be added. Note that the threshold could be chosen to be 0, but that the resulting sample sizes required might be large. Alternatively, it is possible that the modeling proposed in Sect. 4 above can provide information concerning which units to add to the screening sample. We continue applying the procedure until all the units that are connected by the linking mechanism are exhausted.

The collection of the sampled units (created by an adaptive sampling procedure) with spore concentrations greater than the threshold value is called a *network*. We could have multiple networks of various sizes in the building. If a unit with concentration of spores above the threshold is selected in the initial sample and is not connected to any other unit with concentration above the threshold by the established linking mechanism, then this unit would be considered as a network by itself. We can have k networks where k does not exceed the initial sample size n.

All grid units in the k networks are units with above the threshold concentration of spores and should be decontaminated. After the initial remediation based on the findings of the first round of adaptive sampling procedure, another sample of units could be selected randomly from the area that was not covered in the first round of networking. The adaptive sampling procedures would be applied in a similar fashion as before. If units with concentration of spores above threshold are found in this round, they are decontaminated and further sampling would be done of the area not covered in the previous two rounds. If no unit above the threshold is found in the second round, we stop the sampling. In other words, if no unit in the initial sample contains a concentration of spores above the threshold in the first round, one might consider declaring that there is no evidence that the site is not clear and stop further environmental sampling. Alternatively, a second sample could be chosen for further verification of the initial findings.

After decontaminating those units with levels above the threshold, as found in the successive adaptive sampling procedures, a new sample from the whole building should be taken. At this point a nonparametric statistical approach or asymptotic distribution theory could be used to test the hypothesis whether spore concentrations in the decontaminated areas are elevated relative to the whole building sample. In this case it should be noted that the remaining contaminants no longer have the distribution they had before remediation. From the concentration field previously generated, draw a "screening" or initial sample of grid units of sufficient size to estimate the proportion of spatial grid units that contain lethal spore concentrations with acceptable precision and reliability. For example, to determine the number of grid units to be initially considered, one might use the criterion that we want to be 98% confident that estimated contaminant concentrations will have an absolute error of 1% or less. If the sampling results are binary (indicating only whether spores are present or absent), then an appropriate criterion might be that the fraction of grid units (or the total number of grid units) with spores present should be estimated with 98% confidence. It should be noted that the method used in assigning selection probabilities to the grid units is an important part of this approach and that confidence levels will be greater for subsequent samples.

Katzoff et al. [KWG03a] have used computer simulation procedures to study adaptive sampling procedures, assuming a critical spore count per unit was required. They found that the final sample of units with lethal spore counts was orders of magnitude greater for adaptive procedures than for the corresponding nonadaptive procedures. Their results also showed that consideration of the sources of contamination was important, as was the importance of an understanding of the airflow in and between the various subunits.

A sampling scheme to characterize the extent of site decontamination need not be adaptive, but could be based on expert judgment or probabilistic models as discussed in Sects. 3 and 4 above. By subtracting expected values of concentration of the contaminant (from CFD and MZ models) from observed values measured using monitors, residuals may be obtained for examination. These could be modeled to obtain information about model goodness of fit and covariance structure of the observations. If the number of samples is sufficiently large, geostatistial techniques such as variogram estimation may also be used to model the information in these residuals. The aim of this process is the approximate validation of the deterministic approaches, followed by the improvement of these approximations by statistical means. This will ensure that the distribution of contaminants in the building environment is understood.

Environmental sampling thus has important uses:

- Identification of areas of high concentrations of contaminant.
- Comparison of results from adaptive sampling with results from CFD and MZ models.

Successful modeling of precleanup data may also help identify the location of the sources of the contaminant, if that is unknown. This can be useful both for the cleanup operation and for law enforcement.

6 Modeling Contaminant Dispersion: Proposed Investigations

Experimental data may be used to validate and improve CFD models of contaminant concentrations as described in Sect. 5. Contaminant transport experiments in buildings can validate and improve MZ modeling. Then, the numerical models can be used to validate and enhance the statistical approaches such as adaptive sampling and kriging. Adaptive sampling is one of a family of techniques, including kriging, that estimate a field at a spatial resolution higher than what is initially known. CFD may be useful to investigate the accuracy of such estimates, through its high-resolution rendering of fields. These may be viewed as populations, from which samples may be drawn for the adaptive sampling process whose estimates can then be compared to the population. Also, CFD shares with MZ the prediction of transport. Whereas an adaptive sampling routine might look neutrally in all directions from a local maximum to find other high values, the numerical airflow model would predict that higher values occur downwind of a source. In the case of an MZ model of a multiroom system, adjacent rooms fed by different air handlers may have very different concentrations, whereas a neutral adaptive routine would assume a spatial correlation.

The following is a more detailed look at preliminary experiments in a single room in the ventilation laboratory, performed for the purpose of generating a concentration field to compare with CFD. An outline of the proposed experiment is shown in Fig. 3.

The experimental layout and geometry of the ventilation chamber are shown in Fig. 4. As indicated, concentrations of tracer gas are measured in real time at both locations A and B. Distances between measuring points and velocities at airflow boundaries are also shown in Fig. 4.

Results of measurements taken in the ventilation chamber are shown in Table 2 and Fig. 5.

For this example, measurements taken at locations A and B were fit to a negative exponential model of the form

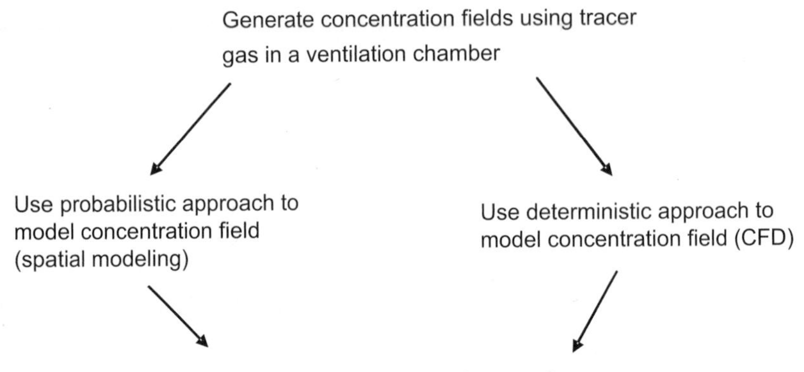

Take a sample of results from each experiment/model above for a given location

Use statistical adaptive sampling algorithm to determine locations of high levels of contaminant for further environmental sampling

Fig. 3. Outline of proposed experiment.

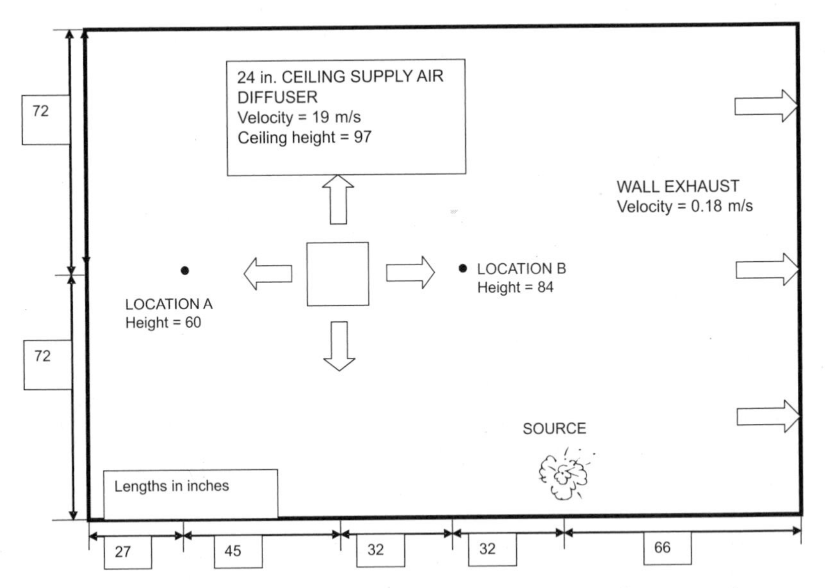

Fig. 4. Experimental layout and CFD geometry viewed from top of room.

$$C(t) = \beta_0(1 - \exp(-\beta_1 t)), \tag{7}$$

where $C(t)$ is concentration at time t, to generate Fig. 5. Notice that the range of measured concentrations is larger at location A than at location B, and that concentration at stationarity is greater at location A (29.2 mg/m^3 in the fitted model, versus 28.1 mg/m^3 for location B), and that it took longer to reach this value at location A (118 seconds versus 66 seconds at location B). Since location A is farther from the source, a long-term diffusion action may be responsible for the longer time to stationarity. The main and more rapid transport mechanism in this room under mixing ventilation is convection. A range of concentrations was measured at both locations.

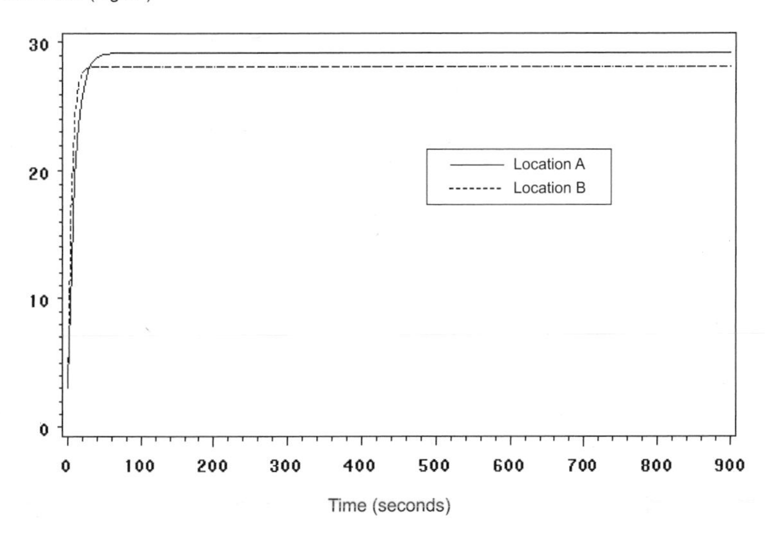

Fig. 5. Predicted concentration* versus time at two monitoring locations. Concentrations predicted under a negative exponential model (7).

Table 2. Results of fitting observed data to negative exponential growth curve

	Location B	Location A
Observed concentrations (mg/m^3)	13.6 – 33.9	16.8 – 31.3
Fitted model		
Parameter β_0	29.2	28.1
Parameter β_1	0.11	0.20
Time to reach β_0 (seconds)	118	66

Experimental measurements taken in the ventilation chamber may be modeled using an equation of the form of (7), but one in which error and autocorrelation of measurements would also be taken into account. Predictions from such a model could be compared to concentrations from CFD predictions. Once the CFD predictions are validated, the entire field produced by CFD can be used as a source for adaptive sampling. The predictions obtained from adaptive sampling from the CFD grid of values will be compared to CFD mean values. If the use of adaptive sampling can be demonstrated successfully, CFD will be used to generate grids for a variety of contaminant situations, on which further adaptive sampling will be carried out and compared to CFD predictions. CFD calculations will be performed using Fluent 6 software. Fluent 6 is a commercial code that has been widely used in academia, government, and industry for many years [Flu98].

6.1 Modeling Dispersion of Tracer Gas in a Room — Limitations and Further Research

As mentioned earlier, the tracer gas experiment has limitations that include the following issues:

- Location: The point source(s) in the tracer gas experiment are known and fixed, whereas source location may be an unknown in a real attack.
- Agent: The tracer is a gas measured in air, whereas the contaminant of interest may be an aerosol in air or deposited on surfaces.
- Generality: Can agreement of this particular experiment with CFD modeling say anything about expected agreement in other rooms, building floors, and whole buildings?

The question of whether experimental validation of the CFD techniques in one situation says anything about accuracy in another situation deserves some attention. It is not proposed to use a CFD solution for the test room as information for the wider field of contaminants and potentially contaminated environments. Rather, the specifics of the process of arriving at the CFD solution, when validated, provide information about how that same process would perform for airflow in another occupied space. To require laboratory validation for every new situation is an unnecessary burden. The aim is to first validate CFD in a situation that is feasibly measurable, then to also apply CFD in the situation of interest not amenable to direct validation. CFD is capable of tracking particles of any reasonable aerodynamic diameter from their source to their fate on a surface. Its ability to accurately predict a tracer gas field in a test room says much about the fidelity of computed particle paths in a building. On the other hand, CFD/MZ methods may provide imperfect predictions in a real-world workplace. Perhaps by combining data and CFD/MZ values, predictions of concentrations of contaminants in unsampled units may be obtained that are better than those produced by either geostatistical methods or CFD/MZ methods by themselves. Since there is also no guarantee that the assumptions required for second-order stationarity and ergodicity (which ensure convergence of the average of the sample values to the mean for large samples) will be met, it may be that alternative modeling approaches including transformation of data or considerations of lognormality may be needed. Furthermore, even if residuals in the modeling process appear to be appropriate, systematic components may still not be completely taken into account.

For instance, if our initial sample can be viewed as a grid, then we could perhaps combine median polish kriging (which adjusts the mean for rows and columns in the data) with estimates from the CFD/MZ data in one statistical model to provide smoothed estimates of mean concentrations at the sampled locations. To the extent that the modeled CFD/MZ data explain all variation, the additional contribution from the rows and columns might be unnecessary. Another plus of this approach would be that it does not seem to require stationary data. If the explanatory components from the model adjust for all changes in the mean, then the residuals will be stationary [Cre93]. Variograms can be fit to the residuals and these can be combined with the mean structure (from the model for the fitted mean) to obtain the full model for prediction of unsampled locations, in order to determine sampling locations for the second phase of sampling. Another advantage of using the median polish method is the use of medians protects against outliers. A somewhat different approach using medians has been discussed by Kafadar and Morris [KM02].

7 Understanding Contaminant Fields: The Need for New Multidisciplinary Approaches

Understanding the distribution and dispersion characteristics of contaminants that might be released during a terrorist attack poses great challenges. Not only are the types of contaminants varied, including chemical, biological, as well as radiological agents, but standard techniques may not be applicable and appropriate ones must be developed.

An important issue is how and when to carry out probability sampling for further characterization of results obtained following initial judgmental sampling (first response), and for determination of the extent of site decontamination. It may be that at some facilities, where anthrax (or some other toxic substance) will very likely be found, expert judgment alone can be used to choose a sampling location that yields a positive result, such as on the surface of a return air grille. However, for other locations, where the presence of the substance is less likely, the need for probability sampling is greater. The question is how to sample and how many samples to take, to accurately assess the risk to workers. Epidemiological information such as where an index case spent time can be helpful, but limiting sampling to a specific area is often hard to justify, due to the variety of pathways for contaminant transport. The same is true for the influence of the ventilation system. The field of inquiry quickly expands then to the entire space where contamination is possible. Therefore, response teams initially rely on necessarily sparse sampling of large spaces.

Although initial environmental samples might be considered to be taken at random locations in a given area, other kinds of probability samples can be taken later to characterize the concentration field — for instance, systematic samples. One approach might be to set up a uniform grid in the area under question and designate sample locations on the grid. Modeling of the contaminant concentration field using a deterministic approach such as CFD or a probabilistic spatial analysis approach could help to determine appropriate sampling locations on the grid. Some researchers [Tho02] have demonstrated the utility of systematically sampling on a uniform grid to determine contaminant concentrations, using results to determine locations for further sampling in an adaptive procedure. In actual practice there might be a rule such as, take one sample from every room in a building.

The approach presented here for modeling the concentration field to direct further environmental sampling represents the marriage of two complementary disciplines, fluid dynamics and statistical science. Such an analytic approach not only adds validity and worth to each, but may also serve as a model for analyses of other aerosols and/or agents. It is also an example of the need for multidisciplinary approaches in understanding contaminant distributions and subsequent environmental sampling and decontamination efforts. CFD has been used in combination with other epidemiological findings to model airborne transmission of the Severe Acute Respiratory Syndrome virus (SARS), for example Yu et al. [YLW04].

While CFD can be further developed to account for patterns of airflow, deposition, and resuspension of particles around furniture, office or postal sorting machines, it is not clear that it is the most efficient method, since deterministic models get very complicated (expensive) as the details reach ever smaller resolved scales. Furthermore, accounting for the contaminant transport induced by normal human activity may be unknowable in a deterministic sense. While MZ is more practical for large buildings, it is a less complete deterministic model. In view of the limitation of both numerical techniques, statistical methods of generating the concentration field for a contaminant may be helpful in understanding contaminant distribution for site decontamination. The use of the lognormal distribution to characterize contaminant concentrations was introduced in Sect. 3. Kriging and geostatistical analyses represent alternative approaches to modeling the concentration field of a contaminant. Another approach for dynamic modeling of transport of airflows (air contaminants) in buildings is the method of Markov chain models or use of stochastic differential equations. Markov chains can be used to model turbulent diffusion and advection of indoor contaminants [Nic01]. Ideally the numerical/deterministic and the statistical/empirical methodologies (such as geostatistical displays) can complement each other in setting appropriate locations to apply adaptive sampling techniques.

Conclusions concerning concentrations of contaminants require sampling. Adaptive sampling provides a novel means of using spatial correlation to identify locations of high concentrations of lethal substances. The use of CFD/MZ and geostatistical methods to identify potential locations at which to sample can enhance adaptive sampling through identification of metaspatial or non-proximal correlations based on airflow and containment patterns.

Such multidisciplinary approaches will be useful in understanding the concentration field of contaminants and in determining methods most appropriate to site characterization and remediation. They will also be of value in other potentially hazardous situations such as characterizing the quality of air in offices or the extent of lead dust or respirable silica in construction/demolition activities.

Acknowledgments

The authors would like to acknowledge the thoughtful review and suggestions by Wayne T. Sanderson, Max Kiefer, Marty Petersen, and one anonymous reviewer. Stanley Shulman expresses his gratitude to his family, Jane, Meryl, and Richard, as well as to his sisters, Sherri and Lisa, for their support over the years.

References

[BCS03] Bennett, J. S., K. G. Crouch, and S. A. Shulman. 2003. "Control of wake-induced exposure using an interrupted oscillating jet." *AIHA Journal* 64:24–29.

[BFK03] Bennett, J. S., C. E. Feigley, J. Khan, and M. H. Hosni. 2003. "Comparison of emission models with computational fluid dynamic simulation and a proposed improved model." *AIHA Journal* 64:739–754.

[BB02] Brookmeyer, R., and N. Blades. 2002. "Prevention of inhalational anthrax in the U. S. outbreak." *Science* 295:1861.

[BB03] Brookmeyer, R., and N. Blades. 2003. "Statistical models and bioterrorism: Application to the U. S. anthrax outbreak." *Journal of the American Statistical Association* 98 (464): 781–788.

[CDC02a] Centers for Disease Control and Prevention. 2002. Basic diagnostic testing protocols for level A laboratories for the presumptive identification of bacillus anthracis. http://www.bt.cdc.gov/agent/anthrax/lab-testing/.

[CDC02b] Centers for Disease Control and Prevention. Revised April 2002. Comprehensive procedures for collecting environmental samples for culturing bacillus anthracis. http://www.bt.cdc.gov/agent/anthrax/environmental-sampling-apr2002.pdf.

[CDC02c] Centers for Disease Control and Prevention. 2002. "Notice to readers: Protecting building environments from airborne chemical, biological, or radiological attacks." *Morbidity and Mortality Weekly Report* 51 (35): 789.

[Cre93] Cressie, N. A. C. 1993. *Statistics for spatial data*, rev. ed. New York: Wiley.

[DW02] Dols, W. S., and G. N. Walton. 2002. "CONTAMW 2.0 user manual; Multizone airflow and contaminant transport software." NISTIR 6921, U. S. Department of Commerce, Technology Administration, National Institute of Standards and Technology, Building and Fire Research Laboratory.

[EH94] Englund, E. J., and N. Heravi. 1994. "Phased sampling for soil remediation." *Environmental and Ecological Statistics* 1 (3): 247–263.

[Flu98] *Fluent 5 user's guide*. 1998. Lebanon, NH.

[Gui99] Guillemin, J. 1999. *Anthrax: The investigation of a lethal outbreak*. Berkeley: University of California Press.

[IOH02] Inglesby, T. V., T. O'Toole, D. A. Henderson, J. G. Bartlett, M. S. Ascher, E. Eitzen, J. Friedlander, J. Gerberding, J. Hauer, J. Hughes, J. McDade, M. T. Osterholm, G. Parker, T. M. Perl, and K. Tonat. 2002. "Anthrax as a biological weapon, 2002: Updated recommendations for management." *Journal of the American Medical Association* 287 (17): 2236–2250.

[JE01] Johnson-Winegar, A., and D.W. Evans. 2001, August. "Anthrax and other vaccines: Use in the U. S. military." *Proceedings of the Joint Statistical Meetings*. American Statistical Association.

[KM02] Kafadar, K., and M. D. Morris. 2002. "Nonlinear smoothers in two dimensions for environmental data." *Chemometrics and Intelligent Laboratory Systems* 60:113–125.

[KWG03a] Katzoff, M. J., A. Wouhib, and J. F. Gonzalez. 2003. "Refinements of adaptive sampling in site decontamination." *Proceedings of the Joint Statistical Meetings*. American Statistical Association, 2092–2096.

[KWG03b] Katzoff, M. J., A. Wouhib, and J. F. Gonzalez. 2003. "Using design-based adaptive sampling procedures in site decontamination." *35th Symposium on the Interface: Computing Science and Statistics, Security, and Infrastructure Protection, Salt Lake City, UT*.

[LBL77] Leidel, N. A., K. A. Busch, and J. R. Lynch. 1977. "Occupational exposure sampling strategy manual." Publication 77-173, U. S. Department of Health, Education, and Welfare, National Institute for Occupational Safety and Health.

[MGH94] Meselson, M., J. Guillemin, M. Hugh-Jones, A. Langmuir, I. Popova, A. Shelokov, and O. Yampolskaya. 1994. "The Sverdlovsk anthrax outbreak of 1979." *Science* 266:1202–1208.

[NIO02] National Institute for Occupational Safety and Health. 2003. "Guidance for protecting building environments from airborne chemical, biological, or radiological attacks." Technical Report DHHS-NIOSH, No. 2002-139, U. S. Department of Health and Human Services, CDC, NIOSH.

[Nic01] Nicas, M. 2001. "Modeling turbulent diffusion and advection of indoor air contaminants by Markov chains." *American Industrial Hygiene Journal* 62 (2): 149–158.

[OT02] Oberkampf, W. L., and T. G. Trucano. 2002. "Verification and validation in computational fluid dynamics." Sandia report SAND2002-0529, Sandia National Laboratories, Albuquerque, NM.

[Rap91] Rappaport, S. M. 1991. "Assessment of long-term exposures to toxic substances in air." *Annals of Occupational Hygiene* 35 (1): 61–121.

[Roa98] Roache, P. J. 1998. *Verification and validation in computational science and engineering.* Albuquerque, NM: Hermosa.

[SHT02] Sanderson, W. T., M. J. Hein, L. Taylor, B. D. Curwin, G. M. Kinnes, T. A. Seitz, T. Popovic, H. T. Holmes, M. E. Kellum, S. K. McAllister, D. N. Whaley, E. A. Tupin, T. Walker, J. A. Freed, D. S. Small, B. Klusaritz, and J. H. Bridges. 2002. "Surface sampling methods for bacillus anthracis spore contamination." *Emerging Infectious Diseases* 8 (10): 1145–1151.

[SNL03] Sandia National Laboratories. 2003. "Terrorism: How vulnerable are we?" *Sandia Technology* 5 (2): 8–12.

[SLS02] Sextro, R. G., D. M. Lorenzetti, M. D. Sohn, and T. L. Thatcher. 2002. "Modeling the spread of anthrax in buildings." *Proceedings: Indoor air 2002 Conference, Monterey, CA.*

[SMW03] Sieber, W. K., K. Martinez, and B. Wackerman. 2003, May. Anthrax sampling and an investigative database. *Second Conference on Statistics and Counterterrorism*, National Academy of Sciences.

[Tho02] Thompson, S. K. 2002. *Sampling*, 2nd ed. New York: Wiley.

[WE92] Weber, D., and E. Englund. 1992. "Evaluation and comparison of spatial interpolators." *Mathematical Geology* 24 (4): 381–391.

[YLW04] Yu, I. T. S., Y. Li, T. W. Wong, W. Tam, A. T. Chan, J. H. W. Lee, D. Y. C. Leung, and T. Ho. 2004. "Evidence of airborne transmission of the severe acute respiratory syndrome virus." *New England Journal of Medicine* 350 (17): 1731–1739.

Secure Statistical Analysis of Distributed Databases

Alan F. Karr[1], Xiaodong Lin[2], Ashish P. Sanil[3], and Jerome P. Reiter[4]

[1] National Institute of Statistical Sciences, karr@niss.org
[2] Statistical and Applied Mathematical Sciences Institute and Department of Mathematical Sciences, University of Cincinnati, lindx@samsi.info
[3] National Institute of Statistical Sciences and Bristol-Myers Squibb, ashish@niss.org
[4] Institute of Statistics and Decision Sciences, Duke University, jerry@stat.duke.edu

1 Introduction

A continuing need in the contexts of homeland security, national defense, and counterterrorism is for statistical analyses that "integrate" data stored in multiple, distributed databases. There is some belief, for example, that integration of data from flight schools, airlines, credit card issuers, immigration records, and other sources might have prevented the terrorist attacks of September 11, 2001, or might be able to prevent recurrences.

In addition to significant technical obstacles, not the least of which is poor data quality [KSS01, KSB05], proposals for large-scale integration of multiple databases have engendered significant public opposition. Indeed, the outcry has been so strong that some plans have been modified or even abandoned. The political opposition to "mining" distributed databases centers on deep, if not entirely precise, concerns about the privacy of database subjects and, to a lesser extent, database owners. The latter is an issue, for example, for databases of credit card transactions or airline ticket purchases. Integrating the data without protecting ownership could be problematic for all parties; the companies would be revealing who their customers are, and where a person is a customer would also be revealed.

For many analyses, however, it is not necessary actually to integrate the data. Instead, as we show in this paper, using techniques from computer science known generically as secure multiparty computation, the database holders can share analysis-specific sufficient statistics anonymously, but in a way that the desired analysis can be performed in a principled manner. If the sole concern is protecting the source rather than the content of data elements, it is even possible to share the data themselves, in which case *any* analysis can be performed.

The same need arises in nonsecurity settings as well, especially scientific and policy investigations. For example, a regression analysis on integrated state databases about factors influencing student performance would be more insightful than individual analyses, or complementary to them. Yet another setting is proprietary data; pharmaceutical companies might all benefit, for example, from a statistical analysis of their combined chemical libraries, but do not wish to reveal which chemicals are in the libraries [KFL05].

The barriers to integrating databases are numerous. One is confidentiality; the database holders — we term them "agencies" — almost always wish to protect the identities of their data subjects. Another is regulation; agencies such as the Census Bureau (CB) and Bureau of Labor Statistics (BLS) are largely forbidden by law to share their data, even with each other, let alone with a trusted third party. A third is scale; despite advances in networking technology, there are few ways to move a terabyte of data from point A today to point B tomorrow.

In this paper we focus on linear regression and related analyses. The regression setting is important because of its prediction aspect; for example, vulnerable critical infrastructure components might be identified using a regression model. We begin in Sect. 2 with background on data confidentiality and on secure multiparty computation. Linear regression is treated for "horizontally partitioned data" in Sect. 3 and for "vertically partitioned data" in Sect. 4. Two methods for secure data integration and an application to secure contingency tables appear in Sect. 5, and conclusions are given in Sect. 6.

Various assumptions are possible about the participating parties, for example, whether they use "correct" values in the computations, follow computational protocols, or collude against one another. The setting in this paper is that of agencies wishing to cooperate but to preserve the privacy of their individual databases. While each agency can "subtract" its own contribution from integrated computations, it should not be able to identify the other agencies' contributions. Thus, for example, if data are pooled, an agency can of course recognize data elements that are not its own, but should not be able to determine which other agency owns them. In addition, we assume that the agencies are "semihonest;" each follows the agreed-on computational protocols, but may retain the results of intermediate computations.

2 Background

In this section we present background from statistics (Sect. 2.1) and computer science (Sect. 2.2).

2.1 Data Confidentiality

From a statistical perspective, the problem we treat lies historically in the domain of data confidentiality or, in the context of official statistics, statistical disclosure limitation (SDL) [DJD93, WD96, WD01]. The fundamental dilemma is that government statistical agencies are charged with the inherently conflicting missions of both protecting the confidentiality of their data subjects and disseminating useful information derived from their data, to Congress, other federal agencies, the public, and researchers.

In broad terms, two kinds of disclosures are possible from a database of records containing attributes of individuals or establishments. An "identity disclosure" occurs when a record in the database can be associated with the individual or establishment that it describes even if the record does not contain explicit identifiers. An "attribute disclosure" occurs if the value of a sensitive attribute, such as income or health status, is disclosed. This may be an issue even without identity disclosure; for instance, if a doctor is known to specialize in treating AIDS, then attribute disclosure may occur for his or her patients. Attribute disclosure is often inferential in nature, and may not be entirely certain. It is also highly domain-dependent.

To prevent identity disclosures, agencies remove explicit identifiers such as name and address or Social Security number, as well as implicit identifiers, such as "Occupation = Mayor of New York." Often, however, this is not enough. Technology poses new threats, through the proliferation of databases and software to link records across databases. Record linkage, which is shown pictorially in Fig. 1, produces identity disclosures by matching a record in the database to a record in another database containing some of the same attributes as well as identifiers. In one well-known example [Swe97], only three attributes — date of birth, 5-digit zip code of residence, and gender — produced identity disclosures from a medical records database by linkage to public voter registration data.

Identity disclosure can also occur by means of rare or extreme attribute values. For example, female Korean dentists in North Dakota are rare, and an intruder — the generic term for a person attempting to break confidentiality — could recognize such a record, or a family member may recognize another family member from household characteristics, or an employer could recognize an employee from salary, tenure, and geography. Establishments (typically, corporations and other organizations) are especially vulnerable in data at high geographical resolution. The largest employer in a county is almost always widely known, so that county-level reporting of both numbers of employees and health benefits expenditures does not protect the latter.

There is a wealth of techniques [DLT01, FCS94, FW98, WD96, WD01] for "preventing" disclosure. In general, these techniques preserve low-dimensional statistical characteristics of the data, but distort disclosure-inducing, high-dimensional characteristics. *Aggregation* — especially geographical aggregation [KLS01, LHK01] — is a principal strategy to reduce identity disclosures. The CB and several other federal agencies do not release data at aggregations less than 100,000. Another is *top-coding*; for example, all incomes exceeding $10,000,000 could be lumped into a single category. *Cell suppression* is the outright refusal to release risky — usually, small count — entries in tabular data. *Data swapping* interchanges the values of one or more attributes, such as geography, between data records. *Jittering* adds random noise to values of attributes such as income. *Microaggregation* groups numerical data records into small clusters and replaces all elements of each cluster by their (componentwise) average [DA95, DN93]. Even entirely *synthetic databases* may be created, which preserve some characteristics of the original data, but whose records simply do not correspond to real individuals or establishments [DK01, RRR03, Rei03a]. Analysis servers [GKR05], which disseminate analyses of data rather than data themselves, are another alternative, as is the approach described in this paper.

Much current research focuses on explicit disclosure risk — data utility formulations for SDL problems [DFK02, DKS03, DKS04, DS04, GKS05, KKO05, Tro03]. These enable agencies to make explicit trade-offs between risk and utility.

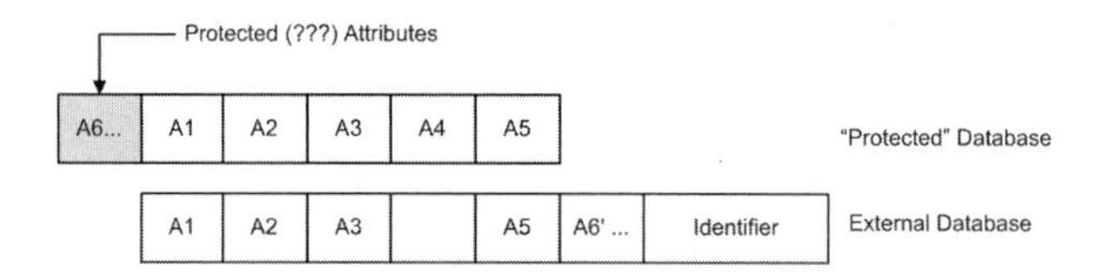

Fig. 1. Pictorial representation of record linkage. The upper record, in the purported protected database, is linked to a record in the external database that has the same values of attributes A1, A2, A3, and A5, but also contains an identifier. If only one record in the external database matches, then the value of A6 is known for the subject of that record. In practice, surprisingly few attributes are needed.

2.2 Secure Multiparty Computation

The generic secure multiparty computation problem [GMW87, Gol97, Yao82] concerns agencies $1, \ldots, K$ with values $v_1, \ldots, v_K$ that wish to compute a known function $f(v_1, \ldots, v_K)$ in such a manner that no agency j learns no more about the other agencies' values than can be determined from v_j and $f(v_1, \ldots, v_K)$. In practice, absolute security may not be possible, so some techniques for secure multiparty computation rely on heuristics [DZ02] or randomization.

The simplest secure, multiparty computation, and the one used in Sect. 3 for secure regression, is to sum values v_j held by the agencies: $f(v_1, \ldots, v_K) = \sum_{j=1}^{K} v_j$. Let v denote the sum. The secure summation protocol [Ben87], which is depicted graphically in Fig. 2, is straightforward in principle, although a "production quality" implementation presents many challenges. Number the agencies $1, \ldots, K$. Agency 1 generates a very large random integer R, adds R to its value v_1, and sends the sum to agency 2. Since R is random, agency 2 learns effectively nothing about v_1. Agency 2 adds its value v_2 to $R + v_1$, sends the result to agency 3, and so on. Finally, agency 1 receives $R+v_1+\cdots+v_K = R+v$ from agency K, subtracts R, and shares the result v with the other agencies. Here is where cooperation matters. Agency 1 is obliged to share v with the other agencies.

Figure 2 contains an extra layer of protection. Suppose that v is known to lie in the range $[0, m)$, where m is a very large number, say 2^{100}, that is known to all the agencies. Then R can be chosen randomly from $\{0, \ldots, m-1\}$ and all computations performed modulo m.

To illustrate, suppose that the agencies have income data and wish to compute the global average income. Let n_j be the number of records in agency j's database and I_j be the sum of their incomes. The quantity to be computed is

$$\bar{I} = \frac{\sum_j I_j}{\sum_j n_j},$$

whose numerator can be computed using secure summation on the I_j's, and whose denominator can be computed using secure summation on the n_j's.

This method for secure summation faces an obvious problem if, contrary to our assumption, some agencies were to collude. For example, agencies $j-1$ and $j+1$ can together compare the values they send and receive to determine the exact value of v_j. Secure summation can be extended to work for an honest majority; each agency divides v_j into shares, and secure summation is used to calculate the sum for each share individually. However, the path used is altered for each share so that no agency has the same neighbor twice. To compute v_j, the neighbors of agency j from every iteration would have to collude.

3 Horizontally Partitioned Data

As the name connotes, this is the case where the agencies have the same attributes on disjoint sets of data subjects [KLR04a, KLR05]. Examples include state-level drivers license databases and data on individuals held by their countries of citizenship.

3.1 The Computations

We assume that there are $K > 2$ agencies, each with the same numerical data on its own n_j data subjects — p predictors X^j and a response y^j — and that the agencies wish to fit the usual linear model

$$y = X\beta + \varepsilon \tag{1}$$

to the "global" data

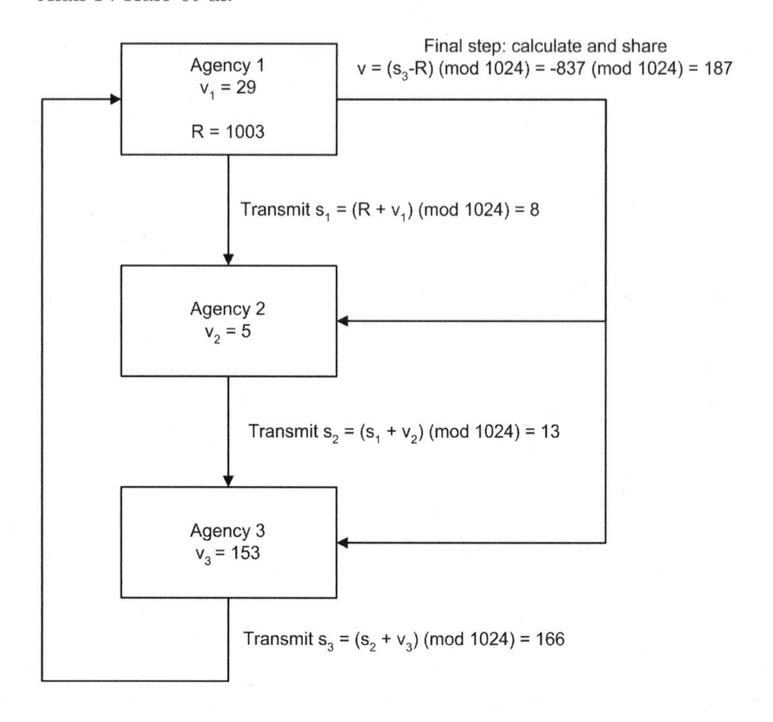

Fig. 2. Values computed at each agency during secure computation of a sum initiated by agency 1. Here $v_1 = 29$, $v_2 = 5$, $v_3 = 152$, and $v = 187$. All arithmetic is modulo $m = 1024$.

$$X = \begin{bmatrix} X^1 \\ \vdots \\ X^K \end{bmatrix} \qquad \text{and} \qquad y = \begin{bmatrix} y^1 \\ \vdots \\ y^K \end{bmatrix}. \tag{2}$$

Figure 3 shows such horizontal partitioning for $K = 3$ agencies. Each X^j is $n_j \times p$.

We embed the constant term of the regression in the first predictor: $X_1^j \equiv 1$ for all j. To illustrate the subtleties associated with distributed data, the usual strategy of centering the predictors and response at their means does not work directly, at least not without another round of secure computation. The means needed are the global, not the local, means, which are not available.[5]

Under the condition that

$$\text{Cov}(\varepsilon) = \sigma^2 I, \tag{3}$$

the least-squares estimator for $\boldsymbol{\beta}$ is of course

[5] They could, of course, be computed using secure summation, as in the average income example in Sect. 2.2.

$$\hat{\boldsymbol{\beta}} = (X^T X)^{-1} X^T \mathbf{y}. \tag{4}$$

To compute $\hat{\boldsymbol{\beta}}$ without data integration, it is necessary to compute $X^T X$ and $X^T \mathbf{y}$. Because of the horizontal partitioning of the data in (2),

$$X^T X = \sum_{j=1}^{K} (X^j)^T X^j. \tag{5}$$

Therefore, agency j simply computes its own $(X^j)^T X^j$, a local sufficient statistic that has dimensions $p \times p$, where p is the number of predictors, and these are combined entrywise using secure summation. This computation is illustrated with $K = 3$ in Fig. 3. Of course, because of symmetry, only $\binom{p}{2} + p$ secure summations are needed. Similarly, $X^T \mathbf{y}$ can be computed by secure, entrywise summation of the $(X^j)^T y^j$.

Finally, each agency can calculate $\hat{\boldsymbol{\beta}}$ from the shared values of $X^T X$ and $X^T \mathbf{y}$. Note that no agency learns any other agency's $(X^j)^T X^j$ or $(X^j)^T y^j$, but only the sum of these over all the other agencies.

The least-squares estimator S^2 of σ^2 in (3) also can be computed securely. Since

$$S^2 = \frac{(\mathbf{y} - X\hat{\boldsymbol{\beta}})^T (\mathbf{y} - X\hat{\boldsymbol{\beta}})}{n - p}, \tag{6}$$

and $X^T X$ and $\hat{\boldsymbol{\beta}}$ have been computed securely, the only thing left is to compute n and $\mathbf{y}^T \mathbf{y}$ using secure summation.

With this method for secure regression, each agency j learns the global $X^T X$ and $X^T \mathbf{y}$. This creates a unilateral incentive to "cheat"; if j contributes a false $(X^j)^T X^j$ and $(X^j)^T y^j$ but every other agency uses its real data, then j can recover

$$\sum_{i \neq j} (X^i)^T X^i$$

and

$$\sum_{i \neq j} (X^i)^T y^i,$$

and thereby the regression for the other agencies, correctly. Every other agency, by contrast, ends up with an incorrect regression. Research on means of preventing this is under way at the National Institute of Statistical Sciences (NISS). Exactly what is learned about an agency's database from one regression — and whether that regression compromises individual data elements — requires additional research.

Virtually the same technique can be applied to any model for which "sufficient statistics" are additive over the agencies. One such example is generalized linear models of the form (1), but with $\Sigma = \mathrm{Cov}(\varepsilon)$ not a diagonal matrix. The least-squares estimator for $\boldsymbol{\beta}$ in the GLM is

$$\boldsymbol{\beta}^* = (X^T \Sigma^{-1} X)^{-1} X^T \Sigma^{-1} \mathbf{y},$$

which can be computed using secure summation, provided that Σ is known to all the agencies. Exactly how Σ would be known to all the agencies is less clear.

Another example is linear discriminant analysis [HTF01]; extension to other classification techniques also remains a topic for future research.

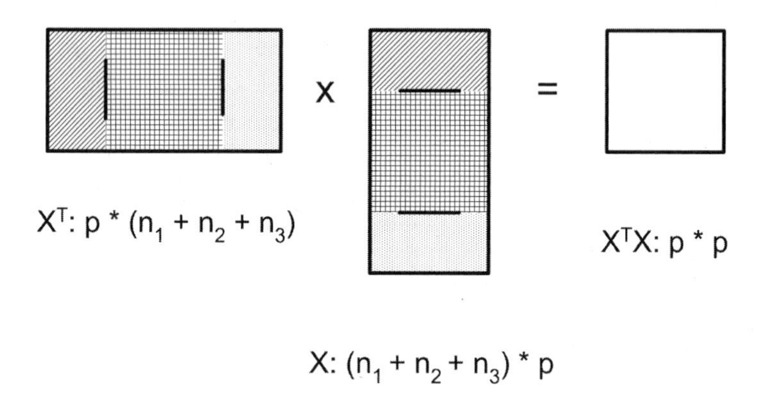

Fig. 3. Pictorial representation of the secure regression protocol for horizontally partitioned data. The dimensions of various matrices are shown.

3.2 Example

We illustrate the secure regression protocol of Sect. 3.1 using the "Boston housing data" [HR78]. There are 506 data cases, representing towns around Boston, which we partitioned, purely for illustrative purposes, among $K = 3$ agencies representing, for example, regional governmental authorities. An alternative, more complicated partition of chemical databases occurs in [KFL05].

The database sizes are comparable: $n_1 = 172$, $n_2 = 182$, and $n_3 = 152$. The response y is median housing value, and three predictors were selected: X_1 = CRIME per capita, X_2 = IND[USTRIALIZATION], the proportion of nonretail business acres, and X_3 = DIST[ANCE], a weighted sum of distances to five Boston employment centers.

Figure 4 shows the results of the computations of their respective $(X^j)^T X^j$ and $(X^j)^T y^j$ performed by the three agencies. The agencies then use the secure regression protocol to produce the global values

$$X^T X = (X^1)^T X^1 + (X^2)^T X^2 + (X^3)^T X^3$$

$$= \begin{bmatrix} 506.00 & 1828.44 & 5635.21 & 1920.29 \\ 1828.44 & 43970.34 & 32479.10 & 3466.28 \\ 5635.21 & 32479.10 & 86525.63 & 16220.67 \\ 1920.29 & 3466.28 & 16220.67 & 9526.77 \end{bmatrix}$$

and

$$X^T\mathbf{y} = (X^1)^T y^1 + (X^2)^T y^2 + (X^3)^T y^3$$
$$= \begin{bmatrix} 11401.60 \\ 25687.10 \\ 111564.08 \\ 45713.87 \end{bmatrix}.$$

These global objects are shared among the three agencies, each of which can then calculate the estimated values of the regression coefficients.

Figure 5 contains these estimators, as well as the estimators for the three agency-specific local regressions. The intercept is $\hat{\beta}_{\text{CONST}}$, the coefficient corresponding to the constant predictor X_1. Each agency j ends up knowing both, but only, the global coefficients and its own local coefficients. To the extent that these differ, it can infer some information about the other agencies' regressions collectively, but not individually. In this example, agency 2 can detect that its regression differs from the global one, but is not able to determine that agency 1 is the primary cause for the difference. Agency 3 is unaware that the regressions of both agency 1 and agency 2 differ from the global regression.

Agency j	n_j	$(X^j)^T X^j$	$(X^j)^T y^j$
1	172	$\begin{bmatrix} 172.00 & 49.03 & 1581.19 & 781.52 \\ 49.03 & 40.42 & 556.29 & 180.95 \\ 1581.19 & 556.29 & 23448.60 & 5631.35 \\ 781.52 & 180.95 & 5631.35 & 4186.07 \end{bmatrix}$	$\begin{bmatrix} 4057.90 \\ 909.24 \\ 32227.19 \\ 18996.12 \end{bmatrix}$
2	182	$\begin{bmatrix} 182.00 & 94.47 & 1563.50 & 746.12 \\ 94.47 & 160.90 & 1433.20 & 231.87 \\ 1563.50 & 1433.20 & 18970.98 & 5224.19 \\ 746.12 & 231.87 & 5224.19 & 3882.02 \end{bmatrix}$	$\begin{bmatrix} 4691.10 \\ 2299.13 \\ 37949.83 \\ 19193.18 \end{bmatrix}$
3	152	$\begin{bmatrix} 152.00 & 1684.95 & 2490.52 & 392.64 \\ 1684.95 & 43769.02 & 30489.61 & 3053.46 \\ 2490.52 & 30489.61 & 44106.05 & 5365.14 \\ 392.64 & 3053.46 & 5365.14 & 1458.68 \end{bmatrix}$	$\begin{bmatrix} 2652.60 \\ 22478.73 \\ 41387.06 \\ 7524.57 \end{bmatrix}$

Fig. 4. Illustration of the secure regression protocol for horizontally partitioned data using the "Boston housing data." As discussed in the text, there are three agencies, each of which computes its local $(X^j)^T X^j$ and $(X^j)^T y^j$. These are combined entrywise using secure summation to produce shared global values $X^T X$ and $X^T\mathbf{y}$, from which each agency calculates the global regression coefficients.

Regression	$\hat{\beta}_{\text{CONST}}$	$\hat{\beta}_{\text{CRIME}}$	$\hat{\beta}_{\text{IND}}$	$\hat{\beta}_{\text{DIST}}$
Global	35.505	-0.273	-0.730	-1.016
Agency 1	39.362	-8.792	-0.720	-1.462
Agency 2	35.611	2.587	-0.896	-0.849
Agency 3	34.028	-0.241	-0.708	-0.893

Fig. 5. Estimated global and agency-specific regression coefficients for the partitioned Boston housing data. The intercept is $\hat{\beta}_{\text{CONST}}$.

3.3 Model Diagnostics

In the absence of model diagnostics, secure regression loses appeal to statisticians. We describe briefly two strategies for producing informative diagnostics. The first is to use quantities that can be computed using secure summation from corresponding local statistics. The second uses secure data integration protocols from Sect. 5 to share synthetic residuals.

A number of diagnostics are computable by secure summation. These include:

1. The coefficient of determination R^2.
2. The least-squares estimate S^2 of the error variance σ^2, which was noted in (6).
3. Correlations between predictors and residuals.
4. The hat matrix $H = X(X^TX)^{-1}X^T$, which can be used to identify X-outliers.

For diagnosing some types of assumption violations, only patterns in relationships among the residuals and predictors suggestive of model misspecification are needed, rather than exact values of the residuals and predictors. Such diagnostics can be produced for the global database using secure data integration protocols (Sect. 5) to share synthetic diagnostics. The synthetic diagnostics are generated in three steps [Rei03b]. First, each agency simulates values of its predictors. Second, using the global regression coefficients, each agency simulates residuals associated with these synthetic predictors in a way — and this is the hard part — that mimics the relationships between the predictors and residuals in its own data. Finally, the agencies share their synthetic predictors and residuals using secure data integration.

4 Vertically Partitioned Data

Vertically partitioned databases contain different sets of attributes for the same data subjects. For example, one government agency might have employment information, another health data, and a third information about education, but all for the same individuals.

In this section, we show how to perform regression analyses on vertically partitioned data. One approach (Sect. 4.1) assumes that the database owners are willing to share sample means and covariances, which allows them to perform much richer sets of analyses than mere coefficient estimation, including inference for the coefficients, model diagnostics, and model selection. The second approach (Sect. 4.2) solves directly the quadratic optimization problem associated with computation of least-squares estimators. It entails less sharing of information, but requires that all agencies have access to the response attribute.

Two assumptions underlie this section. First, we assume that the agencies know that they have data on the same subjects, or that there is a secure method for determining which subjects are common to all their databases. The second, and stronger, assumption is that agencies can link records without error. Operationally, this requires in effect that the databases have a common primary key, such as Social Security number. How realistic this assumption is varies by context. For administrative and financial records, it may be sensible, but it becomes problematic in situations where error-prone keys such as name or address must be used.

For the remainder of the section, we assume that the agencies have aligned their common data subjects in the same order.

4.1 Secure Matrix Products

This method [KLR04b], which is in the spirit of Du et al. [DHC04], computes the off-diagonal blocks of the full data covariance matrix securely.

Since each such block involves only two agencies, we restrict attention to two database owners, labeled agency A and agency B, that possess disjoint sets of attributes for the same n data subjects. Let agency A possess n p-dimensional data elements $X_1^A, \ldots, X_n^A$, and let agency B possess n q-dimensional data elements $X_1^B, \ldots, X_n^B$, so that the full data matrix is

$$[X^A \quad X^B] = \begin{bmatrix} X_{11}^A & \cdots & X_{1p}^A & X_{11}^B & \cdots & X_{1q}^B \\ \vdots & & \vdots & \vdots & & \vdots \\ X_{n1}^A & \cdots & X_{np}^A & X_{n1}^B & \cdots & X_{nq}^B \end{bmatrix}. \tag{7}$$

We assume the two data matrices are of full rank; if not, the agencies remove linearly dependent columns.

The agencies wish to compute securely and share the $p \times q$-dimensional matrix $(X^A)^T X^B$. Assuming that they also share "diagonal blocks" of the covariance matrix, as we describe below, once they have done so, each possesses the "full data" covariance matrix and may perform a variety of statistical analyses of the integrated data.

Computation of Secure Matrix Products

An optimal computational protocol ensures that neither agency learns more about the other's data by using the protocol than it would learn if an omniscient third party were to tell it the result. From the perspective of fairness, the protocol should be symmetric in the amount of information exchanged. A protocol that achieves both of these goals, at least approximately, is:

1. Agency A generates a set of $g = \lfloor (n-p)/2 \rfloor$ orthonormal vectors $Z_1, Z_2, \ldots, Z_g \in \mathbb{R}^n$ such that $Z_i^T X_j^A = 0$ for all i and j, and sends the matrix $Z = [Z_1 Z_2 \cdots Z_g]$ to agency B.
2. Agency B computes
$$W = (I - ZZ^T)X^B,$$
where I is an identity matrix, and sends W to agency A.
3. Agency A calculates, and shares with agency B,
$$(X^A)^T W = (X^A)^T (I - ZZ^T) X^B = (X^A)^T X^B.$$
The latter equality holds since $(X_j^A)^T Z_i = 0$ for all i and j.

A method for generating Z is presented in [KLR04b].

It might appear that agency B's data can be learned exactly since agency A knows both W and Z. However, W has rank $(n-g) = (n-2p)/2$, so that agency A cannot invert it to obtain X^B.

To assess the degree of protection afforded by this protocol, we note that for any matrix product protocol where $(X^A)^T X^B$ is learned by both agencies, including protocols that involve trusted third parties, at a minimum each agency knows pq constraints on the other's data, one for each element of $(X^A)^T X^B$. In realistic settings, the number of data subjects is much greater than the number of terms in the cross-product matrix: $n \gg pq$. Thus, the knowledge of agency A about X^B consists of pq constraints implied by $(X^A)^T X^B$, and that the X_i^B lie in the $g \approx n/2$-dimensional subspace given by $W = (I - ZZ^T)X^B$. Thus, agency A has a total of $g + pq$ constraints on X^B. Assuming $n \gg pq$, we can say that agency A knows the approximately $n/2$-dimensional subspace that the X_i^B lie in. For large n, agency B's data may be considered safe.

Correspondingly, agency B knows pq constraints on X^A implied by $(X^A)^T X^B$, and that the X_i lie in the $(n-g) \approx n/2$-dimensional subspace orthogonal to Z. Thus, agency B has a total of $n - g + pq$ constraints on X^A. Assuming $n \gg pq$ and that $g \approx n/2$, we can say that agency B knows the approximately $n/2$-dimensional subspace that the X_i^A lie in. For large n, agency A's data may be considered safe.

Since agency A and agency B can each place the other's data in an approximately $n/2$-dimensional subspace, the protocol is symmetric in the information exchanged. At higher levels, though, symmetry can break down. For example, if agency A holds the response, but none of its other attributes is a good predictor, whereas the attributes held by agency B are good predictors, then arguably A learns more about B's data than vice versa.

The protocol is not optimal in the sense of each agency's learning as little as possible about the other's data. From $(X^A)^T X^B$ alone, agency A has only pq constraints on X^B, rather than the approximately $n/2$ constraints described above. The symmetry, however, implies a minimax form of optimality; the total amount of information that must be exchanged is n (consider the extreme case that agency A transmits its data to agency B, which computes $(X^A)^T X^B$ and returns the result to A), and so each agency's transmitting $n/2$ constraints on its data minimizes the maximum information transferred.

Nor is the protocol immune to breaches of confidentiality if the agencies do not use their real data. Moreover, disclosures might be generated because of the values of the attributes themselves. A related problem occurs if one agency has attributes that are nearly linear combinations of the other agency's attributes. When this happens, accurate predictions of the data subjects' values can be obtained from linear regressions built from the securely computed matrix products.

Application to Secure Regression

Application of the secure matrix product protocol to perform secure linear regression analyses is straightforward. Altering notation for simplicity, let the matrix of all variables in the possession of the agencies be $D = [D_1, \ldots, D_p]$, with

$$D_i = \begin{bmatrix} d_{i1} \\ \vdots \\ d_{in} \end{bmatrix}, \quad 1 \leq i \leq p \ . \tag{8}$$

The data matrix D is distributed among agencies $A_1, A_2, \ldots, A_K$. Each agency A_j possesses its own p_j columns of D, where $\sum_{j=1}^{K} p_j = p$.

A regression model of some response attribute, say $D_i \in D$, on a collection of the other attributes, say $D^0 \subseteq D \setminus \{D_i\}$, is of the form

$$D_i = D^0 \boldsymbol{\beta} + \varepsilon, \tag{9}$$

where $\varepsilon \sim N(0, \sigma^2)$. As in Sect. 3, an intercept term is achieved by including a column of ones in D^0, which, without loss of generality, we assume is owned by agency A_1.

The goal is to regress any D_i on some arbitrary subset D^0 using secure computation. For simplicity, we suppress dependence of $\boldsymbol{\beta}$, ε, and σ^2 on D^0. The maximum likelihood estimates of $\boldsymbol{\beta}$ and σ^2, as well as the standard errors of the estimated coefficients, can be obtained from the sample covariance matrix of D, using for example the sweep algorithm [Bea64, Sch03]. Hence, the agencies need only the elements of the sample covariance matrix of D to perform the regression. Each agency computes and shares the on-diagonal blocks of the matrix corresponding to its variables, and the agencies use secure matrix computations as described above to compute the off-diagonal blocks.

The types of diagnostic measures available in vertically partitioned data settings depend on the amount of information the agencies are willing to share. Diagnostics based on residuals require the predicted values, $D^0\hat{\boldsymbol{\beta}}$. These can be obtained using the secure matrix product protocol, since

$$D^0\hat{\boldsymbol{\beta}} = D^0\left[(D^0)^T D^0\right]^{-1}(D^0)^T D_i.$$

Alternatively, once the $\hat{\boldsymbol{\beta}}$ is shared, each agency could compute the portion of $D^0\hat{\boldsymbol{\beta}}$ based on the attributes in its possession, and these vectors can be summed across agencies using secure summation.

Once the predicted values are known, the agency with the response D_i can calculate the residuals $E_0 = D_i - D^0\hat{\boldsymbol{\beta}}$. If that agency is willing to share the residuals with the other agencies, each agency can plot residuals versus its predictors and report the nature of any lack of fit to the other agencies. Sharing E_0 also enables all agencies to obtain Cook's distance measures, since these are functions of E_0 and the diagonal elements of $H = D^0[(D^0)^T D^0]^{-1}(D^0)^T$, which can be computed securely, as noted in Sect. 3.

The agency with D_i may be unwilling to share E_0 with the other agencies, since sharing could reveal the values of D_i itself. In this case, one option is to compute the correlations of the residuals with the independent variables using the secure matrix product protocol. When the model fits poorly, these correlations will be far from zero, suggesting model misspecification. Additionally, the agency with D_i can make a plot of E_0 versus $D^0\hat{\boldsymbol{\beta}}$, and a normal quantile plot of E_0, and report any evidence of model violations to the other agencies. The number of residuals exceeding certain thresholds, i.e., outliers, also can be reported.

Variations of linear regression can be performed using the secure matrix product protocol. For example, to perform weighted least-squares regression, the agencies first securely premultiply their variables by $T^{1/2}$, where T is the matrix of weights, and then apply the secure matrix protocol to the transformed variables. To run semiautomatic model selection procedures such as stepwise regression, the agencies can calculate the covariance matrix securely, then select models based on criteria that are functions of it, such as the F-statistic or the Akaike information criterion.

It is also possible to perform ridge regression [HK70] securely. Ridge regression shrinks the estimated regression coefficients away from the maximum likelihood estimates by imposing a penalty on their magnitude. Written in matrix form, ridge regression seeks the $\hat{\boldsymbol{\beta}}$ that minimizes

$$\text{Ridge}(\lambda) = (D_i - D^0\boldsymbol{\beta})^T(D_i - D^0\boldsymbol{\beta}) + \lambda\boldsymbol{\beta}^T\boldsymbol{\beta}, \tag{10}$$

where λ is a specified constant. The ridge regression estimate of the coefficients is

$$\hat{\boldsymbol{\beta}}_R = \left[(D^0)^T D^0 + \lambda I\right]^{-1}(D^0)^T D_i. \tag{11}$$

Since $(D^0)^T D^0$ can be computed using the secure matrix product protocol, $[(D^0)^T D^0 + \lambda I]^{-1}$ can be obtained and shared among the agencies. The agencies also can share $(D^0)^T D_i$ securely, which enables calculation of the estimated ridge regression coefficients.

4.2 Secure Least Squares

A second approach to vertically partitioned data entails less sharing of information than for the secure matrix product protocol of Sect. 4.1, but requires that all agencies possess the response attribute $\mathbf{y}$. If this were not the case, the agency holding $\mathbf{y}$ would be required to share it with the others, which poses obvious disclosure risks.

We assume the model of (1), and that (3) holds. The least-squares estimates $\hat{\boldsymbol{\beta}}$ of (4) are, by definition, the solution of the quadratic optimization problem

$$\hat{\boldsymbol{\beta}} = \arg\min_{\boldsymbol{\beta}} (\mathbf{y} - X\boldsymbol{\beta})^T(\mathbf{y} - X\boldsymbol{\beta}). \tag{12}$$

Denote by I_j the predictors held by agency A_j and assume that the I_j are disjoint. If there were overlaps, the agencies would decide in advance which one "owns" shared attribute. For a vector $\mathbf{u}$, we write u_{I_j} for $\{u_i\}_{i \in I_j}$. The total number of attributes — predictors and response — remains p.

As in other protocols for secure multiparty computation, one agency must assume a lead role in initiating and coordinating the process. This is a purely administrative role and does not imply any information advantage or disadvantage. We assume that agency 1 is the designated leader.

Powell's Algorithm

The basis of the computational protocol is *Powell's method* [Pow64] for solution of quadratic optimization problems with calculating, which in practice means approximating numerically, derivatives. We will use it to calculate $\hat{\boldsymbol{\beta}}$ in (12) directly.

Powell's method is a derivative-free numerical minimization method that solves the multidimensional minimization problem by solving a series of one-dimensional ("line search") minimization problems. A high-level description of the algorithm is as follows:

1. Start with a suitably chosen set of p vectors in $\mathbb{R}^p$ that serve as "search directions."
2. Start at an arbitrary starting point in $\mathbb{R}^p$ and determine the step size δ along the first search direction $s^{(1)}$ that minimizes the objective function.
3. Move distance δ along $s^{(1)}$.
4. Move an optimal step in the second search direction $s^{(2)}$, and so on until all the search directions are exhausted.
5. Make appropriate updates to the set of search directions and continue until the minimum is obtained.

Specifically, the procedure for finding the minimizer of the function $f(\boldsymbol{\beta})$ consists of an initialization step and an iteration block as described below.

Initialization: Select an arbitrary[6] orthogonal basis $\mathbf{s}^{(1)}, \ldots, \mathbf{s}^{(p)}$ for $\mathbb{R}^p$. Also, pick an arbitrary starting point $\tilde{\boldsymbol{\beta}} \in \mathbb{R}^p$.

Iteration: Repeat the following block of steps p times.

- Set $\boldsymbol{\beta} \leftarrow \tilde{\boldsymbol{\beta}}$.
- For $i = 1, \ldots, p$, find δ that minimizes $f(\boldsymbol{\beta} + \delta \mathbf{s}^{(i)})$, and then set $\boldsymbol{\beta} \leftarrow \boldsymbol{\beta} + \delta \mathbf{s}^{(i)}$.
- For $i = 1, \ldots, (p-1)$, set $\mathbf{s}^{(i)} \leftarrow \mathbf{s}^{(i+1)}$.
- Set $\mathbf{s}^{(p)} \leftarrow \boldsymbol{\beta} - \tilde{\boldsymbol{\beta}}$.
- Find δ that minimizes $f(\boldsymbol{\beta} + \delta \mathbf{s}^{(p)})$, and set $\tilde{\boldsymbol{\beta}} \leftarrow \boldsymbol{\beta} + \delta \mathbf{s}^{(p)}$.

Note that each iteration of the iteration block involves solving $(p+1)$ one-dimensional minimization problems, to determine the δ's.

Powell [Pow64] established the remarkable result that if f is a quadratic function, then p iterations of the iteration block yield the *exact minimizer* of f! That is, solving $p(p+1)$ one-dimensional minimization produces the minimizer of a quadratic function.

Application to Secure Regression

The gist of our approach [SKL04] is to apply Powell's method to

$$f(\boldsymbol{\beta}) = (\mathbf{y} - X\boldsymbol{\beta})^T (\mathbf{y} - X\boldsymbol{\beta}),$$

in order to solve (12). The complication, of course, is that no agency possesses all of the data. The details are as follows.

1. Let $\mathbf{s}^{(1)}, \ldots, \mathbf{s}^{(p)} \in \mathbb{R}^p$ be p-dimensional vectors that will serve as a set of search directions in $\mathbb{R}^p$, to be used for finding the optimal estimate $\hat{\boldsymbol{\beta}}$. The $\mathbf{s}^{(r)}$ will be initially chosen and later updated in such a manner that agency A_j knows only the $s^{(r)}_{I_j}$ components of each $\mathbf{s}^{(r)}$.
2. Initially, $\mathbf{s}^{(r)}$ are chosen as follows. Each A_j picks an orthogonal basis $\{\mathbf{v}^{(r)}\}_{r \in I_j}$ for $\mathbb{R}^{d_j}$. Then for $r \in I_j$ let $s^{(r)}_{I_j} = \mathbf{v}^{(r)}$, and $s^{(r)}_l = 0$ for $l \notin I_j$. Each agency should pick its basis at random so that the other agencies cannot guess it.
3. Let $\tilde{\boldsymbol{\beta}} = \left(\tilde{\beta}_{I_1}, \ldots, \tilde{\beta}_{I_k}\right) \in \mathbb{R}^p$ be the initial starting value of $\boldsymbol{\beta}$ obtained by each A_j picking $\tilde{\boldsymbol{\beta}}_{I_j}$ arbitrarily.
4. Perform the *Basic Iteration Block* below p times. The final value of $\tilde{\boldsymbol{\beta}}$ will be the least-squares estimators $\hat{\boldsymbol{\beta}}$.

The *Basic Iteration Block* is:

1. Each A_j sets $\beta_{I_j} \leftarrow \tilde{\beta}_{I_j}$.
2. For $r = 1, \ldots, p$:

[6] Powell's original algorithm used the coordinate axis vectors as the basis, but any orthogonal basis also suffices [Bre73].

a) Each A_j computes $X_{I_j}\beta_{I_j}$ and $X_{I_j}s_{I_j}^{(r)}$.
b) The agencies use secure summation to compute

$$\mathbf{z} = \mathbf{y} - X\beta = \mathbf{y} - \sum_{j=1}^{K} X_{I_j}\beta_{I_j} \quad .$$

and

$$w = X\mathbf{s}^{(r)} = \sum_{j=1}^{K} X_{I_j}s_{I_j}^{(r)}.$$

In (only) the first iteration of this block, for a given r, $X_{I_j}s_{I_j}^{(r)}$ is nonzero only for the agency that owns x_r. Revealing this to all agencies would be too risky, so only that particular agency, say A_r, will compute $\mathbf{w}$, but not reveal it to the others.
c) All agencies compute

$$\delta = \mathbf{z}^T\mathbf{w}/\mathbf{w}^T\mathbf{w}.$$

In the first iteration, A_r computes this and announces it to the other agencies.
d) Each A_j updates $\beta_{I_j} \leftarrow \beta_{I_j} + \delta s_{I_j}^{(r)}$.

3. For $r = 1, \ldots, (p-1)$, each A_j updates $s_{I_j}^{(r)} \leftarrow s_{I_j}^{(r+1)}$.
4. Each A_j updates $s_{I_j}^{(p)} \leftarrow \beta_{I_j} - \tilde{\beta}_{I_j}$.
5. $\mathbf{z}$, $\mathbf{w}$, and δ are computed as before, and each A_j updates $\beta_{I_j} \leftarrow \beta_{I_j} + \delta s_{I_j}^{(p)}$.

After the regression coefficients are calculated and shared, the agencies learn at least three useful quantities. The first of these, of course, is the global coefficients $\hat{\boldsymbol{\beta}}$, enabling each agency to assess the effect of its variables on the response variable after accounting for the effects of the other agencies' variables. Agencies can also assess the size of effects of the other agencies' variables. If an agency obtains a complete record for some individual, the global regression equation can also be used for prediction of the response value. A comparison of the globally obtained coefficients with the coefficients of the local regression (i.e., the regression of y on X_{I_j}) could also be informative.

Agencies also learn the vector of residuals $\mathbf{e} = \mathbf{y} - X\hat{\boldsymbol{\beta}}$, which is equal to the final $\mathbf{z}$ in our iterative procedure. The residuals permit agencies to perform diagnostic tests to determine if the linear regression model is appropriate. The agencies can perform formal statistical tests or use simple visual diagnostics [SKL04]. Finally, agencies can compute the coefficient of determination

$$R^2 = \frac{\mathbf{y}^T\mathbf{y} - \mathbf{e}^T\mathbf{e}}{\mathbf{y}^T\mathbf{y}}. \tag{13}$$

To assess what is revealed by this protocol, consider any one step of the iteration. The only information exchanged by the agencies is the $\mathbf{z}$ and $\mathbf{w}$ vectors. The actual risk to the data x is less since there is some masking with components of the $\mathbf{s}$ vectors. Specifically, the vulnerability is highest in the first step of the iteration since — because of the way we have chosen the initial $\mathbf{s}$ — only one agency contributes to the sum $\mathbf{w}$ at each round of the basic iteration block. We can reduce risk of disclosure by having the contributing agency compute δ privately and announce it to the others. If we assume that the agencies select their initial bases randomly, so that it is impossible for the others to guess them, and if the summation is performed using the secure summation protocol, then no private information is revealed if only $\mathbf{z}$ and $\mathbf{w}$ are common knowledge.

If iterations were independent, then clearly the procedure would be secure. However, the values that each agency contributes to the sum are functionally related from one iteration to the next. The relationship is complex and difficult to express, however, so that this complexity combined with the nature of the secure sum protocol will make it impossible in practice for malicious agencies to exploit the iteration-to-iteration dependency of the values to compromise data privacy.

Whether the approach is feasible computationally has not been established.

5 Secure Data Integration

The procedures described in Sects. 3 and 4 are tailored to regressions, or more generally to statistical analyses for which there exist sufficient statistics that are additive over the agencies. This makes the protocols efficient, but obviously every time a new kind of analysis is needed, so are new algorithms.

If the agencies are concerned primarily with protecting which one holds which data elements, then it is possible to construct an integrated database that can be shared among the agencies, and on which any kind of analysis is possible. There are, however, at least two problematic aspects of this. First, it requires sharing individual data values, with attendant disclosure risks to the data subjects. Second, secure data integration does not work in situations when data values themselves are informative about their source. For instance, it would not work with state-held databases containing zip codes. Nor would it work, for example, for hospital databases containing income when the patient populations have drastically different incomes.

Consider $K > 2$ agencies wishing to share the integrated data among themselves without revealing the origin of any record, and without use of mechanisms such as a trusted third party. We present two algorithms for doing this, neither of which provides any confidentiality protection for data subjects beyond what may already have been imposed by the agencies.

5.1 Algorithm 1

Algorithm 1 passes a continually growing integrated database among the agencies in a known round-robin order, and in this sense is similar to secure summation, although multiple rounds are required. To protect the sources of individual records, agencies are allowed or required to insert both real and "synthetic" records. The synthetic data may be produced by procedures similar to those for construction of synthetic residuals (see Sect. 3.3), by drawing from predictive distributions fit to the data [KLR05], or by some other means. Once all real data have been placed in the integrated database, each agency recognizes and removes its synthetic data, leaving the integrated database.

The steps in Algorithm 1 are:

1. Initialization: Order the agencies by number 1 through K.
2. Round 1: Agency 1 initiates the integrated database by adding *only synthetic data*. Every other agency puts in a mixture of at least 5% of its real data and, optionally, synthetic data, and then randomly permutes the current set of records. The value of 5% is arbitrary, and serves to ensure that the process terminates in at most 21 rounds. Permutation thwarts attempts to identify the source of records from their position in the database.
3. Rounds $2, \ldots, 20$: Each agency puts in at least 5% of its real data or all real data that it has left, and then randomly permutes the current set of records.
4. Round 21: Agency 1, if it has data left, adds them, and removes its synthetic records. In turn, each other agency $2, \ldots, K$ removes its synthetic data.
5. Sharing: The integrated data are shared after all synthetic data have been removed.

The role of synthetic data is analogous to that of the random number R in secure summation. Without it, agency 2 would receive only real data from agency 1 in round 1. However, synthetic data do not protect the agencies completely. In round 1, agency 3 receives a combination of synthetic data from agency 1 and a mixture of synthetic and real data from agency 2. By retaining this intermediate version of the integrated database, which semihonesty allows, and comparing it with the final version, which contains only real data, agency 2 can determine which records are synthetic (they are absent from the final version) and thus identify agency 2 as the source of some real records. The problem propagates, but with decreasing severity. For example, what agency 4 receives in round 1 is a mixture of synthetic data from agency 1, synthetic and real data from agency 2, and synthetic and real data from agency 3. By *ex post facto* removal of the synthetic data, agency 4 is left with real data that it knows to have come from either agency 2 or agency 3, although it does not know which. There are also corresponding vulnerabilities in the last round.

Algorithm 1 is rather clearly vulnerable to poorly synthesized data. For example, if the synthetic data produced by agencies 1 and 2 are readily detectable, then even without retaining intermediate versions of the database, agency 3 can identify the real data received from agency 2 in round 1. There is no guaranteed way to eliminate risks associated with retained intermediate computations in Algorithm 1, other than the agencies' agreeing not to retain intermediate versions of the integrated database. Alternatively, the agencies may simply accept the risks, since only a controllably small fraction of the data is compromised. Given the "at least 5% of real data" requirement in Algorithm 1, agency 2 would be revealing 5% of its data to agency 3, agencies 2 and 3 would reveal collectively 5% of their data to agency 4, and so on. Reducing 5% to a smaller value would reduce this risk, but at the expense of requiring more rounds.

5.2 Algorithm 2

Algorithm 2 is more secure than Algorithm 1, but it is also much more complex. In particular, while the algorithm will terminate in a finite number of stages, there is no fixed upper bound on this number. By randomizing the order in which agencies add data not only are the risks reduced but also the need for synthetic data is almost obviated. In addition to a growing integrated database, Algorithm 2 requires transmission of a binary vector $\mathbf{d} = (d_1, \ldots, d_K)$, in which $d_j = 1$ indicates that agency j has not yet contributed all of its data and $d_j = 0$ indicates that it has.

Steps in Algorithm 2 are:

1. Initialization: A randomly chosen agency is designated as the *stage 1 agency* a_1.
2. Stage 1: The stage 1 agency a_1 initializes the integrated database with some synthetic data and at least one real data record and permutes the order of the records. If a_1 has exhausted its data, it sets $d_{a_1} = 0$. Then, a_1 picks a *stage 2 agency* a_2 randomly from the set of agencies j, other than itself, for which $d_j = 1$, and sends the integrated database and the vector $\mathbf{d}$ to a_2.
3. Intermediate stages $2, \ldots$: As long as more than two agencies have data left, the stage ℓ agency a_ℓ adds at least one real data record and, optionally, as many synthetic data records as it wishes, to the integrated database, and then permutes the order of the records. If its own data are exhausted, it sets $d_{a_\ell} = 0$. It then selects the stage $\ell + 1$ agency $a_{\ell+1}$ randomly from the set of agencies j, other than itself, for which $d_j = 1$ and sends the integrated database and the vector $\mathbf{d}$ to $a_{\ell+1}$.
4. Final round: Each agency removes its synthetic data.

The attractive feature of Algorithm 2 is that because of the randomization of the "next stage agency," no agency can be sure which other agencies other than possibly the agency from which it received the current integrated database have contributed real data to it. The number and order of previous contributors to the growing integrated database cannot be determined. Nor, if it comes from the stage 1 agency, is there even certainty that the database contains real data.

In fact, to a significant extent, Algorithm 2 does not even need synthetic data. The one possible exception is stage 1. If only real data were used, an agency that receives data from the stage 1 agency knows that with probability $1/(K-1)$ that it is the stage 2 agency, and would, even with this low probability, be able to associate them with the stage 1 agency, which is presumed to be known to all agencies. The variant of Algorithm 2 that uses synthetic data at stage 1 and only real data thereafter seems completely workable.

5.3 Application: Secure Contingency Tables

The algorithms for secure data integration have both direct uses, to do data integration, and indirect applications. Here we illustrate the latter, using secure data integration to construct contingency tables containing counts.

Let D be a database containing only categorical variables $V_1, \ldots, V_J$. The associated contingency table is the J-dimensional array T defined by

$$T(v_1, \ldots, v_J) = \#\{r \in D : r_1 = v_1, \ldots, r_J = v_J\}, \tag{14}$$

where each v_i is a possible value of the categorical variable V_i,[7] $\#\{\cdot\}$ denotes "cardinality of $\cdot$," and r_i is the ith attribute of record i. The J-tuple $(v_1, \ldots, v_J)$ is called the cell coordinates. More generally, contingency tables may contain sums of numerical variables rather than counts; in fact the procedure described below works in either case. The table T is a near-universal sufficient statistic, for example for fitting log-linear models [BFH75].

While (14) defines a table as an array, this is not a feasible data structure for large tables (with many cells, which are invariably sparse) with relatively few cells having nonzero counts. For example, the table associated with the CB "long form," which contains 52 questions, has more than 10^{15} cells (1 gigabyte $= 10^9$) but at most approximately 10^8 (the number of households in the USA) of these are nonzero. The *sparse representation* of a table is the data structure of (cell coordinate, cell count) pairs

$$\Big(v_1, \ldots, v_J, T(v_1, \ldots, v_J)\Big),$$

for only those cells for which $T(v_1, \ldots, v_J) \neq 0$. Algorithms that use the sparse representation data structure have been developed for virtually all important table operations.

Consider now the problem of securely building a contingency table from agency databases $D_1, \ldots, D_K$ containing the same categorical attributes for disjoint sets of data subjects. Given the tools described in Sects. 3, 5.1, and 5.2, this process is straightforward. The steps:

1. **List of Nonzero Cells:** Use secure data integration (either protocol) to build the list L of cells with nonzero counts. The "databases" being integrated in this case are the agencies' individual lists of cells with nonzero

[7] For example, if V_1 is gender, then possible values of v_1 are "female" and "male."

counts. The protocols in Sects. 5.1 and 5.2 allow each agency not to reveal in which cells it has data.

2. **Nonzero Cell Counts:** For each cell in L, use secure summation to determine the associated count (or sum).

6 Discussion

In this paper we have presented a framework for secure linear regression and other statistical analyses in a cooperative environment, under various forms of data partitioning.

A huge number of variations is possible. For example, in the case of horizontally partitioned data, to give the agencies flexibility, it may be important to allow them to withdraw from the computation when the perceived risk becomes too great. Ideally, this should be possible without first performing the regression. To illustrate, agency j may wish to withdraw if its sample size n_j is too large relative to the global sample size $n = \sum_{i=1}^{K} n_i$, which is the classical p-rule in the statistical disclosure limitation (SDL) literature [WD01]. But, n can be computed using secure summation, and so agencies may "opt out" according to whatever criteria they wish to employ, prior to any other computations. It is even possible, under a scenario that the process does not proceed if any one of the agencies opts out, to allow the opting out itself to be anonymous. Opting out in the case of vertically partitioned data does not make sense, however.

There are also more complex partitioning schemes. For example, initial approaches for databases that combine features of the horizontally and vertically partitioned cases are outlined in [RKK04]. Both data subjects and attributes may be spread among agencies, and there may be many missing data elements, necessitating expectation-maximization (EM)-algorithm-like methods. Additional issues arise, however, that require both new abstractions and new methods. For example, is there a way to protect the knowledge of which agencies hold which attributes on which data subjects? This information may be very important in the context of counterterrorism if it would compromise sources of information or reveal that data subjects are survey respondents.

Perhaps the most important issue is that the techniques discussed in this paper protect database holders, but not necessarily database subjects. Even when only data summaries are shared, there may be substantial disclosure risks. Consequently, privacy concerns about data mining in the name of counterterrorism might be attenuated, but would not be eliminated, by use of the techniques described here. Indeed, while it seems almost self-evident that disclosure risk is reduced by our techniques, this is not guaranteed, especially for vertically partitioned data. Nor is there any clear way to assess disclosure risk without actually performing the analyses, at which point it is arguably "too late." Research on techniques such as those in Sects. 3–5 from this "traditional" SDL perspective is currently under way at NISS.

Acknowledgments

This research was supported by the National Science Foundation (NSF) grant EIA–0131884 to the National Institute of Statistical Sciences (NISS). Any opinions, findings, and conclusions or recommendations expressed in this publication are those of the authors and do not necessarily reflect the views of the NSF. We thank Max Buot, Christine Kohnen, and Michael Larsen for useful discussion and comments.

References

[Bea64] Beaton, A. E. 1964. "The use of special matrix operations in statistical calculus." *Educational Testing Service Research Bulletin*, vol. RB-64-51.

[Ben87] Benaloh, J. 1987. "Secret sharing homomorphisms: Keeping shares of a secret sharing." In *Advances in cryptology: Proceedings of CRYPTO '86*, edited by G. Goos and J. Hartmanis, Volume 263 of *Lecture Notes in Computer Science*, 251–260. New York: Springer-Verlag.

[BFH75] Bishop, Y. M. M., S. E. Fienberg, and P. W. Holland. 1975. *Discrete multivariate analysis: Theory and practice.* Cambridge, MA: MIT Press.

[Bre73] Brent, R. P. 1973. *Algorithms for minimization without derivatives.* Englewood Cliffs, NJ: Prentice-Hall.

[DA95] Defays, D., and N. Anwar. 1995. "Micro-aggregation: A generic method." *Proceedings of the 2nd International Symposium on Statistical Confidentiality.* Luxembourg: Office for Official Publications of the European Community, 69–78.

[DN93] Defays, D., and P. Nanopoulos. 1993. "Panels of enterprises and confidentiality: The small aggregates method." *Proceedings of the 92 Symposium on Design and Analysis of Longitudinal Surveys.* Ottawa: Statistics Canada, 195–204.

[DFK02] Dobra, A., S. E. Fienberg, A. F. Karr, and A. P. Sanil. 2002. "Software systems for tabular data releases." *International Journal of Uncertainty, Fuzziness, and Knowledge Based Systems* 10 (5): 529–544.

[DKS03] Dobra, A., A. F. Karr, and A. P. Sanil. 2003. "Preserving confidentiality of high-dimensional tabular data: Statistical and computational issues." *Statistics and Computing* 13 (4): 363–370.

[DLT01] Doyle, P., J. Lane, J. J. M. Theeuwes, and L. V. Zayatz. 2001. *Confidentiality, disclosure, and data access: Theory and practical application for statistical agencies.* Amsterdam: Elsevier.

[DHC04] Du, W., Y. Han, and S. Chen. 2004. "Privacy-preserving multivariate statistical analysis: Linear regression and classification." *Proceedings 4th SIAM International Conference on Data Mining.* 222–233.

[DZ02] Du, W., and Z. Zhan. 2002. "A practical approach to solve secure multiparty computation problems." *New Security Paradigms Workshop 2002.* New York: ACM Press, 127–135.

[DJD93] Duncan, G. T., T. B. Jabine, and V. A. de Wolf. 1993. *Private lives and public policies: Confidentiality and accessibility of government statistics.* Washington, DC: National Academies Press. Panel on Confidentiality and Data Access.

[DK01] Duncan, G. T., and S. A. Keller-McNulty. 2001. "Mask or impute?" *Proceedings of ISBA 2000.*

[DKS04] Duncan, G. T., S. A. Keller-McNulty, and S. L. Stokes. 2004. "Disclosure risk vs. data utility: The R-U confidentiality map." Under revision for *Management Science.*

[DS04] Duncan, G. T., and L. Stokes. 2004. "Disclosure risk vs. data utility: The R-U confidentiality map as applied to topcoding." *Chance* 17 (3): 16–20.

[FCS94] Federal Committee on Statistical Methodology. 1994. *Report on statistical disclosure limitation methodology.* Washington, DC: U. S. Office of Management and Budget.

[FW98] Fienberg, S. E., and L. C. R. J. Willenborg, eds. 1998. *Special issue on disclosure limitation methods for protecting the confidentiality of statistical data.* Volume 14(4). *Journal of Official Statistics.*

[GMW87] Goldreich, O., S. Micali, and A. Wigderson. 1987. "How to play any mental game." *Proceedings of the 19th Annual ACM Symposium on Theory of Computing.* 218–229.

[Gol97] Goldwasser, S. 1997. "Multi-party computations: Past and present." *Proceedings 16th Annual ACM Symposium on Principles of Distributed Computing.* New York: ACM Press, 1–6.

[GKR05] Gomatam, S., A. F. Karr, J. P. Reiter, and A. P. Sanil. 2005. "Data dissemination and disclosure limitation in a world without microdata: A risk-utility framework for remote access analysis servers." *Statistical Sciences* 20 (2): 163–177. http://www.niss.org/dgii/technicalreports.html.

[GKS05] Gomatam, S., A. F. Karr, and A. P. Sanil. January 2004. "Data swapping as a decision problem." *Journal of Official Statistics*, (Revised October 2004), http://www.niss.org/dgii/technicalreports.html.

[HR78] Harrison, D., and D. L. Rubinfeld. 1978. "Hedonic prices and the demand for clean air." *Journal Environmental Economics Management* 5:81–102.

[HTF01] Hastie, T., R. Tibshirani, and J. Friedman. 2001. *The elements of statistical learning: Data mining, inference, and prediction.* New York: Springer-Verlag.

[HK70] Hoerl, A. E., and R. Kennard. 1970. "Ridge regression: Biased estimation for nonorthogonal problems." *Technometrics* 12:55–67.

[KFL05] Karr, A. F., J. Feng, X. Lin, J. P. Reiter, A. P. Sanil, and S. S. Young. 2005. "Secure analysis of distributed chemical databases without data integration." *Journal Computer-Aided Molecular Design*, pp. 1–9. http://www.niss.org/dgii/technicalreports.html.

[KKO05] Karr, A. F., C. N. Kohnen, A. Oganian, J. P. Reiter, and A. P. Sanil. June 2005. "A framework for evaluating the utility of data altered to protect confidentiality." *The American Statistician*, http://www.niss.org/dgii/technicalreports.html.

[KLS01] Karr, A. F., J. Lee, A. P. Sanil, J. Hernandez, S. Karimi, and K. Litwin. 2001. "Disseminating information but protecting confidentiality." *IEEE Computer* 34 (2): 36–37.

[KLR04a] Karr, A. F., X. Lin, J. P. Reiter, and A. P. Sanil. 2004. "Analysis of integrated data without data integration." *Chance* 17 (3): 26–29.

[KLR04b] Karr, A. F., X. Lin, J. P. Reiter, and A. P. Sanil. 2004. "Privacy preserving analysis of vertically partitioned data using secure matrix products." Submitted to *Journal of Official Statistics*, http://www.niss.org/dgii/technicalreports.html.

[KLR05] Karr, A. F., X. Lin, J. P. Reiter, and A. P. Sanil. 2005. "Secure regression on distributed databases." *Journal of Computational and Graphical Statistics* 14 (2): 263–279.

[KSB05] Karr, A. F., A. P. Sanil, and D. L. Banks. 2006. "Data quality: A statistical perspective." *Statistical Methodology*, 3(2):137–173 http://www.niss.org/dgii/technicalreports.html.

[KSS01] Karr, A. F., A. P. Sanil, J. Sacks, and E. Elmagarmid. 2001. "Affiliates workshop on data quality." Workshop report, National Institute of Statistical Sciences. http://www.niss.org/affiliates/dqworkshop/report/dq-report.pdf.

[LHK01] Lee, J., C. Holloman, A. F. Karr, and A. P. Sanil. 2001. "Analysis of aggregated data in survey sampling with application to fertilizer/pesticide usage surveys." *Research in Official Statistics* 4:101–116.

[Pow64] Powell, M. J. D. 1964. "An efficient method for finding the minimum of a function of several variables without calculating derivatives." *Computer Journal* 7:152–162.

[RRR03] Raghunathan, T. E., J. P. Reiter, and D. B. Rubin. 2003. "Multiple imputation for statistical disclosure limitation." *Journal of Official Statistics* 19:1–16.

[Rei03a] Reiter, J. P. 2003. "Inference for partially synthetic, public use microdata sets." *Survey Methodology* 29:181–188.

[Rei03b] Reiter, J. P. 2003. "Model diagnostics for remote access regression servers." *Statistics and Computing* 13:371–380.

[RKK04] Reiter, J. P., A. F. Karr, C. N. Kohnen, X. Lin, and A. P. Sanil. 2004. "Secure regression for vertically partitioned, partially overlapping data." *Proceedings of the Joint Statistical Meetings.* American Statistical Association.

[SKL04] Sanil, A. P., A. F. Karr, X. Lin, and J. P. Reiter. 2004. "Privacy preserving regression modelling via distributed computation." *Proceedings Tenth ACM SIGKDD International Conference on Knowledge Discovery and Data Mining.* New York: ACM Press, 677–682.

[Sch03] Schafer, J. L. 2003. *Analysis of incomplete multivariate data.* London: Chapman & Hall.

[Swe97] Sweeney, L. 1997. "Computational disclosure control for medical microdata: The Datafly system." *Record Linkage Techniques 1997: Proceedings of an International Workshop and Exposition.* 442–453.

[Tro03] Trottini, M. 2003. "Decision models for data disclosure limitation." Ph.D. diss., Carnegie Mellon University. http://www.niss.org/dgii/TR/Thesis-Trottini-final.pdf.

[WD96] Willenborg, L. C. R. J., and T. de Waal. 1996. *Statistical disclosure control in practice.* New York: Springer-Verlag.

[WD01] Willenborg, L. C. R. J., and T. de Waal. 2001. *Elements of statistical disclosure control.* New York: Springer-Verlag.

[Yao82] Yao, A. C. 1982. "Protocols for secure computations." *Proceedings of the 23rd Annual IEEE Symposium on Foundations of Computer Science.* New York: ACM Press, 160–164.

Statistical Evaluation of the Impact of Background Suppression on the Sensitivity of Passive Radiation Detectors

Tom Burr[1], James Gattiker[2], Mark Mullen[3], and George Tompkins[4]

[1] Statistical Sciences Group, Los Alamos National Laboratory, tburr@lanl.gov
[2] Statistical Sciences Group, Los Alamos National Laboratory, gatt@lanl.gov
[3] Safeguards Systems Group, Los Alamos National Laboratory, mmullen@lanl.gov
[4] Stockpile Complex Modeling and Analysis, Los Alamos National Laboratory, tompkins@lanl.gov

1 Introduction

Following the 9/11/2001, terrorist attack on the United States, new counter-terrorism measures have been implemented, and others have been considered. Many of these measures aim to monitor for chemical, biological, or radiological and nuclear weapons. Here we consider detectors that monitor cargo or passenger vehicles for radioactive material. Potential weapons include nuclear explosive devices and radiological dispersal devices ("dirty bombs") that could spread harmful radiation. Several detection options are potentially available, but here we focus on those currently deployed that passively (without using penetrating radiation to actively interrogate the vehicle) detect neutrons and gamma and gamma rays.

Passive detectors are generally less expensive and more accepted than active detectors, and debates regarding the appropriate resources for each of the many newly considered threats should always include cost/benefit analyses. Although a cost/benefit analysis of candidate vehicle-screening systems is beyond our scope, we offer an initial assessment of passive detectors in the context of a challenging "background suppression" phenomenon in which the vehicles suppress the natural background, potentially lowering system sensitivity.

The paper is organized as follows. Section 2 gives additional background. Section 3 describes issues involved in determining the threat scenarios (type of material, cargo, and shielding) most appropriate for deployed systems to be expected to detect with high probability. A description of statistical issues involved in the choice of detector characteristics and mode of operation is given in Sect. 4. Our main topic is background suppression, which we describe in detail in Sect. 5 in the context of the threat scenarios and operating mode discussion. Here we evaluate real detector data (sanitized to eliminate the possibility of revealing system performance) in the presence of this challenging phenomenon in which the vehicles suppress the natural background, lowering system sensitivity. Methods to estimate the typical suppression shape and magnitude and accommodate the resulting patterns in residuals are described, illustrating how statistical methods are currently being developed and applied.

2 Background

Ideally, vehicle monitors deployed to protect against radioactive and nuclear material would alarm only on materials that pose a significant threat, but in practice the situation is complex. Complications include: the existence of natural background radiation; the suppression of the natural background by the vehicle; widespread transport of naturally occurring radioactive material (NORM), such as that arising from cat litter, soils, vegetables, granite, concrete, and medical isotopes used in radiophamaceuticals [BF86]; and the fact that the threat list is large with a huge range of expected threat signatures, especially in the presence of background suppression. In addition, in deployed systems, the radiation signatures of radioactive materials (NORM and threat) are highly variable depending on strength, shielding, vehicle speed and proximity, and signals from neighboring traffic lanes ("cross talk").

Several candidate alarm rules should be considered, including one-at-a-time thresholding and trend monitoring with sequential tests, or monitoring for specific signatures using specific linear combinations of counts or forecast errors. However, the complications mentioned imply that assessing and minimizing the false-positive rate for a given alarm rule or set of rules, which includes nuisance alarms that occur due to NORM, will be difficult.

3 Characterization of Threat Scenarios

It is possible to broadly discuss threat scenarios without revealing sensitive information. As an example, weapons-grade plutonium (WGPu) has two kinds of radiation signatures that might in principle be detectable [BF86]. WGPu generates neutrons via spontaneous fission of 240Pu, and those neutrons might be detected. It also emits gamma radiation at many different energies, for example, energies near 414 keV (414-keV gamma rays at these energies are emitted when 239Pu decays). Of course, the detector size, vehicle speed, and proximity and shielding characteristics determine the count rates of the WGPu signal, and the background count rates must also be characterized.

Another potential threat is highly enriched uranium (HEU), which does not passively emit neutrons, but does emit gamma radiation at several energies (186 keV and 1001 keV, for example). However, the 186-keV gamma radiation is less penetrating than the 414-keV gammas from WGPu and is therefore more difficult to detect through shielding. Furthermore, gamma spectra (the relative count rates as a function of energy, with energy bins ranging typically from 1 to 512, depending on detector type and resolution) are impacted in difficult-to-predict ways by interactions of gammas with matter before or while they interact with the detector. In general, HEU is relatively difficult to detect using passive detectors. In addition to WGPu and HEU, there are many other radioactive sources that could lead to major damage if released as a "dirty bomb" using conventional explosives. Methods to detect these threats include simple thresholding of gross count rates and count rate ratios using counts in specific energy bins. Ideally, most NORM events would not trigger the alarm rules, and statistical alarms (count rates reaching the alarm threshold due simply to variation in detected count rates) would also be rare.

A threat scenario specifies how the radioactive or nuclear material is distributed in the vehicle (single "point" source, or spatially distributed for example) and how the vehicle is packed (geometry and contents, because lighter materials have more impact on neutrons while heavy materials such as lead have more impact on the gammas). It is not our intent to exhaustively list all threat scenarios or even all threat materials. However, it is important to realize that sensitivity (the false-pass rate) and specificity (the false-fail rate) are impacted by the scenario details in the library of events to be considered on the threat list.

Assuming that subject experts reach agreement on the threat list, there are typically two complementary approaches to system design and evaluation. First, laboratory and field experiments can be conducted using real (or, more commonly, surrogate) threat items loaded on different vehicles carrying a range of cargo types, and transported through detectors to characterize detector response to a range of source strengths and shapes (empirical approach). Second, computer models that have been calibrated to reproduce (to within statistical uncertainty) count rates from controlled experiments can simulate laboratory or field experiments (model approach).

Computer model uncertainty [STC04] clearly plays a role here. Although real experimental data is typically favored because of model uncertainty, the value of computer models is becoming more widely recognized. Sources of model uncertainty include: (a) uncertainty in input parameters and (b) discrepancies between the model and the physical system being evaluated. MCNP (Monte Carlo N-Particle code) [MCNP] is one of the most commonly used computer codes, and "perturbation cards" allow users to vary the value of input parameters to assess the impact of uncertainty in the inputs. In some cases, model predictions can be compared to measurements to partially assess the impact of uncertainty sources of type (b), although generally total model uncertainty is underestimated because of the difficulty of assessing type (b) error sources. Much more remains to be done regarding computer model uncertainty before MCNP-based threat and background suppression simulations can provide a suitable testbed for alarm rule evaluation. Therefore, in Sect.

5 we focus on methods to lower the alarm threshold without evaluating the false-pass rates for specified threats.

4 Statistical Issues in the Choice of Detector Characteristics and Mode of Operation

Vehicle monitors are similar to airport passenger security lines in that a primary lane screens all vehicles as they drive slowly through passive detectors. Vehicle profiles such as those shown in Fig. 1 (low-energy gamma counts) are collected during typically a few seconds to tens of seconds. Vehicles that cause a primary alarm are sent for a more thorough secondary screening, where additional hand-held gamma spectrometry equipment, or perhaps imaging capability and/or manual inspections are used. Figure 2 is similar to Fig. 1, but is for the neutron counts, which are much lower than the gamma counts, having many 0-valued and 1-valued counts prior to rescaling. The large fraction of 0 and 1 counts impacts the shape of the smooth curves.

There are two curves in each plot in Figs. 1 and 2, with differing smoothness. The degree of smoothing is important and a practical criterion we use is to select a degree of smoothing (strong smoothing means to use large windows, for example, in moving average smoothers) that gives rise to the expected Poisson variation in the residuals. Specifically, we compute a candidate smooth fit and the associated residuals. We then bin the smoothed value into 10 bins and compute the residual variance in each bin. When a fit of the residual variance to the average smoothed value in the corresponding bin indicates that the variance is essentially equal to the mean, the candidate smoother is accepted. Using this "variance approximately equal to mean" criterion for Fig. 1, the first smoother (using smooth.spline in S-PLUS®[Ins03] with 7 degrees of freedom (df) and weights equal to 1/count) is preferred over the second (which has 3 df). Because vehicle suppression should be much smaller for neutrons, a simple and reasonable criterion for neutron profile smoothers is that they exhibit no or relatively small features (such as maxima or minima) in most profiles. The features in the smooth curve using 7 df in Fig. 2 are difficult to interpret and because we focus here on the gamma counts, we will not pursue an explanation. However, it is possible that the 3-df smoother is more appropriate for these neutron profiles.

The background gamma counts arise primarily from naturally occurring uranium, thorium, and potassium in soil, concrete, and asphalt, and are partially shielded by each vehicle. Notice in Fig. 1 that the number of time indices ranges from approximately 20 to 200, which implies that profile lengths differ among vehicles, due to varying speeds. Also, the top left plot exhibits a broad minimum, then a maximum, then another minimum. This pattern is explained by variation in the amount of background shielding as a function of the locations of the vehicle's tires and vehicle shape. When no vehicle is

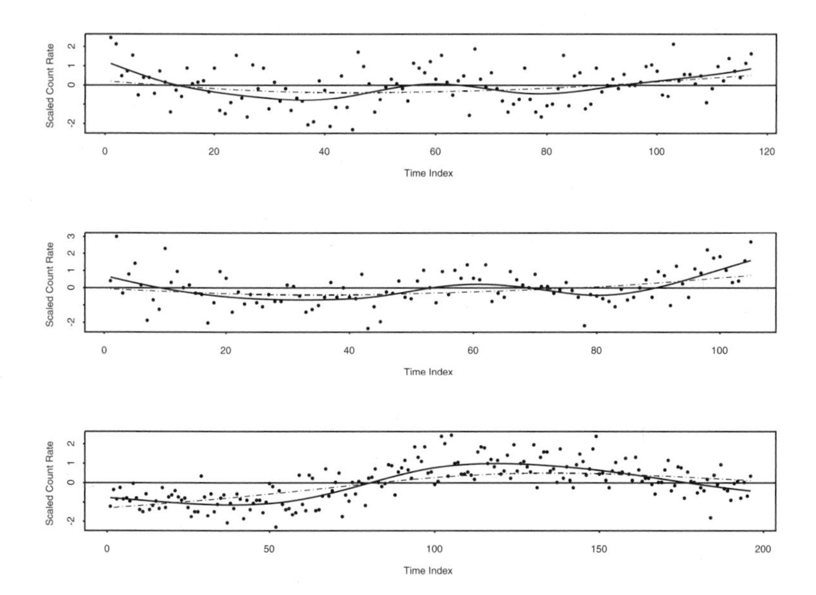

Fig. 1. Three examples of vehicle profiles (low-energy gamma counts), each showing scaled raw count rates (unspecified units) and two smooth curves (the solid one with 7 df and the dotted one with 3 df) fit to the raw profiles.

present, the background is more nearly constant over tens of minutes. Typically, a periodic background reading without any vehicle impacting count rates can be used to establish a baseline. Alternatively, a few counts at the beginning and end of each profile may serve as approximate backgrounds, although there may still be a discernible effect of the vehicle on such pre- and postsamples. In the case we consider below, this pre- and postbackground is approximately 5% smaller than the periodic background, which is taken on longer time periods (and which is therefore less noisy) and is taken without any vehicle present. Position and motion sensors indicate when each vehicle approaches the detector, triggering the collection of pre- and postcounts. The 5% difference between the periodic background and the pre/postbackground indicates that there is some vehicle suppression in these pre/postcounts.

Regardless of whether the empirical or computer-model approach is used to analyze threats, each threat scenario is characterized by the additional count rate (if any) caused by the radioactive or nuclear material, such as shown in Fig. 3 for a generic source. This generic detector exhibits both background suppression and a signal from the material; note that alarm rules based on simply thresholding the count rate would probably not alarm for this example threat. Background suppression is caused by the fact that much of the natural

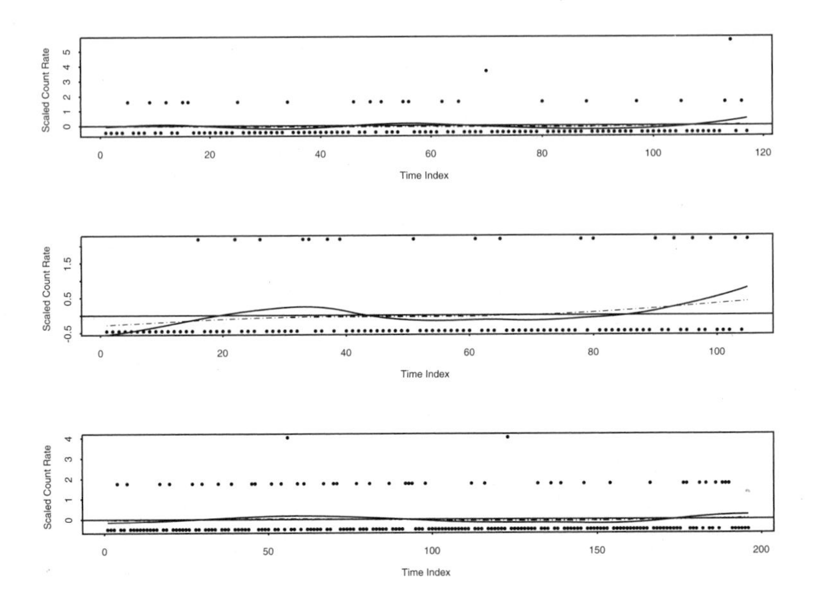

Fig. 2. Three examples of vehicle profiles (neutron counts), each showing scaled raw count rates (unspecified units) and two smooth curves (the solid one with 7 df and the dotted one with 3 df) fit to the raw profiles.

background arises from the nearby ground (asphalt or concrete for example) that is partially shielded by the vehicle while it is in the detector view.

Detector characteristics include: (1) the efficiency (the fraction of impinging radiation that is detected), (2) size (the field of view), and (3) the energy resolution. Resolution (see Fig. 4) is a measure of the width of an energy peak at a specific energy (the width is often defined as the width where the count rate is one-half of its maximum in that region). High-resolution detectors are generally more costly, but they can more clearly separate key peaks, and therefore better characterize radioactive material. Choice of detectors thus entails trade-offs between cost and performance, as well as other factors. Also, in some contexts, lower resolution is better; one obvious reason is the "bias–variance" trade-off in which fewer bins lead to better counting statistics (reduced variance) at the expense of bias introduced by grouping counts from multiple energies into one bin. More subtle reasons involve the characteristics of the threat signals and are beyond our scope. Suffice it to say that resolution studies rarely conclude that the highest possible resolution is optimal, although sometimes it is very helpful if the resolution is high enough to distinguish key energy peaks.

The mode of operation describes: (1) how vehicles pass by the detector (speed, proximity, view angle), (2) what objects enter the detector field of view during a typical vehicle count, and (3) whether the declared cargo is used to direct some vehicles to alternate scans, for example, vehicles that declare NORM-containing cargo might be scanned in a special lane using higher-resolution detectors.

It is a large task to develop a threat list library, but given this library, analysts can determine suitable detector characteristics and the associated mode of operation. Our previous discussion (see Fig. 3, for example) should make it clear that the library must consider the background in addition to the library of events to be detected. If an event library is not explicitly developed, then ad hoc procedures determine detector characteristics and mode of operation, implicitly assuming some uncharacterized threat.

Assuming we have the library of events from the threat list coupled with background and NORM events, one approach to choosing detector characteristics and operation modes is to view the task as a pattern recognition application. The most basic method to parameterize threat/background libraries is a collection of event vectors ("filters") that serve as the patterns to be detected. This is typically done by screening the collection of correlations between a given vehicle profile and each event vector (matched filter) for large values. Performance evaluation is complicated due to: (1) variable dimensions, which require some type of stretching/compressing/interpolation, (2) library size due to the coupling of background suppression and threat events, and (3) relationships among coefficient vectors leading to correlations among the correlations with each matched filter.

If derived features such as ratios of counts in different energy bins (such ratios are thought to be useful because NORM events impact some ratios differently than threat events do) are included as input features, the search for effective pattern recognition strategies is made more complicated. Complications include the need to model the behavior of such ratios for each event in the threat/background library and an increase in the dimension of the candidate feature space used to discriminate among patterns. This candidate feature space includes derived features (such as ratios), temporal patterns among counts from multiple sensors, and choice of energy bins (aggregated and/or omitted bins). There are good reasons to consider count ratio-based alarm rules, but we are less optimistic about matched filters, as Sect. 5 shows.

5 The Impact of Background Suppression on System Performance

Recall that nonthreat vehicle traffic contains a mix of nonradioactive cargo and NORM in the presence of natural background. The natural background sources are primarily in the concrete or asphalt road, so the vehicles suppress this background source to varying degrees depending on vehicle speed, size, and density. One main task is to estimate the false-pass rate for threat-carrying vehicles while maintaining a small false-fail rate for nonthreat vehi-

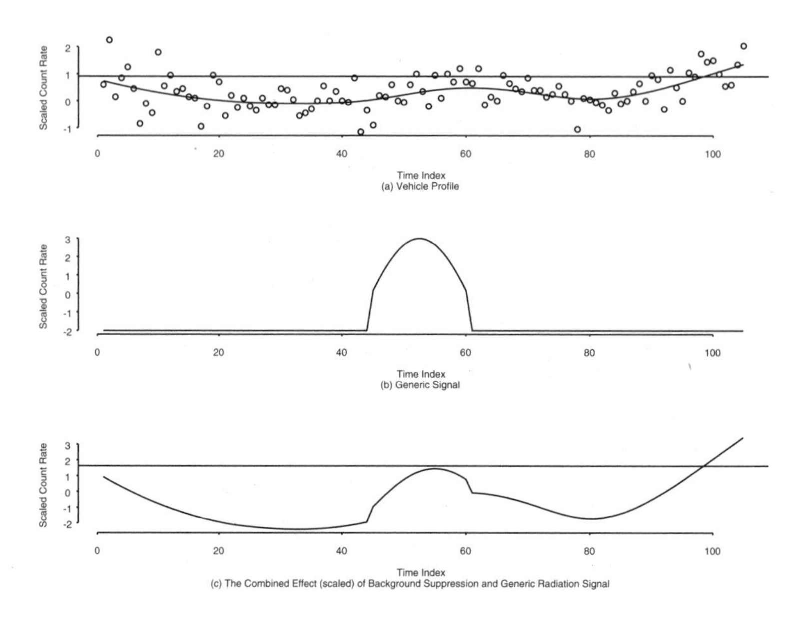

Fig. 3. (a) Example (scaled) raw and smoothed vehicle profile, (b) a generic scaled signal, (c) the combined effect (scaled) of background suppression and radiation signal.

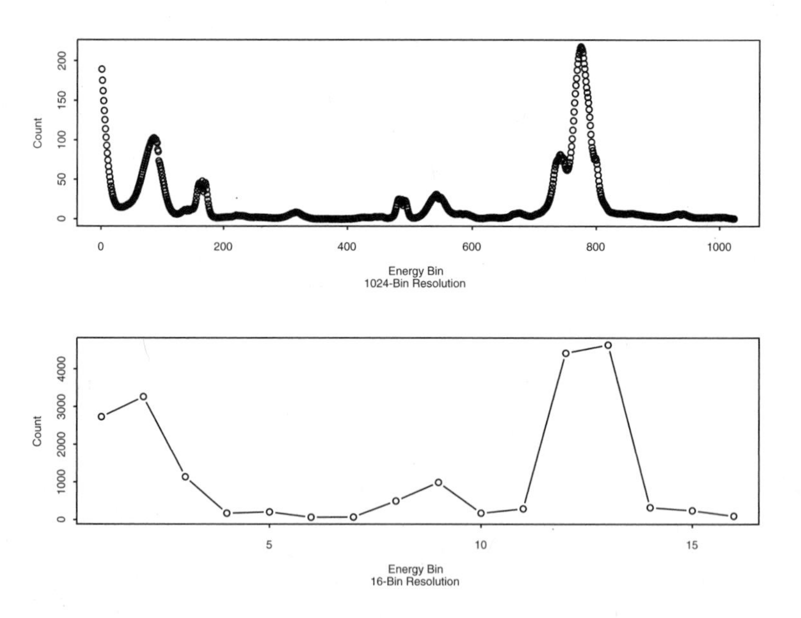

Fig. 4. Example of the same spectra at high (top) and low (bottom) resolution.

cles. To complicate matters, in deployed systems, the radiation signatures of radioactive materials (NORM and threat) are highly variable depending on strength, shielding, vehicle speed and proximity, and signals from neighboring traffic lanes ("cross talk").

Figures 1 and 3 illustrate background suppression of gamma counts and Fig. 3 suggests that there is an infinite number of threats (as characterized by strength, distribution, shielding) and background suppression combinations. In addition, nuisance alarms will occur due to NORM, resulting in a complicated situation for assessing system performance. We want to evaluate alarm rules (false-pass rate for a specified small, false-fail rate) and to support that goal, this section gives quantitative results of the impact of background suppression and a few candidate mitigation strategies.

Background suppression implies: (1) upper control limits should be lower than they would be in the absence of suppression and (2) suppression-corrected count rates make sequential control chart tests more difficult to interpret because of the patterns arising in the residuals.

Concerning (2), note from the threat-scenario discussion that nearly all signals to be detected from the event library involve higher count rates over multiple time periods. Therefore, sequential tests will be more powerful (have a lower false-pass rate) than one-period-at-a-time tests. One question for future investigation is whether the event library is amenable to a custom set of "matched filters" (coefficient vectors that scan the residuals to detect specific patterns with high probability) or is better handled by generic sequential tests such as Page's CuSum [Pag54]. Another question is how to mitigate the anticipated patterns in the residuals that arise due to the fact that not all vehicle profiles exactly match the template suppression (see below). Sequential tests can be applied to residuals having patterns or serial correlation [JB74], but analysis is more difficult, and performance is generally degraded compared to monitoring sequences of residuals that are independent in the zero-signal case.

Concerning (1) and (2), with attention to the impact of the amount of data smoothing, we show that "background suppression" will complicate sequential testing, and in general, will lead to higher false-pass rates for a given, fixed false-fail rate. Figure 5 plots the average aligned scaled (to have unit variance and zero mean) smoothed suppression template (the template can be thought of as an average vehicle profile) over 1210 example vehicle profiles for each of 4 levels of smoothing, and Fig. 6 plots the template (scaled) and the associated 99% upper control limit (UCL) for each. The notation smooth(1) denotes a narrow window that includes the index of interest plus one pre- and post-index, and similarly for smooth(3), smooth(7), and smooth(9). The 99% UCL is computed using, on the square root scale, the template plus 3 times the scalar-valued root mean squared error (RMSE), then transforming back to the original scale, and then rescaling to have zero mean and unit variance. The factor of 3 is justified because the residuals are approximately normally distributed (for the gamma counts) and the use of a scalar-valued RMSE is justified on the square root scale.

Recall that profile lengths range from approximately 20 to over 200, and 150 is the approximate average length of these 1210 vehicles. Therefore, some type of horizontal alignment is required to accommodate the varying profile lengths. Here we use linear stretching or compressing plus interpolation. We also report below that performance using more elaborate horizontal alignment (such as registration via nonlinear mapping to align key features such as minima and maxima [RS97]) has not performed well enough to justify its complexity. In this context, registering to local extrema means that a vehicle profile from start to the first extremum is stretched or compressed to align with the template profile from start to the first extremum. Similarly, the vehicle profile from the first to the second extremum is stretched or compressed to align with the template profile from the first to the second extremum. This process is a nonlinear alignment of the test vehicle's profile with the template profile.

5.1 Performance Measures

One simple performance measure is the average (over all vehicles in the training data) RMSE of the residuals. We define the residual $r_i = C_i - T_i$, where C_i is the count rate (raw or smooth, depending on the context) at index i, and T_i is the prediction (the estimate of the "template") at index i. Note that heavy smoothing suppresses the suppression template, but heavily smoothed data is also expected to exhibit smaller RMSE around its template.

First we describe the best possible (BP) RMSE in a hypothetical situation in which the smooth fit of each vehicle serves as its own template for future profiles of the same vehicle. Physically, this corresponds to hypothetical repeats of the vehicle profile at the same speed, same vehicle lane, same time of day, negligible lane-to-lane cross talk, and same cargo. In this case, the observed variation should be essentially Poisson variation around the template. We would compute each vehicle's smooth profile, calculate the RMSE of the original data around the smooth profile, and average over all vehicles. This is what we will call the BP RMSE (which is unattainable in a realistic setting, but serves as a useful benchmark for comparison).

Alternatively, use each vehicle's smoothed, aligned profile (after some trial and error, we selected length 150, which is approximately the average profile length) to somehow compute an average (template) profile. We have tried principal components and spline basis functions to compute each vehicle's template using vehicle length plus its pre- and postprofile counts to predict its profile. Also, in exploratory mode, we fit the profile itself to principal components or spline basis functions (see Sect. 5.4) but to date have found that simple averaging is as effective as these other two methods. We also consider averaging within groups, where groups can be defined using some type of cluster analysis (see below). For example, vehicle profiles can be clustered via profile length, multidimensional scaling followed by model-based clustering [BR93], or the number of extrema. Again we compute the RMSE for the original (raw) counts around the template.

Figure 7 plots the average RMSE for a naive method, template methods A and B, and the BP method. The naive method uses the average of the pre- and postsamples as the template. Template method B uses the average of the smoothed aligned profile to estimate the template. Template method A is the same as B, but uses the square root of the counts to define the template, and then transforms back to the original scale. The performance of method A is significantly better (both practically and statistically on the basis of t-tests) than that of B. The performance of more sophisticated methods (such as those using clustering to define several classes of background) will fall between A and BP.

Note that the square root transform will approximately stabilize the variance because the variation around the smoothed profile is approximately Poisson. An empirical check that this data is approximately Poisson with a drifting mean was performed using smooth.spline in S-PLUS. In repeated application of smooth.spline, we experimented with the df using simulated Poisson(μ_t) with μ_t obtained from randomly selected smooth fits to real data; and, as with the real data, the choice df = 7 led to the best agreement between variance of residuals and mean value of the smooth fit.

The RMSE of the smoothed data (rather than raw data) around the average template for the four detectors is 7.2, 6.9, 7.3, and 6.8, respectively, which are all significantly smaller than the BP values shown in Fig. 7. In Fig. 6, note that smooth(9) case has a slightly less suppressed (and therefore higher) template, but has much smaller RMSE. However, more smoothing implies less signal, so analyses to estimate the most effective amount of smoothing would require a fully specified signal/background library as previously described (to be presented elsewhere). Our approach here is to fix a reasonable amount of smoothing and always compare methods that have the same amount of smoothing.

5.2 Group-Specific Templates

Next we consider to what extent the UCL can be reduced by defining group-specific templates. To date, we have not significantly reduced the RMSE using clustering, as we indicate below. Figure 8 shows the template for each of the four groups defined by the profile length. The RMSE within each group is 13.3, 13.5, 13.0, and 13.1, respectively, and the groups are defined as: group 1 is all profile lengths, group 4 is lengths 100 to 200, group 7 is length 135 to 165, and group 9 is 145 to 155. Note that these groups overlap, but our intent is to evaluate the reduction in RMSE (if any) when profile lengths are restricted to narrower ranges. We conclude (informally, without a formal test) that restriction to similar-length profiles is not sufficient to achieve significant reduction in RMSE.

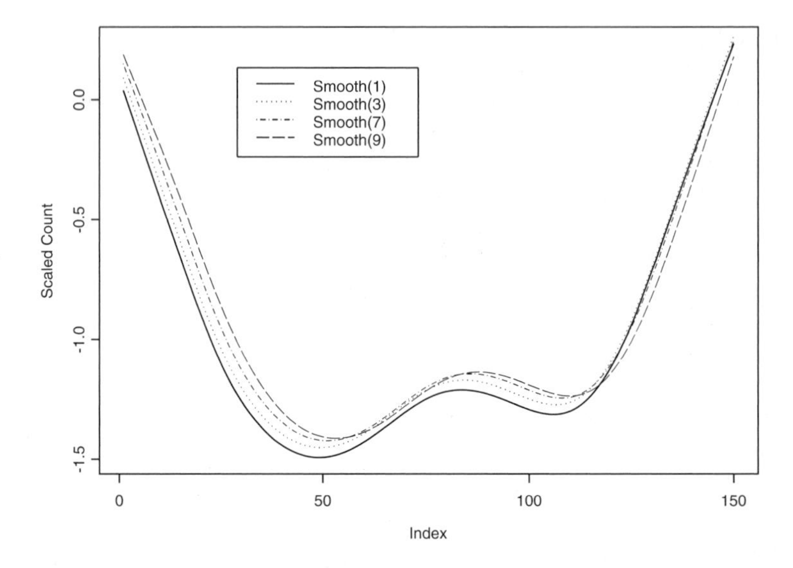

Fig. 5. The average aligned (scaled) suppression resulting from each of four levels of smoothing.

We next use the number of extrema in the smoothed profile to define groups. Figure 9 shows the average aligned profile for vehicles having 1, 2, ..., 7 extrema, and the legend gives the percentage of vehicle profiles among the 1210 having the corresponding number of extrema. More than 50% of profiles have three extrema, in the order minimum, maximum, minimum, which is expected on the basis of current MCNP calculations.

5.3 Sequential Testing

Assuming that we use sequential testing or matched filters to scan for signals from the event library, it will be important that residuals around a template behave approximately as independent, zero mean, constant variance residuals. However, Figs. 10 and 11 illustrate our concern that patterns are likely to be present in the residuals from most vehicle profiles. In Fig. 10, for a profile of length 105, the raw counts, the smooth fit to the counts, the aligned-to-length 150 fit, and the overall template based on all 1210 vehicles are shown. Stretching from 105 to 150 and misfit to the overall template (despite the min, max, min pattern in the vehicle's individual smoothed profile) lead to the pattern shown in the top plot in Fig. 11. The bottom plot in Fig. 11 illustrates that the pattern in the smooth fit to the residuals is not an artifact of smooth.spline in S-PLUS. Figure 11 implies that serial correlation will be

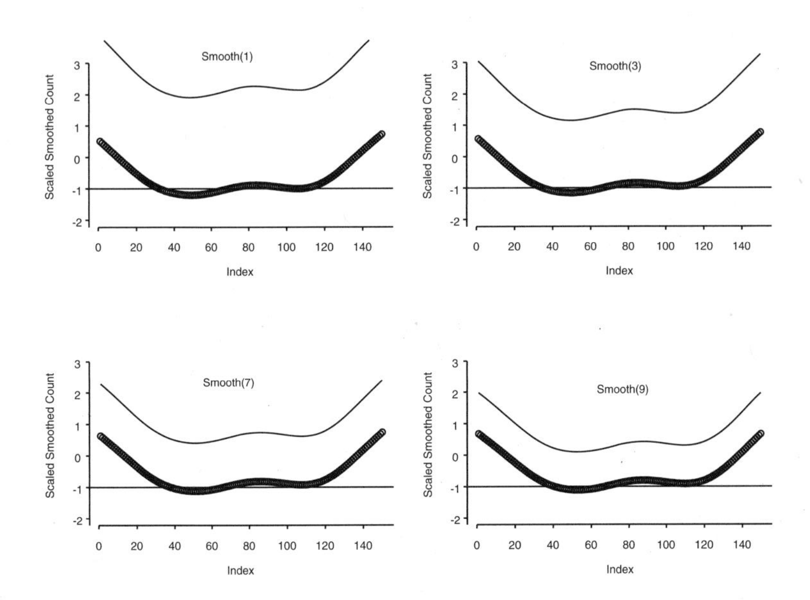

Fig. 6. The template (scaled) and 99% UCL for the same 4 levels of smoothing (top left: 1, top right: 3, bottom left: 7, or bottom right: 9) as in Fig. 5.

present in many of our residual vectors, thus complicating the evaluation of signals from the event/background library. This will be the subject of future work.

It is simple to understand the source of patterns in residuals around an average vehicle template. Although most vehicles (Fig. 8) exhibit the (min, max, min) template, many do not; and among those that do, the distance between extrema is not constant. Therefore, short series of consecutive positive (or negative) residuals are often followed by series of negative (or positive) residuals. These patterns make it difficult to effectively apply sequential tests to detect signals such as the one shown in Fig. 3. Feature registration [RS97] is one area for future research to mitigate these patterns in the residuals; in this context, features will most likely be local extrema. Recall that our initial efforts to align vehicle profiles (nonlinearly) to local extrema have not resulted in improved performance. Also, most of our initial efforts have used smoothed rather than raw counts. Example nonlinear alignment results from a random subset of size 677 from the 1210 vehicles are as follows. If we consider only vehicles that have one extremum (a minimum), then a subset of size 171 from the 677 vehicles had an RMSE of 5.48 unregistered and 4.90 registered. If we consider only vehicles that have three extrema (min, max, min), then a subset of size 506 from the 677 had an RMSE of 5.17 unregistered and 4.87 registered. The combined dataset of size 677 has an RMSE of 5.38 (unregistered). These

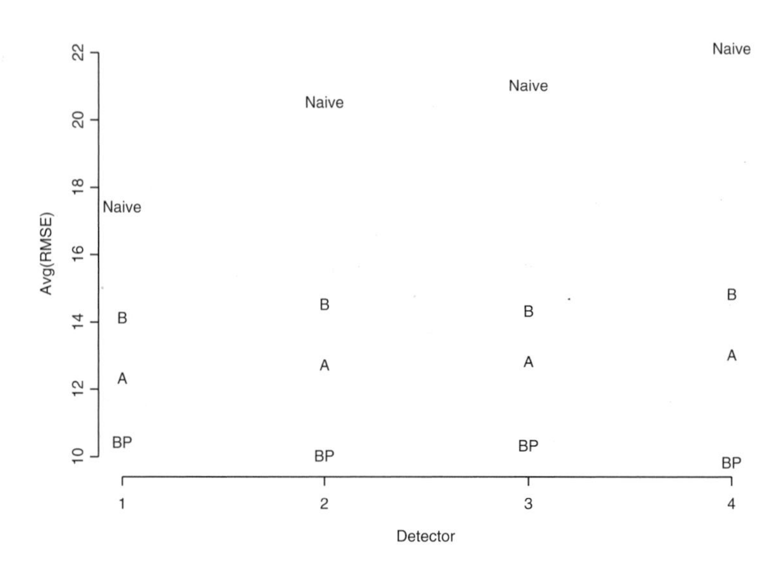

Fig. 7. The average RMSE of the residuals (original scale) for each of four methods, for four different, low-energy gamma detectors, each in a different location in the same vehicle lane. The naive method uses the mean of the pre- and postcounts as the template. The BP method assumes that each vehicle drives repeatedly through the portal with the same speed and loading, and uses the smooth fit for each vehicle as a vehicle-specific template. The resulting average RMSE is due entirely to Poisson variation. Method A transforms to the square root scale to define the average template and then transforms back. Method B uses the original scale throughout.

are small reductions, but perhaps there will be less of a tendency for the residuals to show patterns (to be determined).

5.4 Related Issues

We have suggested that vehicle profiles could be aligned on the basis of locations of minima and maxima. However, this raises the question: how much of the vehicle's profile should be used when either defining its group or investigating its fit to a template? Clearly for the purpose of understanding the types of vehicle profiles to be expected, any such type of exploratory analysis is valid. However, for unknown test vehicles, use of profile features implies that we favor somewhat involved pattern recognition over simple thresholding. Concerns regarding robustness, sensitivity to small threat items coupled with a typical background suppression (such as Fig. 3), and ease-of-use arise.

Also, we have used the mean of pre- and postsamples to estimate the magnitude of the suppression effect, but recall that the pre- and postsam-

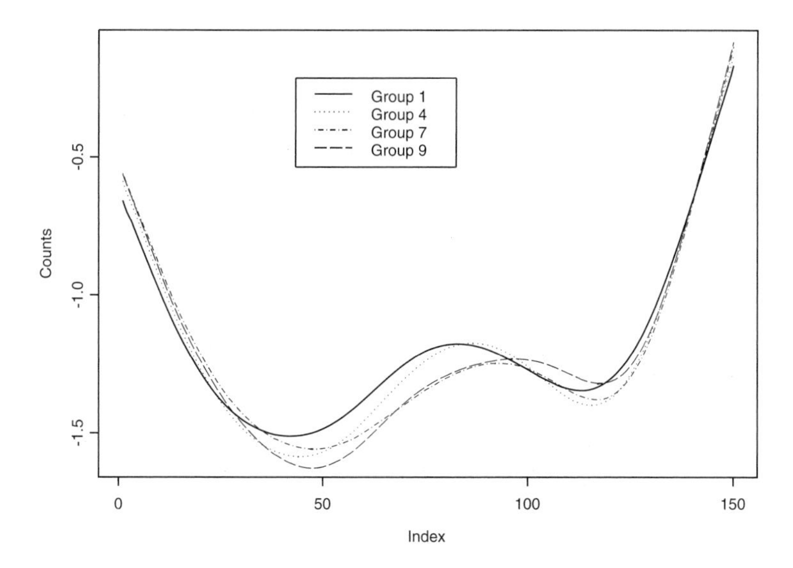

Fig. 8. The template for each of 4 groups defined by vehicle length. Group 1 is all vehicles, group 4 is vehicles whose profile lengths are 100 to 200, group 7 is lengths 135 to 165, and group 9 is lengths 145 to 155.

ples are somewhat suppressed (by approximately 5%) relative to a periodic background count that is much less impacted by vehicle suppression. However, cross talk between lanes can corrupt both the background counts and the profile counts in a given lane. Currently, there is no automated procedure to identify cross talk, but its effect has been noticed using retrospective data analysis.

6 Summary

Physical models and mathematical and statistical sciences have already improved the effectiveness of passive vehicle monitors but more can be done. For example, analytical evaluation of proposed sensors (including active sensors) could facilitate a cost/benefit analysis regarding the merit of using sensors that can interrogate and in principle therefore detect weaker sources. We gave qualitative descriptions of statistical issues involved in passive detector selection and operation, then focused on the "background suppression" issue using real data from existing vehicle monitors. The current alarm criteria are essentially simple thresholds without concern about patterns in the residuals. These thresholds ignore background suppression (which causes approximately a 1.5 standard deviation suppression, averaged over the vehicle profile) and

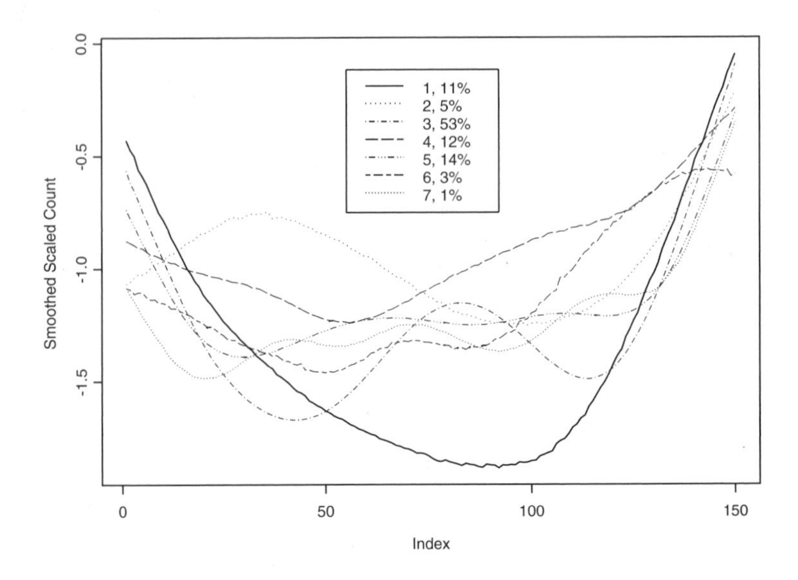

Fig. 9. The aligned (length 150) template (scaled to approximate standard deviation units) for each of the seven groups defined on the basis of the number of extrema. A vehicle belongs to group 1 if its smooth spectrum exhibits one extremum, and similarly for groups 2 to 7. Notice that the group 3 (53% of the 1210 vehicle profiles) spectrum resembles the expected template with the (min, max, min) shape.

therefore are less sensitive than thresholds that adapt to background suppression. If background suppression is to be accounted for, then an open question is whether event/background libraries will be amenable to custom model- and experiment-based matched filters or other pattern recognition tools. Regardless of whether we attempt to correct for background suppression, the presence of background suppression complicates both sequential testing and pattern recognition methods due to the resulting patterns in the residuals. We are experimenting with feature (extrema) registration to mitigate the tendency for the residuals to show patterns, but it is possible (Sect. 5.4) that this would lead to masking of small threat events. In this context, registering to local extrema means that a vehicle profile from start to the first extremum is stretched or compressed to align with the template profile from start to the first extremum. Similarly, the vehicle profile from the first to the second extremum is stretched or compressed to align with the template profile from the first to the second extremum. This process is a nonlinear alignment of the test vehicle's profile with the template profile.

Other issues requiring statistical or mathematical rigor include: evaluation of energy ratio and coincidence algorithms and their sensitivity to the

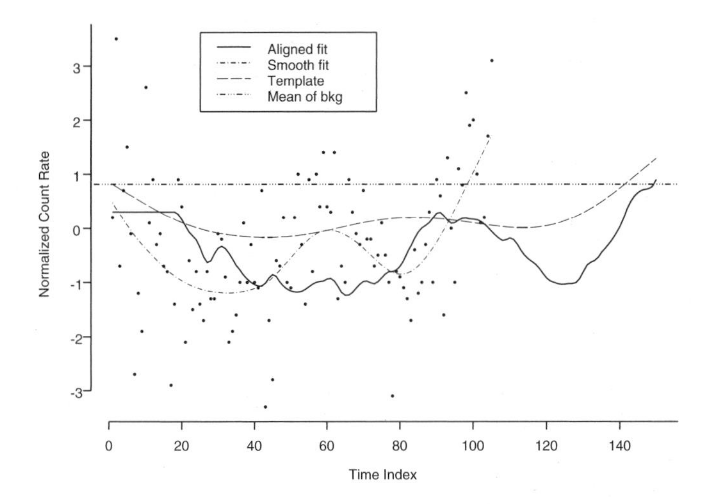

Fig. 10. Example of a vehicle profile of length 105, stretched and aligned to length 150. The template is the average aligned fit of 1210 vehicles, and when it is used to predict the 105 counts for this vehicle, the resulting residuals are shown in Fig. 11.

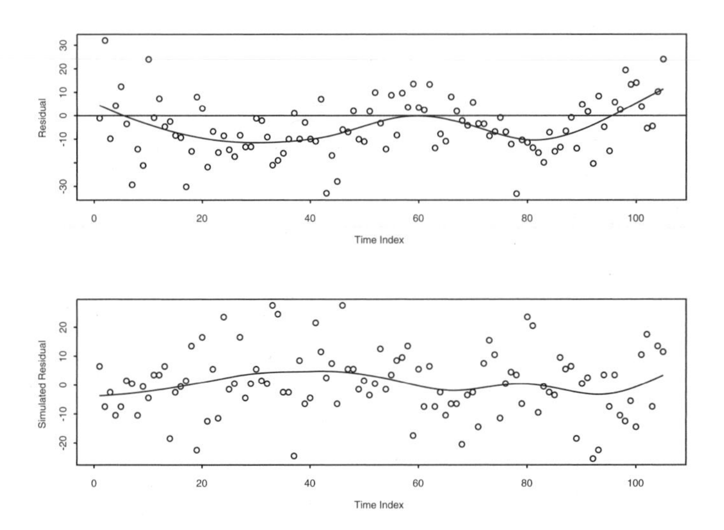

Fig. 11. Residual diagnostics. (Top) The residuals and a smooth fit to the residuals corresponding to the vehicle in Fig. 10. (Bottom) Simulated independent residuals defined by Poisson $(\mu) - \mu$ and a smooth fit to these residuals using the same spline smoother. This illustrates that the pattern in the residuals is not an artifact of the type of smoothing.

threats of interest; spatial and/or temporal profiling algorithms to characterize point sources and perhaps help to resolve and reject NORM events (most NORM would be a distributed source); exploratory data analysis to determine whether there is any value in using axle weight, image data, declared cargo, and neighboring lane sensor data; customization of alarm criteria to individual locations; the impact of model uncertainty on simulated threat signals and visualization of results. Also, alarming vehicles are sent through a more thorough protocol that typically involves more energy channels and better resolution using hand-held detectors, and this second check needs more formal evaluation.

References

[BR93] Banfield, J., and A. Raftery. 1993. "Model-based Gaussian and non-Gaussian clustering." *Biometrics* 49:803–821.

[BF86] Browne, E., and R. Firestone. 1986. *Table of radioactive isotopes.* New York: John Wiley & Sons.

[Ins03] Insightful Corp. 2003. *S-PLUS version 6.2 for Windows.* Seattle, WA: Insightful Corp.

[JB74] Johnson, R., and M. Bagshaw. 1974. "The effect of serial correlation on the performance of CuSum tests." *Technometrics* 16:103–112.

[MCNP] MCNP5, Monte Carlo n-Particle transport code. http://www-xdiv.lanl.gov/x5/MCNP/.

[Pag54] Page, E. S. 1954. "Continuous inspection schemes." *Biometrika* 41:100–115.

[RS97] Ramsay, J., and B. Silverman. 1997. *Functional data analysis.* New York: Springer.

[STC04] Saltelli, A., S. Tarantola, F. Campolongo, and M. Ratto. 2004. *Sensitivity analysis in practice.* West Sussex, UK: John Wiley & Sons.

Index